Stochastic Modeling in Biological System

Stochastic Modeling in Biological System

Editor

Mario Abundo

MDPI

Basel • Beijing • Wuhan • Barcelona • Belgrade • Novi Sad • Cluj • Manchester

Editor
Mario Abundo
University "Tor Vergata"
Rome
Italy

Editorial Office
MDPI
St. Alban-Anlage 66
4052 Basel, Switzerland

This is a reprint of articles from the Special Issue published online in the open access journal *Fractal and Fractional* (ISSN 2504-3110) (available at: https://www.mdpi.com/journal/fractalfract/special_issues/biological_system).

For citation purposes, cite each article independently as indicated on the article page online and as indicated below:

Lastname, A.A.; Lastname, B.B. Article Title. *Journal Name* **Year**, *Volume Number*, Page Range.

ISBN 978-3-7258-1241-7 (Hbk)
ISBN 978-3-7258-1242-4 (PDF)
doi.org/10.3390/books978-3-7258-1242-4

Contents

fractal and fractional

MDPI

Article

Impact of Lévy Noise with Infinite Activity on the Dynamics of Measles Epidemics

Yuqin Song [1] and Peijiang Liu [2,*]

[1] School of Science, Hunan University of Technology, Zhuzhou 412007, China
[2] School of Statistics and Mathematics, Guangdong University of Finance and Economics, Guangzhou 510320, China
* Correspondence: liupj@gdufe.edu.cn

Abstract: This research article investigates the application of Lévy noise to understand the dynamic aspects of measles epidemic modeling and seeks to explain the impact of vaccines on the spread of the disease. After model formulation, the study utilises uniqueness and existence techniques to derive a positive solution to the underlying stochastic model. The Lyapunov function is used to investigate the stability results associated with the proposed stochastic model. The model's dynamic characteristics are analyzed in the vicinity of the infection-free and endemic states of the associated ODEs model. The stochastic threshold $\mathbb{R}_s$ that ensures disease's extinction whenever $\mathbb{R}_s < 1$ is calculated. We utilized data from Pakistan in 2019 to estimate the parameters of the model and conducted simulations to forecast the future behavior of the disease. The results were compared to actual data using standard curve fitting tools.

Keywords: stochastic models; Lévy jump; persistence; parameter estimation; real data; measles in Pakistan

Citation: Song, Y.; Liu, P. Impact of Lévy Noise with Infinite Activity on the Dynamics of Measles Epidemics. *Fractal Fract.* 2023, 7, 434. https://doi.org/10.3390/fractalfract7060434

Academic Editor: Carlo Cattani

Received: 17 April 2023
Revised: 21 May 2023
Accepted: 26 May 2023
Published: 27 May 2023

1. Introduction

The measles infection, also called rubella, is extremely contagious and spreads throughout the globe. It is caused by the virus Morbilli, related to the community of Paramyxoviridae [1,2]. Vaccination is a powerful tool in reducing the impact of the disease over time. However, the force of mortality continues to affect young children under the age of five [3]. Measles infects ten million children each year, and millions of them die due to factors such as the unavailability of a balanced diet, weak digestion, and pneumonia [4]. Disease spread is caused by personal contact, sneezing, coughing, and contact with airborne and nasal droplets. The virus stays active for approximately two hours, and remains highly contagious during this period. The initial symptoms of the disease include nasal discharge, sore throat, and the presence of small white spots on the tongue. In later stages, coughing may occur. The typical incubation period for this infection is about four days before the appearance of the rash, and lasts for nearly five days after the rash appears. The mean incubation time is about fourteen or fifteen days, changing with external factors such as the environment and weather of the surroundings [5]. In real practice, vaccinated people may suffer side-effects, or it may be the case that vaccines are not available. On the other hand, vaccination acts as a preventive measure against measles, reducing the incidence rate by up to 73% over the past eighteen years. According to the World Health Organisation, Measles is present in many developed countries around the world, particularly in Asia and Africa. More than one hundred and fifty thousand people died due to measles in 2018. In the meantime, the death to infection ratio has decreased by about 85 percent [6,7]. According to reports from the WHO, nearly 110,000 people died in 2017 due to measles, with a majority of them being children under six years old. This highlights the need for effective and safe vaccination [8]. The discovery of vaccines for infectious diseases has lowered the ratio of death and infection to a large extent. This process saves about three hundred thousand individuals from death each year. Vaccines produce antibodies that act against reinfection and strengthen the immune system [9]. Specifically, measles can be controlled by the MMR vaccines. These vaccines are considered very safe for both children and the elderly, and can reduce the severity of measles infection to a great extent. According to studies, one dose of the MMR vaccine can prevent disease in around 91% of cases, while two doses can provide approximately 95% protection. Other diseases such as the mumps can be controlled by this vaccine as well [10]. Nigeria experiences periodic outbreaks of measles, as the disease is endemic in the country. The disease spreads in Nigeria during all seasons of the year, and is highly contiguous during dry weather.

In addition, Pakistan is among the countries in the WHO's Eastern Mediterranean Region most highly burdened by measles [11]. Over the past 9 to 12 years, there has been a sharp increase in measles outbreaks throughout Pakistan; some 2845 cases of measles were reported in Pakistan in 2016. This number increased to 6791 in 2017 and 33,007 in 2018, which represents approximately 43%, 21%, and 50% of all cases reported in the Eastern Mediterranean region comprising 23 countries [11]. Nearly 129 children died from measles in 2017, and in 2018 the number increased to about 300 [12].

Mathematical modeling is regarded as a significant and powerful approach for examining and forecasting the dynamic patterns of epidemics [7,12–14]. Most of the models used to date employ the integer order derivative, and their analysis is related to classical theory. The foundation of a mathematical modeling relies heavily on the availability of biological information and data regarding the epidemic under consideration. Mathematical models for the measles have wide application for understanding its spreading dynamics and controlling it. White noise plays a vital role in explaining the dynamics of physical and biological problems. The impact of external environmental factors such as white noise on the dynamics of measles epidemics is significant [15]. Due to the non-homogeneity and nonlinearity of population interactions and other complexities, epidemic prediction cannot be depicted using traditional modeling approaches. Moreover, the dynamics and control of various epidemics may be affected by global environmental factors.

The human population is subject to many complex and random variations in the real world. Therefore, stochastic models are a more appropriate technique for modelling epidemics. It has been demonstrated that stochastic models are more realistic compared to deterministic models. Several researchers have recently focused on perturbations that can capture the true dynamics of the epidemic based on stochastic modelling [13,15,16]. The Lévy white noise is important at various velocities for the spacing of threshold parameters. Adding Lévy noise terms can lead to more realistic and accurate results in stochastically analyzed models for various infectious diseases. The stochastic version of the underlying deterministic mode can be obtained by including this noise in the aforementioned systems. Typically, two types of noise are utilized in such models, namely, Lévy noise and Gaussian noise. The Lévy noise is more suitable than the Gaussian noise because the resulting system is able to model systems with a higher degree of complexity [16–19]. Due to fluctuations in diffusion problems, disturbances cannot be described by the continuous stochastic model; thus, it is important to model these phenomena using jump processes. For this reason, we consider the problem using Lévy noise.

The remainder of this manuscript is structured as follows. In Section 2, we propose a model based on random processes for the transmission dynamics of measles. In Section 3, we present the dynamic features of the globalized positive model's solution. In Sections 4 and 5, we obtain necessary conditions related to disease elimination and persistence of the proposed stochastic model. We optimize the proposed problem using the cases of measles infection in the population of Pakistan for January to October 2019 in Section 6. In Section 7, we verify the theory behind the obtained results qualitatively and quantitatively, and provide numerical simulations. Finally, we conclude our analysis in Section 8 with remarks and recommendations for future research work.

2. Model Formulation

Olumuyiwa et al. [20] recently developed a mathematical model for describing the infection dynamics of the epidemic measles by using a deterministic approach. The authors employed the epidemiological concept of population and divided the entire human population into six classes, including susceptible, vaccinated, exposed, infectious, hospitalised, and recovered individuals, respectively denoted by $\mathbb{S}(t)$, $\mathbb{V}(t)$, $\mathbb{E}(t)$, $\mathbb{I}(t)$, $\mathbb{H}(t)$, and $\mathbb{R}(t)$. The addition to the susceptible population per day is provided by the rate ϕ. The vulnerable class is subject to a vaccination rate τ and loses immunity against the disease at a rate ω, which is commonly known as the "waning rate of vaccination". The rate of infection for the susceptible class is denoted by α. Therefore, the term $\alpha\mathbb{S}\mathbb{I}$ represents the total infection rate per time unit. Additionally, the transition from the exposed class to the infected class is represented by β. Individuals in the infected population are hospitalized at a rate ρ and then recover from measles at a rate γ. The model assumes a constant natural death rate, denoted by μ, for all population classes. In addition, the infection-related death rate is provided by the parameter δ. Here, we do not assume any natural recovery from measles infection. The model chart is illustrated in Figure 1 [21]. The mathematical formulation of the above discussion can be converted into a model of ordinary differential equations (ODEs), represented in symbolic form as follows:

$$
\begin{aligned}
\frac{d\mathbb{S}(t)}{t} &= -\alpha\mathbb{I}(t)\mathbb{S}(t) + \phi - (\mu+\tau)\mathbb{S}(t) + \omega\mathbb{V}(t),\\
\frac{d\mathbb{V}(t)}{t} &= -(\omega+\mu)\mathbb{V}(t) + \tau\mathbb{S}(t),\\
\frac{d\mathbb{E}(t)}{t} &= -(\beta+\mu)\mathbb{E}(t) + \alpha\mathbb{I}(t)\mathbb{S}(t),\\
\frac{d\mathbb{I}(t)}{t} &= -(\mu+\rho+\delta)\mathbb{I}(t) + \beta\mathbb{E}(t),\\
\frac{d\mathbb{H}(t)}{t} &= -(\mu+\gamma+\delta)\mathbb{H}(t) + \rho\mathbb{I}(t),\\
\frac{d\mathbb{R}(t)}{t} &= -\mu\mathbb{R}(t) + \gamma\mathbb{H}(t).
\end{aligned}
\tag{1}
$$

The threshold parameter for the model (1) is obtained using standard techniques, and has the following form:

$$
\mathbb{R}_0^D = \frac{(\mu+\omega)\phi\beta\alpha}{(\mu+\beta)(\mu+\delta+\rho)(\mu+\omega+\tau)\mu}. \tag{2}
$$

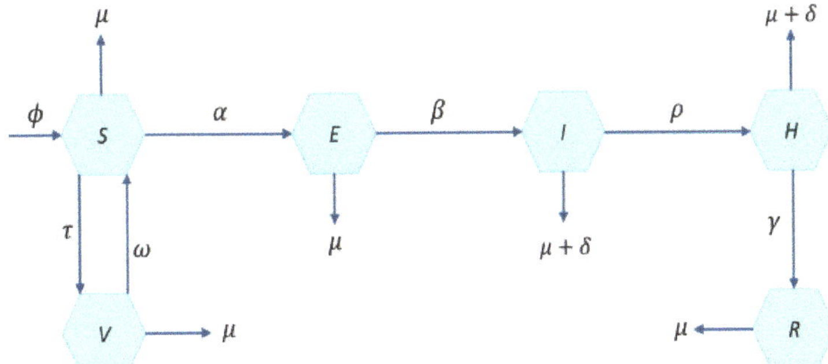

Figure 1. Moments of individuals among the classes in the measles model (1) [21].

The main theme of the present manuscript is to modify model (1) by including white and Lévy noise as well as the incidence rate of nonlinear shapes; the white noise is used for the continuity part and the Lévy noise for the jumping part. The extended form of the deterministic system can be written in stochastic form as follows:

$$\begin{aligned}
d\mathbb{S} &= \left[\phi - \frac{\alpha\mathbb{S}(t)\mathbb{I}(t)}{\mathbb{N}(t)} + \omega\mathbb{V}(t) - (\tau+\mu)\mathbb{S}(t)\right]dt + \xi_1\mathbb{S}(t)d\mathbb{W}_1(t) + \int_Y \mathbb{X}_1(y)\mathbb{S}(t^-)\tilde{\mathbb{N}}(dt,dy),\\
d\mathbb{V} &= \left[\tau\mathbb{S}(t) - (\mu+\omega)\mathbb{V}(t)\right]dt + \xi_2\mathbb{V}d\mathbb{W}_2(t) + \int_Y \mathbb{X}_2(y)\mathbb{V}(t^-)\tilde{\mathbb{N}}(dt,dy),\\
d\mathbb{E} &= \left[\frac{\alpha\mathbb{S}(t)\mathbb{I}(t)}{\mathbb{N}(t)} - (\mu+\beta)\mathbb{E}(t)\right]dt + \xi_3\mathbb{E}(t)d\mathbb{W}_3(t) + \int_Y \mathbb{X}_3(y)\mathbb{E}(t^-)\tilde{\mathbb{N}}(dt,dy),\\
d\mathbb{I} &= \left[\beta\mathbb{E}(t) - (\rho+\delta+\mu)\mathbb{I}(t)\right]dt + \xi_4\mathbb{I}(t)d\mathbb{W}_4(t) + \int_Y \mathbb{X}_4(y)\mathbb{I}(t^-)\tilde{\mathbb{N}}(dt,dy),\\
d\mathbb{H} &= \left[\rho\mathbb{I}(t) - (\gamma+\delta+\mu)\mathbb{H}(t)\right]dt + \xi_5\mathbb{I}(t)d\mathbb{W}_5(t) + \int_Y \mathbb{X}_5(y)\mathbb{H}(t^-)\tilde{\mathbb{N}}(dt,dy),\\
d\mathbb{R} &= \left[\gamma\mathbb{H}(t) - \mu\mathbb{R}(t)\right]dt + \xi_6\mathbb{R}(t)d\mathbb{W}_6(t) + \int_Y \mathbb{X}_6(y)\mathbb{R}(t^-)\tilde{\mathbb{N}}(dt,dy),
\end{aligned} \tag{3}$$

where $\mathbb{W}_i(t)$ for $i = 1,\cdots,6$ is standard Brownian motion defined in a complete probability space $(\Omega,\mathbb{F},\mathbb{P})$ with filtration $\{\mathbb{F}_t\}_{t\geqslant 0}$, satisfying the usual condition $\xi_1,\xi_2,\xi_3,\xi_4,\xi_5$, with ξ_6 representing the intensity of noise, $\mathbb{S}(t^-),\mathbb{V}(t^-),\mathbb{E}(t^-),\mathbb{I}(t^-),\mathbb{H}(t^-)$ and $\mathbb{R}(t^-)$ being the left limit of $\mathbb{S},\mathbb{V},\mathbb{E},\mathbb{I},\mathbb{H}$ and $\mathbb{R}$, respectively, $\tilde{\mathbb{N}} = \mathbb{N}(dt,dy) - v(dy)dt$ and $\mathbb{N}(dy,dt)$ a Poisson counting measure with characteristic measure ν on the measurable subset Y of $[0,\infty)$, and with $\nu(Y) < \infty$ and $\mathbb{X}_i : Z\times\Omega \longrightarrow \mathbb{R}_+$, $(i = 1,2,3,4,5,6)$ representing the effect of random jumps, which are assumed to be bounded and continuous with respect to ν and to be $\mathbb{B}(Y)\times\mathbb{F}_t$-measurable.
With regard to the model (3), in the present work we are particularly interested in answering the following questions:

- Q_1: Does the Lévy noise influence the dynamic properties of measles outbreaks?
- Q_2: How can contaminated vaccinations contribute to the spread of measles, and what measures are in place to prevent such incidents?
- Q_3: What criterion is used to determine the extinction of a disease?
- Q_4: What are the criteria that indicate the persistence of the system?

3. The Existence of a Positive Solution and Its Uniqueness

In this section, we intend to apply the techniques from [16] to establish a proof of the existence of a global and positive solution to the proposed stochastic model. To prove the existence and uniqueness of a solution of model (3), it is important to consider the following two conditions:

(C$_1$). For every $\mathbb{Q} > 0\ \exists\ \mathbb{L}_{\mathbb{Q}} > 0$;

$$\int_Y |A_i(x_1,y) - A_i(x_2,y)|^2 v(dy) \le \mathbb{L}_{\mathbb{Q}}|x_1 - x_2|^2, i = 1,\cdots,6, \tag{4}$$

for $|y_1| \vee |y_2| \le \mathbb{M}$; here,

$$\begin{aligned}
A_1(x,y) &= \mathbb{X}_1(y)x \text{ at } x = \mathbb{S}(t^-),\\
A_2(x,y) &= \mathbb{X}_2(y)x \text{ at } x = \mathbb{V}(t^-),\\
A_3(x,y) &= \mathbb{X}_3(y)x \text{ at } x = \mathbb{E}(t^-),\\
A_4(x,y) &= \mathbb{X}_4(y)x \text{ at } x = \mathbb{I}(t^-),\\
A_5(x,y) &= \mathbb{X}_5(y)x \text{ at } x = \mathbb{H}(t^-),\\
A_6(x,y) &= \mathbb{X}_6(y)x \text{ at } x = \mathbb{R}(t^-).
\end{aligned}$$

(C$_2$). $|\log(\mathbb{X}_i(x))| \le C$ for $\mathbb{X}_i(x) > -1$, $\quad i = 1,\cdots,6$, C is a positive constant.

Theorem 1. *For a given initial value* $(\mathbb{S},\mathbb{V},\mathbb{E},\mathbb{I},\mathbb{H},\mathbb{R})(0)\in\mathbb{R}^6_+$*, system (3) has one global root* $(\mathbb{S},\mathbb{V},\mathbb{E},\mathbb{I},\mathbb{H},\mathbb{R})(t)\in\mathbb{R}^6_+$ *for all* $t\geq 0$ *a.s.*

Proof. Using $(\mathbb{C}_1)$, the results due to drifted and the diffused terms are local Lipschitzian, for every initial conditions $(\mathbb{S},\mathbb{V},\mathbb{E},\mathbb{I},\mathbb{H},\mathbb{R})(0)\in\mathbb{R}^6_+$, $\exists$ a unique localized root $(\mathbb{S}(t),\mathbb{V}(t),\mathbb{E}(t),\mathbb{I}(t),\mathbb{H}(t),\mathbb{R}(t))$ on $t\in[0,\tau_e)$, τ_e is the explosion time. To show that the solution is global, we have to prove that $\tau_e=\infty$ a.s. In the first attempt, we prove the condition that $(\mathbb{S}(t),\mathbb{V}(t),\mathbb{E}(t),\mathbb{I}(t),\mathbb{H}(t),\mathbb{R}(t))$ are not approaching infinity in a finite duration of time. Let $k_0>0$ is too much large such that $(\mathbb{S}(0),\mathbb{V}(0),\mathbb{E}(0),\mathbb{I}(0),\mathbb{H}(0),\mathbb{R}(0))$ lies in $[\frac{1}{k_0},k_0]$. For every integer value $k\leq k_0$, consider the stopping time in the form of

$$\tau_k=\inf\left\{t\in[0,\tau_e)/(\mathbb{S}(t),\mathbb{V}(t),\mathbb{E}(t),\mathbb{I}(t),\mathbb{H}(t),\mathbb{R}(t))\notin\left(\frac{1}{k},k\right)\right\}. \quad (5)$$

Let $\inf\emptyset=\infty$; from this, we can see that $\tau^+\leq\tau_e$, which implies that $\tau^+=\infty$ a.s showing $\tau_e=+\infty$ a.s. Considering that τ^+ is less than ∞, there must exist a number $T>0$ such that $0<\mathbb{P}(\tau^+<T)$.

Next, let us consider the operator $\mathbb{F}:\mathbb{R}^6_+\rightarrow\mathbb{R}_+$ from the C^2 set in the form of

$$\mathbb{F}(\mathbb{S},\mathbb{V},\mathbb{E},\mathbb{I},\mathbb{H},\mathbb{R})=\mathbb{H}+\mathbb{V}+\mathbb{S}+\mathbb{I}+\mathbb{E}+\mathbb{R}-6-(\log\mathbb{S}+\log\mathbb{V}+\log\mathbb{E}+\log\mathbb{I}+\log\mathbb{H}+\log\mathbb{R}). \quad (6)$$

Applying *Itô* formula to $\mathbb{F}$ for all $t\in[0,\tau^+]$, we have

$$\begin{aligned}d\mathbb{F}&=L\mathbb{F}(\mathbb{S},\mathbb{V},\mathbb{E},\mathbb{I},\mathbb{H},\mathbb{R})dt+\xi_1(\mathbb{S}-1)d\mathbb{W}_1(t)+xi_2(\mathbb{V}-1)d\mathbb{W}_2(t)\\&+\xi_3(\mathbb{E}-1)d\mathbb{W}_3(t)+\xi_4(\mathbb{I}-1)d\mathbb{W}_4(t)+\xi_5(\mathbb{H}-1)d\mathbb{W}_5(t)+\xi_6(\mathbb{R}-1)d\mathbb{W}_6(t)\\&+\int_{\mathrm{Y}}[\mathbb{X}_1(y)\mathbb{S}-log(\mathbb{X}_1(y)+1)]\tilde{\mathbb{N}}(d\chi)+\int_{\mathrm{Y}}[\mathbb{X}_2(y)\mathbb{V}-log(1+\mathbb{X}_2(y))]\tilde{\mathbb{N}}(d\chi)\\&+\int_{\mathrm{Y}}[\mathbb{X}_3(y)\mathbb{E}-\log(1+\mathbb{X}_3(y))]\tilde{\mathbb{N}}(d\chi)+\int_{\mathrm{Y}}[\mathbb{X}_4(y)\mathbb{I}-\log(1+\mathbb{X}_4(y))]\tilde{\mathbb{N}}(d\chi),\\&+\int_{\mathrm{Y}}[\mathbb{X}_5(y)\mathbb{H}-\log(1+\mathbb{X}_5(y))]\tilde{\mathbb{N}}(d\chi)+\int_{\mathrm{Y}}[\mathbb{X}_6(y)\mathbb{R}-\log(1+\mathbb{X}_6(y))]\tilde{\mathbb{N}}(d\chi).\end{aligned} \quad (7)$$

In Equation (7), $L\mathbb{F}:\mathbb{R}^6_+\rightarrow\mathbb{R}_+$ is given by using the assumption $\mathbb{C}_2$; thus,

$$\begin{aligned}L\mathbb{F}&\leq\phi+6\mu+\tau+\omega+\beta+\rho+\gamma+2\delta+\frac{\xi_1^2+\xi_2^2+\xi_3^2+\xi_4^2+\xi_5^2+\xi_6^2}{2}\\&+\int_{\mathrm{Y}}[\mathbb{X}_1(y)-\log(1+\mathbb{X}_1(y))]\nu(dy)+\int_{\mathrm{Y}}[\mathbb{X}_2(y)-\log(1+\mathbb{X}_2(y))]\nu(dy)\\&+\int_{\mathrm{Y}}[\mathbb{X}_3(y)-\log(1+\mathbb{X}_3(y))]\nu(dy)+\int_{\mathrm{Y}}[\mathbb{X}_4(y)-\log(1+\mathbb{X}_4(y))]\nu(dy)\\&+\int_{\mathrm{Y}}[\mathbb{X}_5(y)-\log(1+\mathbb{X}_5(y))]\nu(dy)+\int_{\mathrm{Y}}[\mathbb{X}_6(y)-\log(1+\mathbb{X}_6(y))]\nu(dy).\end{aligned} \quad (8)$$

We can refer to Theorem 2.1 of Fatin et al. [16] for the remaining proof, and therefore we skip it here. □

4. Extinction for System (3)

Minimizing the infection effect on any community depends mostly on time and certain useful conditions taken from analysis of the dynamics of the disease in question. In this section, the investigation of the main condition for the vanishing of the infection is discussed through stochastic modeling. To proceed further, we first define the threshold parameter for the deterministic model (3) as follows:

$$\mathbb{R}_0^D=\frac{(\mu+\omega)\phi\beta\alpha}{(\mu+\beta)(\mu+\delta+\rho)(\mu+\omega+\tau)\mu}. \quad (9)$$

Based on the standard techniques for stochastic systems, we are now ready to calculate the threshold number for the stochastic model, which is provided by

$$\mathbb{R}_s = \frac{\alpha}{\left[(\mu+\beta)(\mu+\delta+\rho) + \frac{\xi_3^2}{2} + \frac{\xi_4^2}{2} + \int_y \{(\mathbb{X}_3(y) + \mathbb{X}_4(y)) - \ln(1 + (\mathbb{X}_3(t) + \mathbb{X}_4(y)))\}\nu(dy)\right]}. \tag{10}$$

Subsequently, we introduce the following concept, which is beneficial for the forthcoming discourse.

$$\langle \mathbb{X}(t) \rangle = \frac{1}{t}\int_0^t \mathbb{X}(w)dw.$$

Lemma 1 ((Strong Law) [21,22]). *If an operator fulfills the condition of continuity of process $\mathbb{Z} = \{\mathbb{Z}\}_{0 \leq t}$ within a local Martingale as $t \to 0$, it vanishes; then,*

$$\begin{aligned} &\lim_{t\to\infty} \langle \mathbb{Z}, \mathbb{Z} \rangle_t = \infty, \ a.s., \Rightarrow \lim_{t\to\infty} \frac{\mathbb{Z}_t}{\langle \mathbb{Z}, \mathbb{Z} \rangle_t} = 0, \ a.s. \\ &\limsup_{t\to\infty} \frac{\langle \mathbb{Z}, \mathbb{Z} \rangle_t}{t} < 0, \ a.s., \Rightarrow \lim_{t\to\infty} \frac{\mathbb{Z}_t}{t} = 0, \ a.s. \end{aligned} \tag{11}$$

Theorem 2. *Consider a solution $(\mathbb{S}, \mathbb{V}, \mathbb{E}, \mathbb{I}, \mathbb{H}, \mathbb{R})(t)$ of model (3) with initial condition $(\mathbb{S}, \mathbb{V}, \mathbb{E}, \mathbb{I}, \mathbb{H}, \mathbb{R})(0) \in \mathbb{R}^6$. Next, for $q > \frac{(\xi_1^2 \vee \xi_2^2 \vee \xi_3^2 \vee \xi_4^2 \vee \xi_5^2 \vee \xi_6^2)}{2}$ and $\mathbb{R}_s < 1$, we have*

$$\lim_{t\to\infty} \frac{\log \langle \mathbb{E}(t) \rangle}{t} < 0, \quad and \quad \lim_{t\to\infty} \frac{\log \langle \mathbb{I}(t) \rangle}{t} < 0, a.s.$$

The above inequality implies that the compartments $\mathbb{E}(t)$ and $\mathbb{I}(t)$ approaching 0 a.s., showing the elimination of the disease with unit probability.

Furthermore,

$$\begin{aligned} &\lim_{t\to\infty} \langle \mathbb{S}(t) \rangle = \frac{(\mu+\omega)\phi}{(\mu+\omega+\tau)\mu}, \\ &\lim_{t\to\infty} \langle \mathbb{V}(t) \rangle = \frac{\tau\phi}{\mu(\tau+\omega+\mu)}, \\ &\lim_{t\to\infty} \langle \mathbb{E}(t) \rangle = 0, \\ &\lim_{t\to\infty} \langle \mathbb{I}(t) \rangle = 0, \\ &\lim_{t\to\infty} \langle \mathbb{H}(t) \rangle = 0, \\ &\lim_{t\to\infty} \langle \mathbb{R}(t) \rangle = 0. \end{aligned} \tag{12}$$

Proof. Let us consider a solution $(\mathbb{S}, \mathbb{V}, \mathbb{E}, \mathbb{I}, \mathbb{H}, \mathbb{R})(t)$ of the proposed model (3) associated with the initial condition $(\mathbb{S}, \mathbb{V}, \mathbb{E}, \mathbb{I}, \mathbb{H}, \mathbb{R})(0)$ in the positive cone of $\mathbb{R}_+^6$. Further, let us define

$$G_1(t) = (\beta+\mu)\mathbb{I}(t) + \beta\mathbb{E}(t). \tag{13}$$

If we differentiate relation (13) and then follow the Ito' formula, we have

$$
\begin{aligned}
d(lnG_1(t)) &= \frac{1}{G_1}\left[\frac{\alpha\beta\mathbb{S}\mathbb{I}}{\mathbb{N}} - (\beta+\mu)(\rho+\delta+\mu)\mathbb{I}\right] - \frac{\beta^2\mathbb{E}^2\xi_3^2 + (\mu+\beta)^2\xi_4^2\mathbb{I}^2}{2(G_1)^2} \\
&+ \frac{\beta\xi_3}{[\beta\mathbb{E}(t)+(\mu+\beta)\mathbb{I}]}\mathbb{E}d\mathbb{W}_3(t) + \frac{(\mu+\beta)\xi_4}{[\mathbb{E}+(\mu+\beta)\mathbb{I}]}\mathbb{I}d\mathbb{W}_4(t) \\
&+ \int_y\left\{\ln\left(1+\frac{\beta\mathbb{X}_3(t)\mathbb{E}+(\mu+\beta)\mathbb{X}_4(y)\mathbb{I}}{G_1}\right) - \frac{\beta\mathbb{X}_3(y)\mathbb{E}+(\mu+\beta)\mathbb{X}_4(y)\mathbb{I}}{G_1}\right\}\nu(dy) \\
&+ \int_y \ln\left(1+\frac{\beta\mathbb{X}_3(y)\mathbb{E}(t^-)+(\mu+\beta)\mathbb{X}_4\mathbb{I}(t^-)}{G_1(t^-)}\right)\tilde{\mathbb{N}}(d\chi) \\
&\le \frac{1}{G_1}\left[\alpha\beta\mathbb{I} - (\beta+\mu)(\rho+\delta+\mu)\mathbb{I}\right] - \frac{\beta^2\mathbb{E}^2\xi_3^2}{2(G_1)^2} - \frac{(\mu+\beta)^2\xi_4^2\mathbb{I}^2}{2(G_1)^2} \\
&- \int_y\left\{\frac{\beta\mathbb{X}_3(y)\mathbb{E}+(\mu+\beta)\mathbb{X}_4(y)\mathbb{I}}{G_1} - \ln\left(1+\frac{\beta\mathbb{X}_3(t)\mathbb{E}+(\mu+\beta)\mathbb{X}_4(y)\mathbb{I}}{G_1}\right)\right\}\nu(dy) \\
&+ \frac{\beta\xi_3}{[\beta\mathbb{E}(t)+(\mu+\beta)\mathbb{I}]}\mathbb{E}d\mathbb{W}_3(t) + \frac{(\mu+\beta)\xi_4}{[\mathbb{E}+(\mu+\beta)\mathbb{I}]}\mathbb{I}d\mathbb{W}_4(t) \\
&+ \int_y \ln\left(1+\frac{\beta\mathbb{X}_3(y)\mathbb{E}(t^-)+(\mu+\beta)\mathbb{X}_4\mathbb{I}(t^-)}{G_1(t^-)}\right)\tilde{\mathbb{N}}(d\chi), \ [\because \mathbb{S} \le \mathbb{N}] \\
&\le \frac{1}{(\beta+\mu)}\left[-(\mu+\rho+\delta)(\beta+\mu)+\alpha\right] - \frac{\xi_3^2}{2} - \frac{\xi_4^2}{2} \\
&- \int_y\{(\mathbb{X}_3(y)+\mathbb{X}_4(y)) - \ln(1+(\mathbb{X}_3(t)+\mathbb{X}_4(y)))\}\nu(dy) \\
&+ \frac{\beta\xi_3}{[\beta\mathbb{E}(t)+(\beta+\mu)\mathbb{I}]}\mathbb{E}d\mathbb{W}_3(t) + \frac{(\mu+\beta)\xi_4}{[\mathbb{E}+(\mu+\beta)\mathbb{I}]}\mathbb{I}d\mathbb{W}_4(t) \\
&+ \int_y \ln\left(1+\frac{\beta\mathbb{X}_3(y)\mathbb{E}(t^-)+(\mu+\beta)\mathbb{X}_4\mathbb{I}(t^-)}{G_1(t^-)}\right)\tilde{\mathbb{N}}(d\chi). \ [\because \mathbb{I} \le \mathbb{I} + \frac{\beta\mathbb{E}}{(\mu+\beta)}]
\end{aligned} \tag{14}
$$

We obtain the following result if we integrate both sides of the previous inequality over the interval $[0, t]$:

$$
\begin{aligned}
lnG_1(t) \leq & \frac{1}{(\beta+\mu)}\Bigg\{\alpha - \Bigg[(\mu+\beta)(\rho+\delta+\mu) + \frac{\xi_3^2}{2} + \frac{\xi_4^2}{2} \\
& + \int_y \{(\mathbb{X}_3(y)+\mathbb{X}_4(y)) - \ln(1+(\mathbb{X}_3(t)+\mathbb{X}_4(y)))\}\nu(dy)\Bigg]\Bigg\} \\
& + \int_0^t \frac{\mathbb{E}\xi_3\beta d\mathbb{W}_3(t)}{[(\beta+\mu+)\mathbb{I}+\beta\mathbb{E}(t)]} + \int_0^t \frac{(\beta+\mu)\mathbb{I}\xi_4 d\mathbb{W}_4(t)}{[(\beta+\mu)\mathbb{I}+\mathbb{E}]} \\
& + \int_0^t \int_y \ln\left(1+\frac{\beta\mathbb{X}_3(y)\mathbb{E}(t^-)+(\mu+\beta)\mathbb{X}_4\mathbb{I}(t^-)}{G_1(t^-)}\right)\tilde{\mathbb{N}}(d\chi), \\
\leq & \frac{1}{(\beta+\mu)}\Bigg\{\alpha - \Bigg[(\mu+\rho+\delta)(+\beta+\mu) + \frac{\xi_3^2}{2} + \frac{\xi_4^2}{2} \\
& + \int_y \{(\mathbb{X}_3(y)+\mathbb{X}_4(y)) - \ln(1+(\mathbb{X}_3(t)+\mathbb{X}_4(y)))\}\nu(dy)\Bigg]\Bigg\} \\
& + \int_0^t \frac{\mathbb{E}\xi_3\beta d\mathbb{W}_3(t)}{[(\beta+\mu+)\mathbb{I}+\beta\mathbb{E}(t)]} + \int_0^t \frac{(\beta+\mu)\mathbb{I}\xi_4 d\mathbb{W}_4(t)}{[(\beta+\mu)\mathbb{I}+\mathbb{E}]} \\
& + \int_0^t \int_y \ln\left(1+\frac{\beta\mathbb{X}_3(y)\mathbb{E}(t^-)+(\mu+\beta)\mathbb{X}_4\mathbb{I}(t^-)}{G_1(t^-)}\right)\tilde{\mathbb{N}}(d\chi), \\
\leq & \frac{\left[(\mu+\beta)(\mu+\rho+\delta) + \frac{\xi_3^2}{2} + \frac{\xi_4^2}{2} + \int_y \{(\mathbb{X}_3(y)+\mathbb{X}_4(y)) - \ln(1+(\mathbb{X}_3(t)+\mathbb{X}_4(y)))\}\nu(dy)\right]}{(\mu+\beta)}\Big[\mathbb{R}_s - 1\Big] \\
& + \int_0^t \frac{\mathbb{E}\xi_3\beta d\mathbb{W}_3(t)}{[(\beta+\mu)\mathbb{I}+\beta\mathbb{E}(t)]} + \int_0^t \frac{(\beta+\mu)\mathbb{I}\xi_4 d\mathbb{W}_4(t)}{[(\beta+\mu)\mathbb{I}+\mathbb{E}]} \\
& + \int_0^t \int_y \ln\left(1+\frac{\beta\mathbb{X}_3(y)\mathbb{E}(t^-)+(\mu+\beta)\mathbb{X}_4\mathbb{I}(t^-)}{G_1(t^-)}\right)\tilde{\mathbb{N}}(d\chi).
\end{aligned}
\tag{15}
$$

By taking the superior limit as $t \to \infty$ after dividing Equation (15) by t and using Lemma 1, we obtain

$$
\begin{aligned}
& \limsup_{t\to\infty}(lnG_1(t)) \leq \\
& \frac{\left[(\mu+\rho+\delta)(\beta+\mu) + \frac{\xi_3^2}{2} + \frac{\xi_4^2}{2} + \int_y \{(\mathbb{X}_3(y)+\mathbb{X}_4(y)) - \ln(1+(\mathbb{X}_3(t)+\mathbb{X}_4(y)))\}\nu(dy)\right]}{(\mu+\beta)}\Big[\mathbb{R}_s - 1\Big].
\end{aligned}
\tag{16}
$$

If $1 > \mathbb{R}_s$, then $\lim_{t\to\infty} G_1 = 0$, a.s whenever $\mathbb{R}_s < 1$. As $\mu+\beta$ and β are both positive, per relation (13) we have $\lim_{t\to\infty}[(\beta+\mu)\mathbb{I}(t)+\beta\mathbb{E}(t)] = 0 \Longrightarrow \lim_{t\to\infty}\mathbb{E} = \lim_{t\to\infty}\mathbb{I} = 0$; thus, we reach the conclusion. □

5. Persistence in Mean

The purpose of this section is to perform an analysis of the persistence of the disease and examine the long-term behaviour of the infection. First, we present the mean persistence, as can be seen in [16].

Definition 1 ([19]). *Under the following assumption, model (3) shows the persistence of the infection*

$$\liminf_{t\to\infty}\frac{1}{t}\int_0^t \mathbb{F}(r)dr > 0 \ a.s. \tag{17}$$

For more details on disease persistence, interested readers are referred to the results provided in [16,17].

Lemma 2. *Let $g \in \mathbb{C}(\mathbb{R}_{\geq 0} \times \Omega, \mathbb{R}_{>0})$ and $\mathbb{G} \in \mathbb{C}(\mathbb{R}_{\geq 0} \times \Omega, \mathbb{R})$ such that $\lim_{t\to\infty} \frac{\mathbb{G}(t)}{t} = 0$ a.s. If we assume the following relation to be true for all positive values of t*

$$\log g(t) \geq \lambda_0 t + \mathbb{G}(t) - \lambda \int_0^t g(s)ds, \text{ a.s.}$$

then

$$\liminf_{t\to\infty} \langle g(t) \rangle \geq \frac{\lambda_0}{\lambda} \text{ a.s.},$$

where λ_0 is non-negative and λ is a positive real number.

Next, we provide several mathematical assumptions for the mean persistency of system (3), where the conclusion of the said part is provided by the following theorem.

Theorem 3. *If $\mathbb{R}_0^s > 1$, then for initial approximations $(\mathbb{S}, \mathbb{V}, \mathbb{E}, \mathbb{I}, \mathbb{H}, \mathbb{R})(0) \in \mathbb{R}_+^6$ the disease class $\mathbb{I}(t)$ has the following property:*

$$\liminf_{t\to\infty} \langle \mathbb{I}(t) \rangle \geq \frac{3\phi\left(\sqrt{\mathbb{R}_0^s} - 1\right)}{\mathbb{C}_1 \alpha}, \text{ a.s.}, \tag{18}$$

where $\mathbb{C}_1 = \frac{\phi}{(\tau+\mu+\frac{\xi_1^2}{2}+\int_Y \mathbb{X}_1(y)+\log(1+\mathbb{X}_1(y))\nu(dy))}$ implies that infection is present in the community.
Let us now reproduce the threshold for the stochastic system as

$$\mathbb{R}_0^s = \frac{\alpha\beta}{abc}, \tag{19}$$

where

$$\begin{aligned} a &= \left(\tau + \mu + \frac{\xi_1^2}{2} + \int_Y \mathbb{X}_1(y) + \log(1 + \mathbb{X}_1(y))\nu(dy))\right), \\ b &= \left(\beta + \mu + \frac{\xi_3^2}{2} + \int_Y \mathbb{X}_3(y) + \log(1 + \mathbb{X}_3(y))\nu(dy)\right), \\ c &= \left(\mu + \delta + \rho + \frac{\xi_4^2}{2} + \int_Y \mathbb{X}_4(y) + \log(1 + \mathbb{X}_4(y))\nu(dy)\right). \end{aligned} \tag{20}$$

Proof. Let

$$G_1 = -\mathbb{C}_1 ln\mathbb{S} - \mathbb{C}_2 ln\mathbb{E} - \mathbb{C}_3 ln\mathbb{I}, \tag{21}$$

here, $\mathbb{C}_1, \mathbb{C}_1$ and $\mathbb{C}_3$ are constants, and will be calculated later. Using the *Itô* formula with Equation (21), we obtain

$$\begin{aligned} dG_1 = {} & LG_1 - \mathbb{C}_1\xi_1 d\mathbb{W}_1(t) - \mathbb{C}_2\xi_3 d\mathbb{W}_3(t) - \mathbb{C}_3\xi_4 d\mathbb{W}_4(t) \\ & - \int_Y [\log(1 + \mathbb{X}_1(y))]\tilde{\mathbb{N}}(d\chi) - \int_Y [\log(1 + \mathbb{X}_3(y))]\tilde{\mathbb{N}}(d\chi) - \int_Y [\log(1 + \mathbb{X}_4(y))]\tilde{\mathbb{N}}(d\chi), \end{aligned} \tag{22}$$

where

$$
\begin{aligned}
LG_1 =& -\frac{\mathbb{C}_1\phi}{\mathbb{S}}+\frac{\mathbb{C}_1\alpha\mathbb{I}}{\mathbb{N}}-\frac{\mathbb{C}_1\omega\mathbb{V}}{\mathbb{S}}+\mathbb{C}_1(\tau+\mu)-\mathbb{C}_2\frac{\alpha\mathbb{S}\mathbb{I}}{\mathbb{E}}+\mathbb{C}_2(\mu+\beta)-\mathbb{C}_3\frac{\beta\mathbb{E}}{\mathbb{I}}+\mathbb{C}_3(\mu+\delta+\rho)\\
&+\frac{\mathbb{C}_1\varsigma_1^2}{2}+\frac{\mathbb{C}_2\varsigma_3^2}{2}+\frac{\mathbb{C}_3\varsigma_4^2}{2}+\int_Y \mathbb{C}_1\mathbb{X}_1(y)+\mathbb{C}_1\log(1+\mathbb{X}_1(y))\nu(dy)+\int_Y \mathbb{C}_2\mathbb{X}_3(y)\\
&+\mathbb{C}_2\log(1+\mathbb{X}_3(y))\nu(dy)+\int_Y \mathbb{C}_3\mathbb{X}_4(y)+\mathbb{C}_2\log(1+\mathbb{X}_4(y))\nu(dy),\\
\leq& -\frac{\mathbb{C}_1\phi}{\mathbb{S}}-\mathbb{C}_2\frac{\alpha\mathbb{S}\mathbb{I}}{\mathbb{E}}-\mathbb{C}_3\frac{\beta\mathbb{E}}{\mathbb{I}}+\mathbb{C}_1(\tau+\mu+\frac{\varsigma_1^2}{2}+\int_Y \mathbb{X}_1(y)+\log(1+\mathbb{X}_1(y))\nu(dy))\\
&+\mathbb{C}_2(\mu+\beta+\frac{\varsigma_3^2}{2}+\int_Y \mathbb{X}_3(y)+\log(1+\mathbb{X}_3(y))\nu(dy))\\
&+\mathbb{C}_3(\mu+\delta+\rho+\frac{\varsigma_4^2}{2}+\int_Y \mathbb{X}_4(y)+\log(1+\mathbb{X}_4(y))\nu(dy))+\frac{\mathbb{C}_1\alpha\mathbb{I}}{\mathbb{N}}.
\end{aligned}
\tag{23}
$$

Let

$$
\begin{aligned}
\mathbb{C}_1(\tau+\mu+\frac{\varsigma_1^2}{2}+\int_Y \mathbb{X}_1(y)+\log(1+\mathbb{X}_1(y))\nu(dy)) =&\phi,\\
\mathbb{C}_2(\mu+\beta+\frac{\varsigma_3^2}{2}+\int_Y \mathbb{X}_3(y)+\log(1+\mathbb{X}_3(y))\nu(dy)) =&\phi,\\
\mathbb{C}_2(\mu+\delta+\rho+\frac{\varsigma_4^2}{2}+\int_Y \mathbb{X}_4(y)+\log(1+\mathbb{X}_4(y))\nu(dy)) =&\phi,
\end{aligned}
\tag{24}
$$

namely,

$$
\begin{aligned}
a =& \left(\tau+\mu+\frac{\varsigma_1^2}{2}+\int_Y \mathbb{X}_1(y)+\log(1+\mathbb{X}_1(y))\nu(dy))\right),\\
b =& \left(\mu+\beta+\frac{\varsigma_3^2}{2}+\int_Y \mathbb{X}_3(y)+\log(1+\mathbb{X}_3(y))\nu(dy)\right),\\
c =& \left(\mu+\delta+\rho+\frac{\varsigma_4^2}{2}+\int_Y \mathbb{X}_4(y)+\log(1+\mathbb{X}_4(y))\nu(dy)\right).
\end{aligned}
$$

Then, we can write inequality (23) in the form of

$$
\begin{aligned}
LG_1 \leq& -3\sqrt{\frac{\mathbb{C}_1\phi}{\mathbb{S}}\times\frac{\mathbb{C}_3\alpha\mathbb{S}\mathbb{I}}{\mathbb{E}}\times\frac{\mathbb{C}_3\beta\mathbb{E}}{\mathbb{I}}}\\
&+\mathbb{C}_1(\tau+\mu+\frac{\varsigma_1^2}{2}+\int_Y \mathbb{X}_1(y)+\log(1+\mathbb{X}_1(y))\nu(dy))\\
&+\mathbb{C}_2(\mu+\beta+\frac{\varsigma_3^2}{2}+\int_Y \mathbb{X}_3(y)+\log(1+\mathbb{X}_3(y))\nu(dy))\\
&+\mathbb{C}_3(\mu+\delta+\rho+\frac{\varsigma_4^2}{2}+\int_Y \mathbb{X}_4(y)+\log(1+\mathbb{X}_4(y))\nu(dy))+\mathbb{C}_1\alpha\mathbb{I},\\
=& -3\sqrt{\frac{\phi^3\alpha\beta}{abc}}+3\phi+\mathbb{C}_1\alpha\mathbb{I},\\
=& -3\phi\left(\sqrt{\mathbb{R}_0^s}-1\right)+\mathbb{C}_1\alpha\mathbb{I}.
\end{aligned}
\tag{25}
$$

By putting Equation (25) into Equation (21) and taking the integral of both sides of the stochastic model (3), we have

$$\frac{G_1(\mathbb{S},\mathbb{E},\mathbb{I})(t) - G_1(\mathbb{S},\mathbb{E},\mathbb{I})(0)}{t} \leq -3\phi\left(\sqrt{\mathbb{R}_0^s} - 1\right) + \mathbb{C}_1\alpha\mathbb{I} - \mathbb{C}_1\xi_1 d\mathbb{W}_1(t) - \mathbb{C}_2\xi_3 d\mathbb{W}_3 - \mathbb{C}_3\xi_4 d\mathbb{W}_4 - \int_Y [\log(1+\mathbb{X}_1(y))]\tilde{\mathbb{N}}(d\chi) - \int_Y [\log(1+\mathbb{X}_3(y))]\tilde{\mathbb{N}}(d\chi) - \int_Y [\log(1+\mathbb{X}_4(y))]\tilde{\mathbb{N}}(d\chi). \quad (26)$$

If we define the notion $\Psi(t)$ in the form of

$$\Psi(t) = -\mathbb{C}_1\xi_1 d\mathbb{W}_1(t) - \mathbb{C}_2\xi_3 d\mathbb{W}_3 - \mathbb{C}_3\xi_4 d\mathbb{W}_4 - \int_Y [\log(1+\mathbb{X}_1(y))]\tilde{\mathbb{N}}(d\chi) - \int_Y [\log(1+\mathbb{X}_3(y))]\tilde{\mathbb{N}}(d\chi) - \int_Y [\log(1+\mathbb{X}_4(y))]\tilde{\mathbb{N}}(d\chi). \quad (27)$$

Then, from Strong's law as provided in Lemma 1, we have

$$\lim_{t\to\infty} \Psi(t) = 0. \quad (28)$$

Further, from Equation (27), we have

$$\begin{aligned} C_1\alpha\langle\mathbb{I}(t)\rangle &\geq 3\phi\left(\sqrt{\mathbb{R}_0^s} - 1\right) - \Psi(t) + \frac{G_1(\mathbb{S}(t),\mathbb{E}(t),\mathbb{I}(t)) - G_1(\mathbb{S}(0),\mathbb{E}(0),\mathbb{I}(0))}{t}, \\ \langle\mathbb{I}(t)\rangle &\geq \frac{3\phi\left(\sqrt{\mathbb{R}_0^s} - 1\right)}{C_1\alpha} - \frac{\Psi(t)}{C_1\alpha} - \frac{1}{C_1\alpha}\left(\frac{G_1(\mathbb{S}(t),\mathbb{E}(0),\mathbb{I}(t)) - G_1(\mathbb{S}(0),\mathbb{E}(0),\mathbb{I}(0))}{t}\right). \end{aligned} \quad (29)$$

From Lemma 1 and Equation (28), the superior limit of Equation (5) takes the form

$$\liminf_{t\to\infty} \langle\mathbb{I}(t)\rangle \geq \frac{3\phi\left(\sqrt{\mathbb{R}_0^s} - 1\right)}{C_1\alpha} \geq 0, \text{ a.s.}, \quad (30)$$

showing that $\liminf_{t\to\infty}\langle\mathbb{I}(t)\rangle \geq 0$. □

Thus, the proof of Theorem 3 is concluded.

6. Estimation

Utilizing practical observations to obtain insights into certain missing epidemiological factors is a widely used technique in biological systems analysis. The verification of analytical results pertaining to the measles model (1) and determination of the parameters were performed by considering the measles data presented in Table 1; accordingly, the model was fitted against the data. From the WHO reports for 2018, the continuous rate of fatality μ for a Pakistani individual is 66.5 years (1.253×10^{-4} per month), and with a population of 207,862,518, the inflow rate is estimated to be $\phi \approx 25{,}983$ individuals per month. In [12], it is reported that the Measles vaccine has an efficacy rate of approximately 97%, indicating that the vaccination outcome, denoted by τ, is nearly 0.97. The effects of the other parameters, such as the interactive rate β, the retrieval rate δ, the rate of clinically tested symptoms α, and the rate of vaccine coverage ω, are presented in Table 2 in relation to the calculated numbers. These parameters were estimated, and the modeled predictions using real data are plotted in Figure 2 using the MATLAB software considering the data for the first ten months of 2019. It can be seen from Figure 2 that model (1) provides a good fit, and the actual measles data are almost covered by the curve predicted by model (1). Using the result $\frac{1}{10}\sum_{k=1}^{10}\left|\frac{\varkappa_k^{real} - \varkappa_k^{approximate}}{\varkappa_k^{real}}\right| \approx 1.4685e^{-01}$, we can find the mean relative error of the fitting procedure.

Table 1. Reported measles cases in Pakistan during the first ten months of 2019 [12].

Jan	Feb	Mar	Apr	May	June	July	Aug	Sep	Oct
238	253	398	398	277	169	71	29	24	18

Table 2. Justification and values of the parameters used for simulating the model (1).

Parameter	Description	Source
ϕ	260,479	Estimated
α	1.253133×10^{-3}	Estimated
ω	0.97	Estimated
τ	1.60056×10^{-7}	Fitted
μ	9.3408	Fitted
β	9.2373×10^{-1}	Fitted
δ	5.8306×10^{-1}	Fitted
ρ	0	Estimated
γ	5.8306×10^{-1}	Fitted

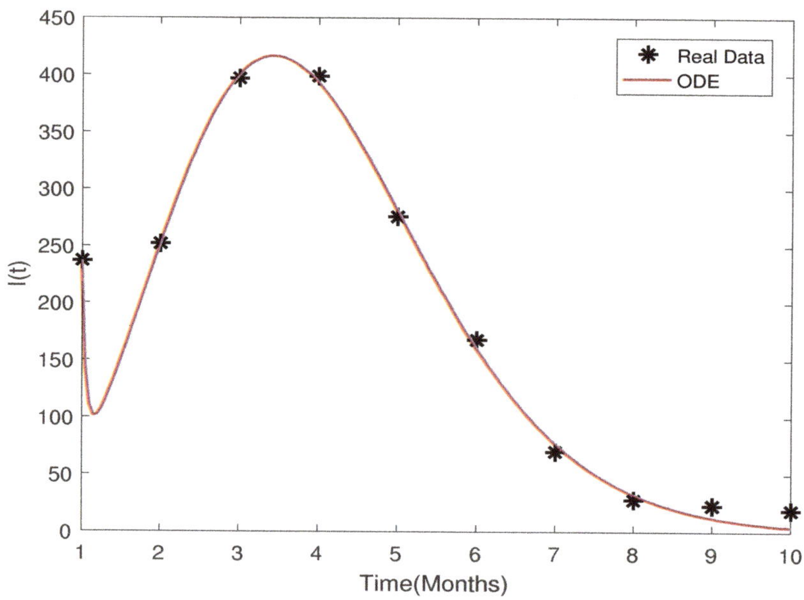

Figure 2. Fitting of the model (1) against reported measles cases from Table 1.

7. Numerical Results and Discussion

In this section, we present a graphical representation of the model dynamics using the available numerical method(s) and values of the parameters discussed in the previous section. For illustrative purposes, we provide the findings of our scheme and compute the approximate paths for the stochastic model (3) and its deterministic counterpart. The desired time interval is $[0, 100]$ and the step size is $\Delta = 0.3$, whereas the initial size of the population is provided by $(\mathbb{S}_0, \mathbb{V}_0, \mathbb{E}_0, \mathbb{I}_0, \mathbb{H}_0, \mathbb{R}_0) = (0.5, 0.4, 0.3, 0.5, 0.2, 0.1)$.

To simulate the model, we utilized two different types of hypothetical data. Initially, we simulated the model by considering the intensity of the noise and the parameter values from Example 1. The figure presented in Figure 3 demonstrates that as $R_0^D < 1$, the solution of system (3) with L'evy jump approaches the disease-free equilibrium point of the corresponding deterministic system, indicating that the disease is in the process of disappearing. Theorem 2 provides the necessary conditions for the extinction of system (3). The graphical representation verifies that the result of Theorem 2 is valid only if $\mathbb{R}_s < 1$.

Example 1. *In the present scenario, we have considered the parameter values as* $\phi = 1.0$, $\tau = 0.3$, $\alpha = 0.3$, $\mu = 0.03$, $\beta = 0.005$, $\delta = 0.4$, $\gamma = 0.02$, $\rho = 0.01$, $\omega = 0.02$, *and the intensities of the white noise as* $\xi_1 = 0.55$, $\xi_2 = 0.40$, $\xi_3 = 0.30$, $\xi_4 = 0.20$, $\xi_5 = 0.10$, $\xi_6 = 0.20$, *and* $\mathbb{X}_i(y) = \frac{-k_i y^2}{1+y^2}$, *with* $y = 0.5$ *and* k_i *equal to* 0.50, 0.50, 0.40, 0.20, 0.20, *and* 0.30, *respectively, for* $i = 1, \cdots, 6$. *The simulations of the model indicate that the epidemic will become extinct, as predicted by Theorem 2. The disappearance of the infection based on the predictions of the stochastic system are depicted in Figure 3a–f.*

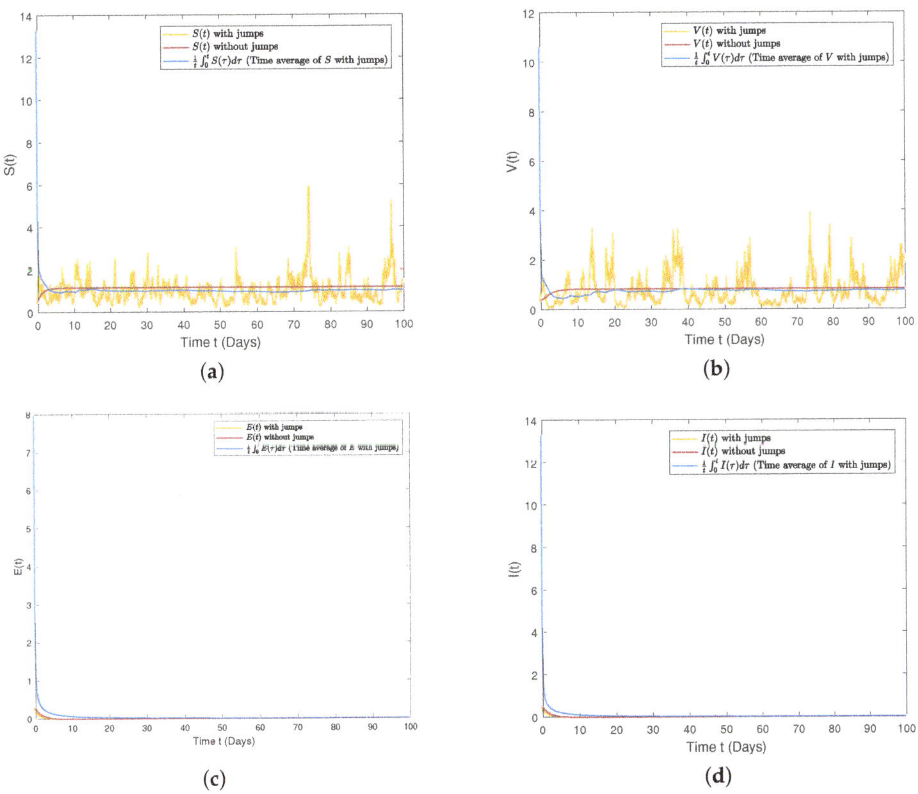

Figure 3. *Cont.*

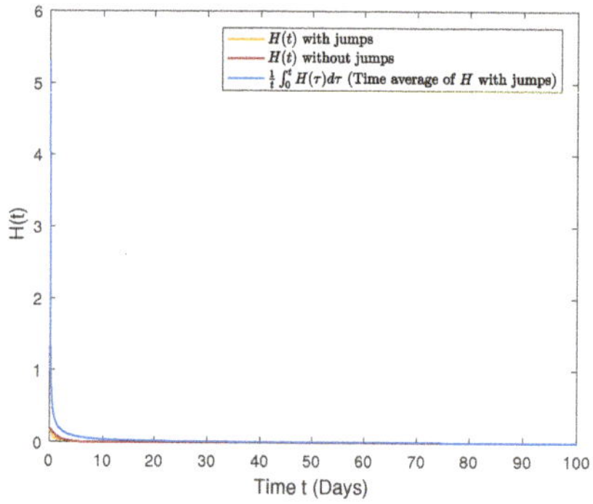

(e)

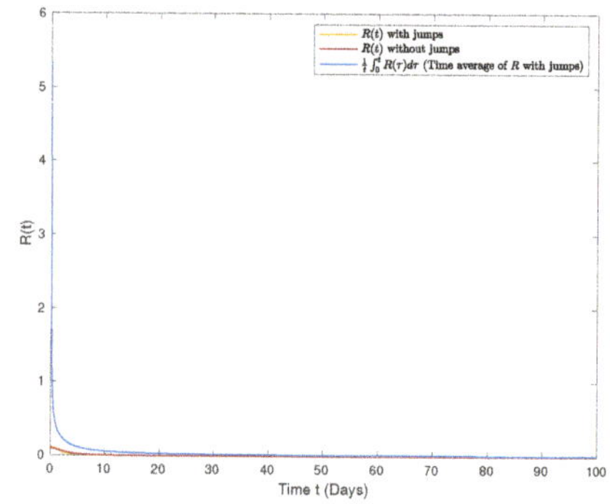

(f)

Figure 3. The plot shows the dynamics of the stochastic system (3) and the associated deterministic system (1) subject to jumps and without jumps. (**a**) The plot shows the $\mathbb{S}(t)-$curve with and without jumps. (**b**) The plot shows the dynamics of $\mathbb{V}(t)$ with and without jumps. (**c**) Three different scenarios for class $\mathbb{I}(t)$. (**d**) The dynamics of the exposed class with and without jumps. (**e**) Time evolution of the number of hospitalized people with and without jumps. (**f**) Three different scenarios for class $\mathbb{R}(t)$.

The findings of Theorem 3 suggest that the disease will continue to exist in the community with the dynamics predicted by model (3) if the value of $\mathbf{R}_0^{\mathbf{D}}$ is greater than 1. The statement $\mathbf{R}_0^{\mathbf{D}} > 1$ indicates that $\mathbf{R}_0^{\mathfrak{s}}$ will be greater than 1 even with low intensities of the noise. This conclusion is supported by the numerical simulations presented in Figures 3c,d and 4c,d, which validate the expected behavior of the epidemic model (3) as per Theorem 3.

Example 2. *In this example, the numerical values of the parameters are* $\phi = 0.9$, $\tau = 0.75$, $\alpha = 0.5$, $\mu = 0.06$, $\beta = 0.05$, $\delta = 0.4$, $\gamma = 0.2$, $\rho = 0.1$, *and* $\omega = 0.2$, *and the intensity of the white noise is* $\xi_1 = 0.45$, $\xi_2 = 0.30$, $\xi_3 = 0.10$, $\xi_4 = 0.20$, $\xi_5 = 0.15$, $\xi_6 = 0.20$, *and* $\mathbb{X}_i(y) = \frac{-k_i y^2}{1+y^2}$, *with* $y = 0.5$ *and where* k_i *equal to* 0.50, 0.35, 0.30, 0.35, 0.10, *and* 0.20, *respectively, for* $i = 1, \cdots, 6$.

For this set of parameters, it is easy to verify that both the stochastic threshold $\mathbf{R}_0^{\mathfrak{s}}$ *and the basic reproduction number* $\mathbf{R}_0^{\mathbf{D}}$ *are greater than one. The simulation results of the mean shown in Figure 4 confirm the persistence of the disease, which supports the conclusion of Theorem 3. Additionally, Figure 4a–f illustrates that the epidemic will continue to persist in the population in the meantime.*

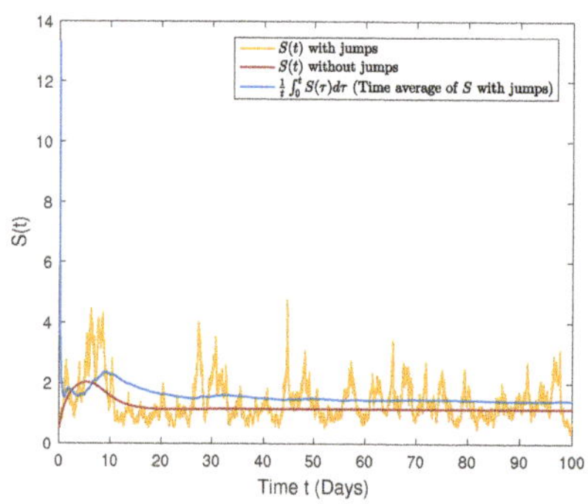

(a)

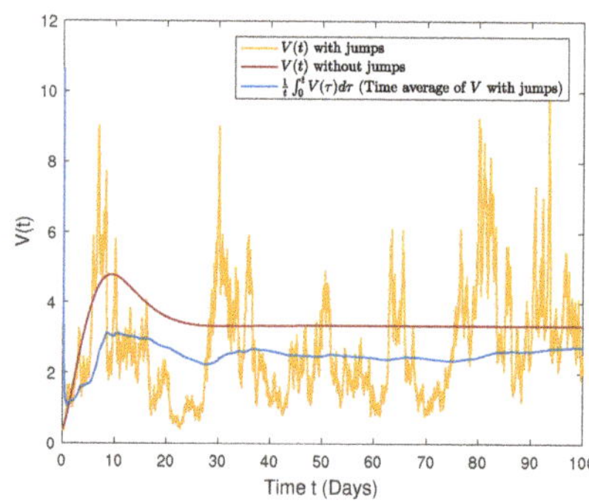

(b)

Figure 4. *Cont.*

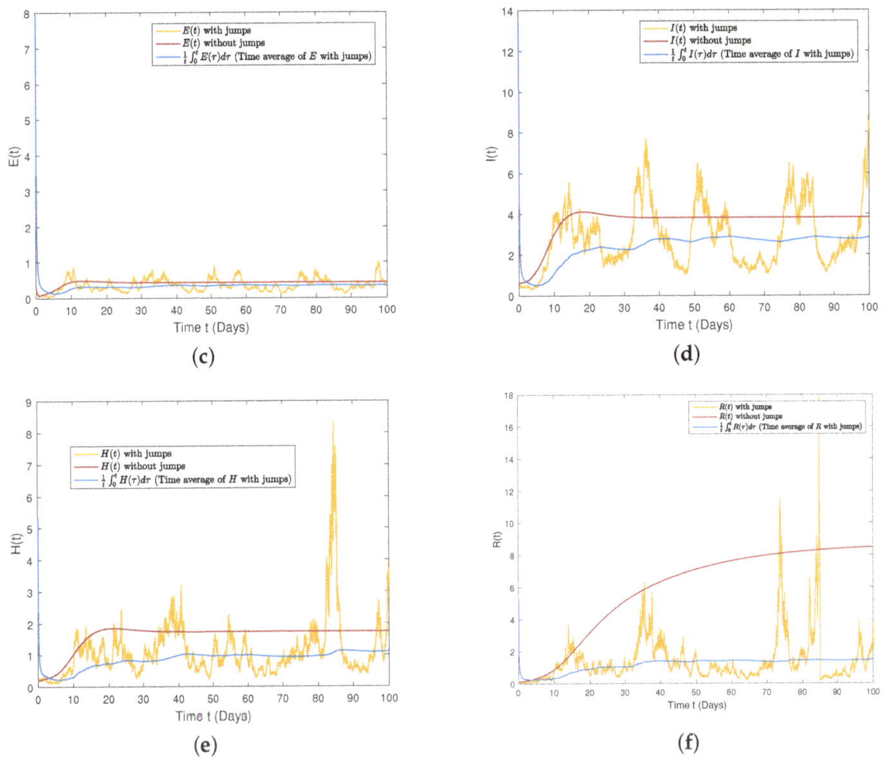

Figure 4. Simulation of the stochastic system (3) and deterministic system (1) with and without jumps for the case when the respective thresholds are greater than one. (**a**) The dynamics of the susceptible class with and without jumps. (**b**) Time evolution of the number of vaccinated people both with and without jumps. (**c**) The plot shows the $\mathbb{E}(t)-$curve with and without jumps. (**d**) The dynamics of the infected class with and without jumps. (**e**) The plot shows the $\mathbb{H}(t)-$curve with and without jumps. (**f**) Time evolution of the number of recovered people when the threshold exceeds one.

The Impact of Lévy Noise on the $\mathbb{I}$ Class

The impact of the intensity of the white noise on class $\mathbb{I}$ corresponding to system (3) is shown in Figure 5a–c. These figures suggest that increasing values of ξ_i for $i = 1, \cdots, 6$ lead towards the extinction of the disease. This means that the size of the infected class approaches zero as the intensity value of the noise increases. Further, this indicates that for small values of the noise intensity the infected class oscillates around the endemic steady state $\mathbb{I}^{\star}$m which confirms the result of Theorem 3. However, if the intensity of the white noise term is sufficiently high, the solution $\mathbb{I}$ may not exhibit oscillations near the endemic equilibrium point. This demonstrates that continuous efforts to increase stochastic disruptions through mass recovery of susceptible individuals and the effective treatment and care of the infected persons can significantly lower the spread of the measles virus in the population.

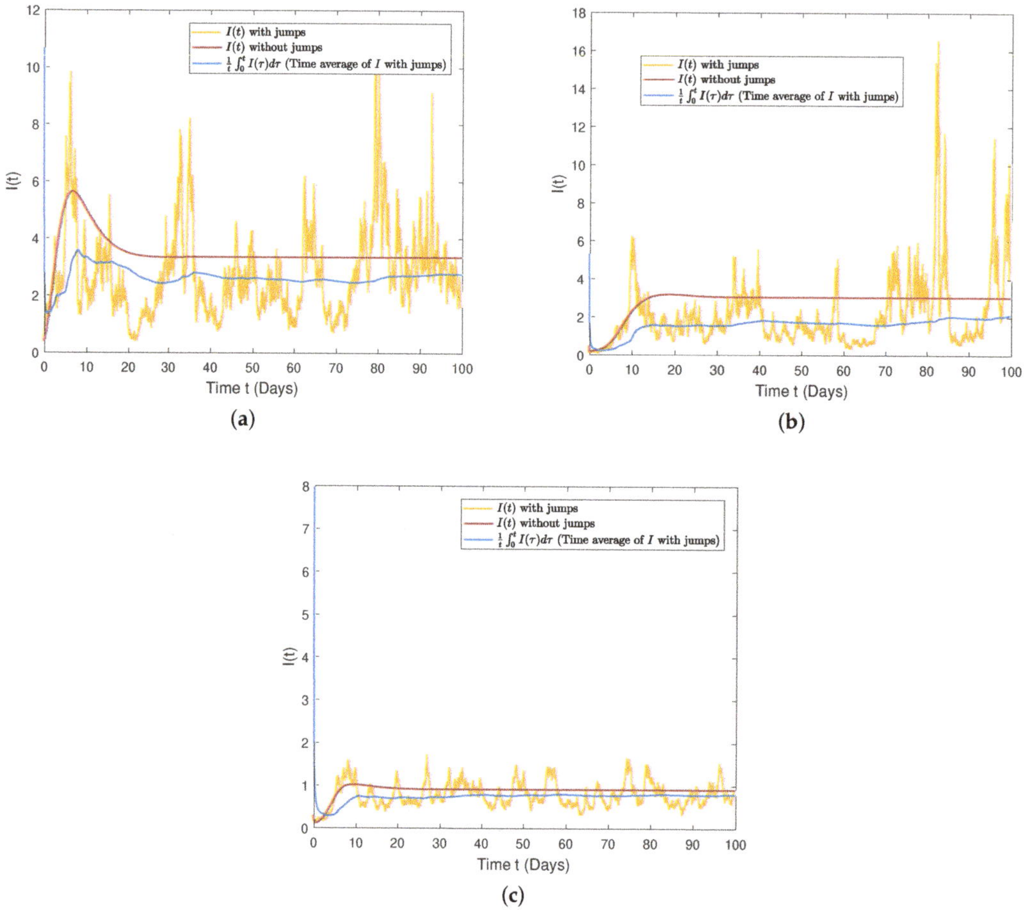

Figure 5. Simulations of $\mathbb{I}(t)$ based on the stochastic and deterministic systems, showing the effect of the intensities on class $\mathbb{I}$ when $\phi = 0.5$, $\tau = 0.90$, $\alpha = 0.5$, $\mu = 0.06$, $\beta = 0.05$, $\delta = 0.4$, $\gamma = 0.2$, $\rho = 0.2$, $\omega = 0.3$, and the noise intensity is assumed as $\xi_1 = 0.50$, $\xi_2 = 0.40$, $\xi_3 = 0.10$, $\xi_4 = 0.20$, $\xi_5 = 0.15$, $\xi_6 = 0.20$, and $\mathbb{X}_i(y) = \frac{-k_i y^2}{1+y^2}$, with $y = 0.5$ and where k_i equals 0.60, 0.34, 0.30, 0.35, 0.10, 0.20, respectively, for $i = 1, \cdots, 6$ and $(\mathbb{S}_0, \mathbb{V}_0, \mathbb{E}_0, \mathbb{I}_0, \mathbb{H}_0, \mathbb{R}_0) = (0.5, 0.4, 0.5, 0.3, 0.2, 0.1)$. (**a**) The dynamics of $\mathbb{I}(t)$ subject to sufficiently large values of noise intensity. (**b**) The plot shows the behavior of the infected class subject to moderate values of noise intensity. (**c**) The dynamic behavior of the infected class subject to low values of noise intensity.

8. Concluding Remarks and Future Directions

In this work, we have considered a stochastic $\mathbb{SVEIHR}$ epidemic model with Lévy noise while considering the nonlinear incidence rate. Our study aimed to investigate the impact of environmental noise on human society, and the results can shed light on the crucial role of noise in disease persistence and extinction. The proposed model assumed six different classes: a healthy class $(\mathbb{S})$, vaccinated class $(\mathbb{V})$, infected class $(\mathbb{I})$, exposed class $(\mathbb{E})$, hospitalized class$(\mathbb{H})$, and class of recovered individuals$(\mathbb{R})$. After model formulation, the study utilized the uniqueness and existence techniques to derive a positive solution for the underlying stochastic model. The stability results of the model were investigated using the Lyapunov function. The model's dynamic characteristics were analyzed in the context of the infection-free and endemic states of the associated ODE model. The stochastic threshold $\mathbb{R}_s$ was calculated to ensure disease extinction whenever $\mathbb{R}_s < 1$. Using data from Pakistan in 2019, we estimated the model parameters and conducted simulations to forecast future disease behavior. The results were compared to actual data using standard curve fitting tools. In our future research, we plan to investigate the impact of regime switching and temporary immunity on system (3).

Author Contributions: Methodology, Y.S.; Software, Y.S.; Formal analysis, P.L.; Investigation, P.L.; Writing—original draft, Y.S.; Writing—review and editing, P.L.; Project administration, Y.S.; Funding acquisition, P.L. All authors have read and agreed to the published version of the manuscript.

Funding: This work was supported by the project of "Research on Key parameters of structure design of petal vortex shaft torsional section" (No. JG523001), Development of Optimization Algorithm for 3D Data of Light Field (No. 3630019001), and the National Natural Science Foundation of China (No. 11901114), and Guangzhou Science and Technology Innovation general project (No. 201904010010), young innovative talents project of the Guangdong Provincial Department of Education (No. 2018KQNCX087).

Institutional Review Board Statement: Not applicable.

Informed Consent Statement: Not applicable.

Data Availability Statement: Not applicable.

Conflicts of Interest: The authors declare that they have no competing interest.

References

1. Mose, O.F.; Sigey, J.K.; Okello, J.A.; Okwoyo, J.M.; Giterere, J. *Mathematical Modeling on the Control of Measles by Vaccination: Case Study of Kisii County, Kenya*; Kenya Education Research Database: Kenya, 2014. Available online: https://kerd.ku.ac.ke/ (accessed on 16 April 2023).
2. Garba, S.M.; Safi, M.A.; Usaini, S. Mathematical model for assessing the impact of vaccination and treatment on measles transmission dynamics. *Math. Methods Appl. Sci.* **2017**, *40*, 6371–6388. [CrossRef]
3. Roberts, M.G.; Tobias, M.I. Predicting and preventing measles epidemic in New Zealand: Application of mathematical model. *Epidem. Infect.* **2000**, *124*, 279–287. [CrossRef] [PubMed]
4. Qureshi, S.; Jan, R. Modeling of measles epidemic with optimized fractional order under Caputo differential operator. *Chaos Solitons Fractals* **2021**, *145*, 110766. [CrossRef]
5. Perry, R.T.; Halsey, N.A. The clinical significance of measles: A review. *J. Infect. Dis.* **2004**, *189* (Suppl. S1), S4–S16. [PubMed]
6. Guo, B.; Khan, A.; Din, A. Numerical Simulation of Nonlinear Stochastic Analysis for Measles Transmission: A Case Study of a Measles Epidemic in Pakistan. *Fractal Fract.* **2023**, *7*, 130. [CrossRef]
7. Mossong, J.; Muller, C.P. Modelling measles re-emergence as a result of waning of immunity in vaccinated populations. *Vaccine* **2003**, *21*, 4597–4603. [CrossRef] [PubMed]
8. Bolarin, G. On the dynamical analysis of a new model for measles infection. *Int. J. Math. Trends Technol.* **2014**, *2*, 144–155. [CrossRef]
9. Taiwo, L.; Idris, S.; Abubakar, A.; Nguku, P.; Nsubuga, P.; Gidado, S. Factors affecting access to information on routine immunization among mothers of under 5 children in Kaduna state Nigeria, 2015. *Pan Afr. Med. J.* **2017**, *27*, 1–8. [CrossRef] [PubMed]
10. Mere, M.O.; Goodson, J.L.; Chandio, A.K.; Rana, M.S.; Hasan, Q.; Teleb, N.; Alexander, J.P., Jr. Progress toward measles elimination—Pakistan, 2000–2018. *Morb. Mortal. Wkly. Rep.* **2019**, *68*, 505. [CrossRef] [PubMed]
11. World Health Organization. *Eastern Mediterranean Vaccine Action Plan 2016–2020: A Framework for Implementation of the Global Vaccine Action Plan (No. WHO-EM/EPI/353/E)*; World Health Organization, Regional Office for the Eastern Mediterranean: Geneva, Switzerland, 2019.
12. Memon, Z.; Qureshi, S.; Memon, B.R. Mathematical analysis for a new nonlinear measles epidemiological system using real incidence data from Pakistan. *Eur. Phys. J. Plus* **2020**, *135*, 378. [CrossRef] [PubMed]
13. Din, A.; Ain, Q.T. Stochastic optimal control analysis of a mathematical model: Theory and application to non-singular kernels. *Fractal Fract.* **2022**, *6*, 279. [CrossRef]
14. Khan, F.M.; Khan, Z.U.; Lv, Y.P.; Yusuf, A.; Din, A. Investigating of fractional order dengue epidemic model with ABC operator. *Results Phys.* **2021**, *24*, 104075. [CrossRef]
15. Liu, P.; Ikram, R.; Khan, A.; Din, A. The measles epidemic model assessment under real statistics: An application of stochastic optimal control theory. *Comput. Methods Biomech. Biomed. Eng.* **2022**, 26, 138–159. [CrossRef] [PubMed]
16. El Fatini, M.; Sekkak, I. Lévy noise impact on a stochastic delayed epidemic model with Crowly–Martin incidence and crowding effect. *Phys. Stat. Mech. Its Appl.* **2020**, *541*, 123315. [CrossRef]

17. Sabbar, Y.; Khan, A.; Din, A.; Tilioua, M. New Method to Investigate the Impact of Independent Quadratic Stable Poisson Jumps on the Dynamics of a Disease under Vaccination Strategy. *Fractal Fract.* **2023**, *7*, 226. [CrossRef]
18. Berrhazi, B.E.; El Fatini, M.; Caraballo Garrido, T.; Pettersson, R. A stochastic SIRI epidemic model with Lévy noise. *Discret. Contin. Dyn.-Syst.-Ser. B* **2018**, *23*, 3645–3661. [CrossRef]
19. Li, X.P.; Din, A.; Zeb, A.; Kumar, S.; Saeed, T. The impact of Lévy noise on a stochastic and fractal-fractional Atangana–Baleanu order hepatitis B model under real statistical data. *Chaos Solitons Fractals* **2022**, *154*, 111623. [CrossRef]
20. James Peter, O.; Ojo, M.M.; Viriyapong, R.; Abiodun Oguntolu, F. Mathematical model of measles transmission dynamics using real data from Nigeria. *J. Differ. Equ. Appl.* **2022**, *28*, 753–770. [CrossRef]
21. Zhao, Y.; Jiang, D. The threshold of a stochastic SIRS epidemic model with saturated incidence. *Appl. Math. Lett.* **2014**, *34*, 90–93. [CrossRef]
22. Zhao, Y.; Jiang, D. The threshold of a stochastic SIS epidemic model with vaccination. *Appl. Math. Comput.* **2014**, *243*, 718–727. [CrossRef]

fractal and fractional

MDPI

Article

First-Passage Times and Optimal Control of Integrated Jump-Diffusion Processes

Mario Lefebvre

Department of Mathematics and Industrial Engineering, Polytechnique Montréal, Montréal, QC H3T 1J4, Canada; mlefebvre@polymtl.ca

Abstract: Let $Y(t)$ be a one-dimensional jump-diffusion process and $X(t)$ be defined by $\mathrm{d}X(t) = \rho[X(t), Y(t)]\,\mathrm{d}t$, where $\rho(\cdot,\cdot)$ is either a strictly positive or negative function. First-passage-time problems for the degenerate two-dimensional process $(X(t), Y(t))$ are considered in the case when the process leaves the continuation region at the latest at the moment of the first jump, and the problem of optimally controlling the process is treated as well. A particular problem, in which $\rho[X(t), Y(t)] = Y(t) - X(t)$ and $Y(t)$ is a standard Brownian motion with jumps, is solved explicitly.

Keywords: Brownian motion; Kolmogorov backward equation; dynamic programming; method of similarity solutions

MSC: 60J75; 93E20

1. Introduction

Diffusion processes are used as models in various applications, in particular in neuroscience to emulate the dynamics of the membrane potential of a neuron [1]. Moreover, to take into account the spikes of the neuron, jump-diffusion processes have been proposed by Jahn et al. [2] and Melanson and Longtin [3], among others.

Now, diffusion and jump-diffusion processes both increase and decrease in any interval, however small. However, in some applications, it is not realistic to assume that the process can decrease or increase. For example, if $X(t)$ represents the wear of a machine at time t, the stochastic process $\{X(t), t \geq 0\}$ should increase with time.

One way to obtain a strictly increasing or decreasing process is to consider degenerate two-dimensional diffusion processes $(X(t), Y(t))$ defined by

$$\mathrm{d}X(t) = \rho[X(t), Y(t)]\,\mathrm{d}t, \tag{1}$$

$$\mathrm{d}Y(t) = f[Y(t)]\,\mathrm{d}t + \{v[Y(t)]\}^{1/2}\mathrm{d}B(t), \tag{2}$$

where $\rho(\cdot,\cdot)$ is either a strictly positive or negative function and $\{B(t), t \geq 0\}$ is a standard Brownian motion. The functions f and v are such that $\{Y(t), t \geq 0\}$ is a diffusion process. When $\rho[X(t), Y(t)] = Y(t)$, the process $\{X(t), t \geq 0\}$ is called an integrated diffusion process. We can of course generalize the definition to the case when $\{Y(t), t \geq 0\}$ is a jump-diffusion process.

The author has published a number of papers on integrated diffusion processes; see, for instance, Lefebvre [4] for a recent one. Other papers on this topic include those by Lachal [5], Makasu [6], Metzler [7], Caravelli et al. [8] and Levy [9].

Next, we define the first-passage time

$$T(x,y) = \inf\{t > 0 : (X(t), Y(t)) \notin C \mid (X(0), Y(0)) = (x,y) \in C\}, \tag{3}$$

where C is a subset of $\mathbb{R}^2$.

Citation: Lefebvre, M. First-Passage Times and Optimal Control of Integrated Jump-Diffusion Processes. *Fractal Fract.* **2023**, *7*, 152. https://doi.org/10.3390/fractalfract7020152

Academic Editors: Carlo Cattani and Mario Abundo

Received: 14 December 2022
Revised: 21 January 2023
Accepted: 30 January 2023
Published: 3 February 2023

Let $\phi(\xi,\eta;x,y)$ be the joint probability density function (pdf) of the random vector $(X(t),Y(t))$, with $(X(0),Y(0)) = (x,y)$. As is well known (see, for example, Cox and Miller [10], p. 247), the function ϕ satisfies the Kolmogorov backward equation

$$\frac{1}{2}v(y)\,\phi_{yy} + f(y)\,\phi_y + \rho(x,y)\,\phi_x = \phi_t \quad \text{for } (x,y) \in C. \tag{4}$$

Moreover, the pdf $g(t;x,y)$ of the random variable $T(x,y)$ satisfies the same partial differential equation (PDE):

$$\frac{1}{2}v(y)\,g_{yy} + f(y)\,g_y + \rho(x,y)\,g_x = g_t \tag{5}$$

(subject to different boundary conditions). It follows that the moment-generating function of the random variable $T(x,y)$, namely

$$M(x,y;\alpha) := E\left[e^{-\alpha T(x,y)}\right], \tag{6}$$

where $\alpha > 0$, is a solution of the following PDE:

$$\frac{1}{2}v(y)\,M_{yy} + f(y)\,M_y + \rho(x,y)\,M_x = \alpha\,M \quad \text{for } (x,y) \in C, \tag{7}$$

where $M_{yy} := \partial^2 M(x,y;\alpha)/\partial y^2$, etc. Furthermore, this equation is subject to the boundary condition

$$M(x,y;\alpha) = 1 \quad \text{for } (x,y) \notin C. \tag{8}$$

We now replace the diffusion process $\{Y(t), t \geq 0\}$ by the jump-diffusion process defined by

$$Y(t) = Y(0) + \int_0^t f[Y(s)]\,\mathrm{d}s + \int_0^t \{v[Y(s)]\}^{1/2}\mathrm{d}B(s) + \sum_{n=1}^{N(t)} Z_n, \tag{9}$$

where $\{N(t), t \geq 0\}$ is a Poisson process with rate λ. The random variables $Z_1, Z_2, \ldots$ are assumed to be independent and identically distributed (i.i.d.), and also independent of the Poisson process. We can state the following proposition.

Proposition 1. *The function $M(x,y;\alpha)$ satisfies the integro-differential equation (dropping the dependence on α from the notation)*

$$\begin{aligned} \alpha\,M(x,y) \quad &= \quad \frac{1}{2}v(y)\,M_{yy}(x,y) + f(y)\,M_y(x,y) + \rho(x,y)\,M_x(x,y) \\ &\quad + \lambda\left\{\int_{-\infty}^{\infty} M(x,y+z)\,f_Z(z)\,\mathrm{d}z - M(x,y)\right\} \end{aligned} \tag{10}$$

for $(x,y) \in C$, where $f_Z(z)$ is the common density function of the Z_ns. As above, the equation is subject to the boundary condition (8).

Proof. This result is obtained by generalizing the *infinitesimal generator* of the jump-diffusion process $\{Y(t), t \geq 0\}$ in Kou and Wang [11] to the case when $f(y)$ and $v(y)$ are not necessarily constant functions. □

Remark 1. *See also the remark after the proof of Proposition 2.*

There are still few explicit solutions to first-passage problems for jump-diffusion processes. Kou and Wang [11] obtained explicit formulae for the Laplace transform of the pdf of the first-passage time τ to a constant boundary for a Wiener process with jumps having a double exponential distribution. This model was generalized or modified in various papers. In Chen et al. [12], τ was the first-exit time from a finite interval, while in Yin et al. [13] the jumps were mixed-exponential random variables. Karnaukh [14] considered the case when the parameters of the Wiener process depend on a finite Markov chain. In Lefebvre [15], the jump sizes were assumed to be uniformly distributed, while in Abundo [16] the jumps (positive and/or negative) were of a constant size and the boundaries were time-dependent. Because obtaining exact analytical solutions to such problems is difficult, some authors presented numerical techniques to obtain the quantities of interest; see, for example, Belkaid and Utzet [17].

Di Crescenzo et al. [18] computed bounds for first-crossing-time probabilities of a Wiener process with random jumps driven by a counting process. Fernández et al. [19] proposed algorithms to compute double-barrier first-passage-time probabilities of a jump-diffusion process with an arbitrary jump size distribution.

D'Onofrio and Lanteri [20] obtained numerical approximations for the density functions of first-passage times in the case of diffusion processes with state-dependent jumps. Finally, in Lefebvre [21], the author was able to obtain exact solutions to optimal control problems for Wiener processes with exponential jumps.

In the current paper, explicit results will be presented for the first-passage time of a *two-dimensional* jump-diffusion process.

In the next section, the special case when the two-dimensional process $(X(t), Y(t))$ is *killed* at the latest at the moment of the first jump, will be considered. We are also interested in the mean value of $T(x,y)$, as well as in the probability that $(X(t), Y(t))$ will leave the continuation region through a given part of its boundary ∂C. In Section 3, the problem of maximizing or minimizing the time the controlled version of the process $(X(t), Y(t))$ spends in the continuation region C will be treated. A particular problem will be solved explicitly in Section 4. Finally, we will conclude with a few remarks in Section 5.

2. Killed Processes

Assume that the random variables $Z_1, Z_2, \ldots$ are such that no overshoot is possible. That is, the degenerate two-dimensional jump-diffusion process $(X(t), Y(t))$ cannot jump over the boundary of the continuation region C. Let $m(x,y) := E[T(x,y)]$ and

$$p(x,y) := P[(X(T), Y(T)) \in \partial C_0 \mid X(0) = x, Y(0) = y], \tag{11}$$

where $\partial C_0 \subset \partial C$. We have the following corollaries.

Corollary 1. *The function $m(x,y)$ satisfies the integro-differential equation*

$$\begin{aligned} -1 &= \frac{1}{2} v(y)\, m_{yy}(x,y) + f(y)\, m_y(x,y) + \rho(x,y)\, m_x(x,y) \\ &\quad + \lambda \left\{ \int_{-\infty}^{\infty} m(x, y+z) f_Z(z)\, \mathrm{d}z - m(x,y) \right\} \end{aligned} \tag{12}$$

for $(x,y) \in C$, subject to the boundary condition

$$m(x,y) = 0 \quad \text{for } (x,y) \notin C. \tag{13}$$

Proof. It follows from the expansion of $M(x,y;\alpha)$ into an infinite series (see Cox and Miller [10]):

$$M(x,y;\alpha) := E\left[e^{-\alpha T(x,y)}\right] = E\left[1 - \alpha T(x,y) + \frac{1}{2}\alpha^2 T^2(x,y) - \ldots\right] \tag{14}$$

$$= 1 - \alpha m(x,y) + \frac{1}{2}\alpha^2 E[T^2(x,y)] - \ldots \tag{15}$$

Notice that in our case, $E[T^n(x,y)]$ will exist for any $n \in \{1, 2, \ldots\}$ because of Equation (18) below. □

Corollary 2. *The probability $p(x,y)$ is a solution of the integro-differential equation*

$$\begin{aligned} 0 &= \frac{1}{2}v(y)\,p_{yy}(x,y) + f(y)\,p_y(x,y) + \rho(x,y)\,p_x(x,y) \\ &\quad + \lambda\left\{\int_{-\infty}^{\infty} p(x,y+z)\,f_Z(z)\,\mathrm{d}z - p(x,y)\right\} \end{aligned} \tag{16}$$

for $(x,y) \in C$. Moreover, the boundary condition is

$$p(x,y) = \begin{cases} 1 & \text{if } (x,y) \in \partial C_0, \\ 0 & \text{if } (x,y) \in \partial D, \end{cases} \tag{17}$$

where $\partial D := \partial C \setminus \partial C_0$.

In this paper, we consider the special case when the random variable Z_1 is such that the process $(X(t), Y(t))$ will leave the continuation region C at the latest at the moment τ_1 of the first event of the Poisson process. Let $T_0(x,y)$ be the random variable that corresponds to $T(x,y)$ when $\lambda = 0$. We can write that

$$T(x,y) = \min\{T_0, \tau_1\}, \tag{18}$$

where τ_1 has an exponential distribution with parameter λ. Furthermore, the sum in Equation (9) can be replaced by $Z_1\,\mathbb{1}_{\{N(t)>0\}}$, where $\mathbb{1}_{\{N(t)>0\}}$ is the indicator variable of the event $\{N(t) > 0\}$, and the equation is valid for $t \in [0, T(x,y)]$. We can say that the process $(X(t), Y(t))$ is *killed* at time $T(x,y)$.

An application of the above problem is the following: as mentioned in Section 1, a more realistic model for the wear of a machine is the degenerate two-dimensional process $(X(t), Y(t))$ defined in Equations (1) and (2), when $\rho(\cdot,\cdot)$ is a strictly increasing function. Rishel [22] proposed this model (in n dimensions) in the context of an optimal control problem. If $X(t)$ denotes the remaining amount of deterioration that a device can undergo before it must be repaired or replaced, then $\rho(\cdot,\cdot)$ should be strictly negative instead. Moreover, the remaining lifetime of the device is the first-passage time to zero or to a level at which it is considered worn out.

Now, many electronic devices, especially mobile phones, are often replaced as soon as they break down, rather than being repaired. A mobile phone failure can be seen as a jump from the current value of $X(t)$ to zero, so that the device is *killed* at the time of the jump. It is also possible that the device will be replaced before a failure occurs, due to normal wear and tear or because it has become obsolete. Thus, *deterioration* could also include the age of the device.

Similarly, in the case of humans, the downward jump to zero could occur during a massive heart attack or stroke.

Because we assume that $(x, y+z) \in \partial C$ for any possible value z of the random variable Z, the integro-differential Equations (10) and (12) become, respectively, the partial differential equations

$$\begin{aligned} \alpha\,M(x,y) &= \frac{1}{2}v(y)\,M_{yy}(x,y) + f(y)\,M_y(x,y) + \rho(x,y)\,M_x(x,y) \\ &\quad + \lambda\,[1 - M(x,y)] \end{aligned} \tag{19}$$

and

$$-1 = \frac{1}{2}v(y)\,m_{yy}(x,y) + f(y)\,m_y(x,y) + \rho(x,y)\,m_x(x,y) - \lambda\,m(x,y). \tag{20}$$

In the case of Equation (16), if $(x, y+z) \in \partial C_0 \; \forall z$, then

$$0 = \frac{1}{2} v(y)\, p_{yy}(x,y) + f(y)\, p_y(x,y) + \rho(x,y)\, p_x(x,y) + \lambda\,[1 - p(x,y)], \tag{21}$$

whereas, we have

$$0 = \frac{1}{2} v(y)\, p_{yy}(x,y) + f(y)\, p_y(x,y) + \rho(x,y)\, p_x(x,y) - \lambda\, p(x,y) \tag{22}$$

if $(x, y+z) \in \partial D \; \forall z$. If $(x, y+z)$ belongs to ∂C_0 for some values of z, and to ∂D for other values of z, then the integral in Equation (16) is replaced by

$$\int_{-\infty}^{\infty} \mathbb{1}_{\{(x,y+z) \in \partial C_0\}} f_Z(z)\, \mathrm{d}z = P[(x, y+Z) \in \partial C_0]. \tag{23}$$

Solving integro-differential equations explicitly and exactly is not an easy task. In Section 4, an example of a problem that we can indeed solve analytically will be presented. As above, the integro-differential equations will be reduced to PDE's, and the method of similarity solutions will be used to transform these PDE's into ordinary differential equations.

3. Optimal Control

In this section, we consider a controlled version of the two-dimensional process $(X(t), Y(t))$:

$$X_u(t) = X_u(0) + \int_0^t \rho[X_u(s), Y_u(s)]\, \mathrm{d}s, \tag{24}$$

$$\begin{aligned} Y_u(t) &= Y_u(0) + \int_0^t b\, u[X_u(s), Y_u(s)]\, \mathrm{d}s + \int_0^t f[Y_u(s)]\, \mathrm{d}s \\ &\quad + \int_0^t \{v[Y_u(s)]\}^{1/2} \mathrm{d}B(s) + \sum_{n=1}^{N(t)} Z_n, \end{aligned} \tag{25}$$

where $u(\cdot,\cdot)$ is the control variable, which is assumed to be a continuous function, and b is a non-zero constant. The aim is to find the value of the control that minimizes the expected value of the cost function

$$J(x,y) := \int_0^{T(x,y)} \left\{ \frac{1}{2} q\, u^2[X_u(t), Y_u(t)] + \theta \right\} \mathrm{d}t, \tag{26}$$

where $q > 0$ and θ are constants. If the parameter θ is positive (respectively negative), the optimizer must try to minimize (respectively maximize) the time spent by the controlled process in the continuation C, taking the quadratic control costs into account. This type of problem is known as a *homing problem*; see Whittle [23] and/or [24].

To solve the above problem, we can make use of dynamic programming. We define the *value function*

$$F(x,y) = \inf_{\substack{u[X_u(t), Y_u(t)] \\ t \in [0, T(x,y))}} E[J(x,y)]. \tag{27}$$

That is, $F(x,y)$ is the expected cost (or reward, if it is negative) obtained by choosing the optimal value of the control variable in the interval $[0, T(x,y))$.

Proposition 2. *The value function $F(x,y)$ satisfies the second-order, non-linear partial integro-differential equation*

$$\begin{aligned} 0 &= \theta - \frac{1}{2}\frac{b^2}{q} F_y^2(x,y) + \rho(x,y)\, F_x(x,y) + f(y)\, F_y(x,y) + \frac{1}{2} v(y)\, F_{yy}(x,y) \\ &\quad + \lambda \left\{ \int_{-\infty}^{\infty} F(x, y+z) f_Z(z)\, \mathrm{d}z - F(x,y) \right\}. \end{aligned} \tag{28}$$

Moreover, we have the boundary condition

$$F(x,y) = 0 \quad \textit{for } (x,y) \notin C. \tag{29}$$

Proof. First, thanks to Bellman's principle of optimality, we can write that

$$\begin{aligned} F(x,y) &= \inf_{\substack{u[X_u(t),Y_u(t)] \\ 0 \le t \le \Delta t}} E\Bigg[\int_0^{\Delta t} \left\{ \frac{1}{2} q u^2[X_u(t), Y_u(t)] + \theta \right\} \mathrm{d}t \\ &\quad + F\Bigg(x + \rho(x,y)\,\Delta t, y + [f(y) + b u(x,y)]\,\Delta t + v^{1/2}(y)\, B(\Delta t) \\ &\quad + \sum_{n=1}^{N(\Delta t)} Z_n \Bigg) + o(\Delta t) \Bigg] \end{aligned} \tag{30}$$

$$\begin{aligned} &= \inf_{\substack{u[X_u(t),Y_u(t)] \\ 0 \le t \le \Delta t}} E\Bigg[\left\{ \frac{1}{2} q u^2(x,y) + \theta \right\} \Delta t \\ &\quad + F\Bigg(x + \rho(x,y)\,\Delta t, y + [f(y) + b u(x,y)]\,\Delta t + v^{1/2}(y)\, B(\Delta t) \\ &\quad + \sum_{n=1}^{N(\Delta t)} Z_n \Bigg) + o(\Delta t) \Bigg]. \end{aligned} \tag{31}$$

Next, let

$$\xi := x + \rho(x,y)\,\Delta t \tag{32}$$

and

$$\eta := y + [f(y) + b u(x,y)]\,\Delta t + v^{1/2}(y)\, B(\Delta t). \tag{33}$$

We have

$$E\left[F\left(\xi, \eta + \sum_{n=1}^{N(\Delta t)} Z_n \right) \right] = E\left[E\left[F\left(\xi, \eta + \sum_{n=1}^{N(\Delta t)} Z_n \right) \,\middle|\, N(\Delta t) \right] \right]. \tag{34}$$

Since $N(\Delta t)$ has a Poisson distribution with parameter $\lambda\,\Delta t$, we can write that

$$P[N(\Delta t) = 1] = \lambda \Delta t e^{-\lambda \Delta t} = \lambda \Delta t + o(\Delta t) \tag{35}$$

and

$$P[N(\Delta t) \ge 2] = o(\Delta t). \tag{36}$$

Hence,

$$E\left[F\left(\xi, \eta + \sum_{n=1}^{N(\Delta t)} Z_n \right) \right] = E[F(\xi, \eta)]\,(1 - \lambda \Delta t) + E[F(\xi, \eta + Z_1)]\,\lambda \Delta t + o(\Delta t). \tag{37}$$

Now, assuming that $F(x,y)$ is twice differentiable with respect to x and to y, making use of Taylor's formula for functions of two variables, we obtain that

$$\begin{aligned}
&F(x+\rho(x,y)\,\Delta t, y+[f(y)+bu(x,y)]\,\Delta t+v^{1/2}(y)\,B(\Delta t)) \\
&= F(x,y)+\rho(x,y)\,\Delta t\,F_x(x,y) \\
&\quad+\left\{[f(y)+bu(x,y)]\,\Delta t+v^{1/2}(y)\,B(\Delta t)\right\}F_y(x,y) \\
&\quad+\frac{1}{2}[\rho(x,y)\,\Delta t]^2 F_{xx}(x,y) \\
&\quad+\frac{1}{2}\left\{[f(y)+bu(x,y)]\,\Delta t+v^{1/2}(y)\,B(\Delta t)\right\}^2 F_{yy}(x,y) \\
&\quad+[\rho(x,y)\,\Delta t]\left\{[f(y)+bu(x,y)]\,\Delta t+v^{1/2}(y)\,B(\Delta t)\right\}F_{xy}(x,y) \\
&\quad+o(\Delta t).
\end{aligned} \tag{38}$$

Furthermore, we have $E[B(\Delta t)]=0$ and $E[B^2(\Delta t)]=V[B(\Delta t)]=\Delta t$, which implies that

$$\begin{aligned}
E[F(\xi,\eta)](1-\lambda\,\Delta t) &= F(x,y)+\rho(x,y)\,\Delta t\,F_x(x,y) \\
&\quad+[f(y)+bu(x,y)]\,\Delta t\,F_y(x,y)+\frac{1}{2}v(y)\,\Delta t\,F_{yy}(x,y) \\
&\quad-F(x,y)\,\lambda\,\Delta t+o(\Delta t).
\end{aligned} \tag{39}$$

Similarly, we find that

$$\begin{aligned}
E[F(\xi,\eta+Z_1)]\,\lambda\,\Delta t &= E[F(x,y+Z_1)]\,\lambda\,\Delta t+o(\Delta t) \qquad (40)\\
&= \lambda\,\Delta t\int_{-\infty}^{\infty}F(x,y+z)\,f_Z(z)\,\mathrm{d}z+o(\Delta t).
\end{aligned}$$

Indeed, by independence, we have

$$E[F(\xi,\eta+Z_1)]=\int_{-\infty}^{\infty}E[F(\xi,\eta+z)]\,f_Z(z)\,\mathrm{d}z. \tag{41}$$

Let $w:=y+z$. We compute

$$\begin{aligned}
&E\Big[F(x+\rho(x,y)\,\Delta t, y+[f(y)+bu(x,y)]\,\Delta t+v^{1/2}(y)\,B(\Delta t)+z)\Big] \\
&= F(x,w)+\rho(x,y)\,\Delta t\,F_x(x,w)+[f(y)+bu(x,y)]\,\Delta t\,F_w(x,w) \\
&\quad+\frac{1}{2}v(y)\,\Delta t\,F_{ww}(x,w)+o(\Delta t),
\end{aligned} \tag{42}$$

so that

$$\begin{aligned}
&E\Big[F(x+\rho(x,y)\,\Delta t, y+[f(y)+bu(x,y)]\,\Delta t+v^{1/2}(y)\,B(\Delta t)+z)\Big]\lambda\,\Delta t \\
&= F(x,w)\,\lambda\,\Delta t+o(\Delta t)=F(x,y+z)\,\lambda\,\Delta t+o(\Delta t).
\end{aligned} \tag{43}$$

Thus,

$$\begin{aligned}
E[F(\xi,\eta+Z_1)]\,\lambda\,\Delta t &= \int_{-\infty}^{\infty}[F(x,y+z)\,\lambda\,\Delta t+o(\Delta t)]\,f_Z(z)\,\mathrm{d}z \\
&= E[F(x,y+Z_1)]\,\lambda\,\Delta t+o(\Delta t).
\end{aligned} \tag{44}$$

From Equation (31) and the above results, we deduce that

$$
\begin{aligned}
0 \;=\; & \inf_{\substack{u[X_u(t),Y_u(t)]\\ 0\le t\le \Delta t}} \left\{ \left[\frac{1}{2} q u^2(x,y) + \theta\right] \Delta t + \rho(x,y)\,\Delta t\, F_x(x,y) \right. \\
& + [f(y) + b u(x,y)]\,\Delta t\, F_y(x,y) + \frac{1}{2} v(y)\,\Delta t\, F_{yy}(x,y) \\
& \left. - F(x,y)\,\lambda\,\Delta t + \lambda\,\Delta t \int_{-\infty}^{\infty} F(x,y+z)\, f_Z(z)\,\mathrm{d}z + o(\Delta t) \right\}.
\end{aligned} \tag{45}
$$

Dividing both sides of the above equation by Δt and letting Δt decrease to zero, we obtain the *dynamic programming equation*

$$
\begin{aligned}
0 \;=\; & \inf_{u(x,y)} \left\{ \frac{1}{2} q u^2(x,y) + \theta + \rho(x,y)\, F_x(x,y) \right. \\
& + [f(y) + b u(x,y)]\, F_y(x,y) + \frac{1}{2} v(y)\, F_{yy}(x,y) \\
& \left. - \lambda F(x,y) + \lambda \int_{-\infty}^{\infty} F(x,y+z)\, f_Z(z)\,\mathrm{d}z \right\}.
\end{aligned} \tag{46}
$$

From the preceding equation, we find that the optimal control $u^*(x,y)$ can be expressed as follows:

$$
u^*(x,y) = -\frac{b}{q} F_y(x,y). \tag{47}
$$

Substituting the optimal control into Equation (46), we obtain Equation (28).

Finally, the boundary condition (29) follows at once from the definition of $F(x,y)$, since $T(x,y) = 0$ if $(x,y) \notin C$. □

Remark 2. *Suppose that we set $u[X_u(s), Y_u(s)]$ equal to zero in Equation (25) and that we replace the cost function $J(x,y)$ defined in Equation (26) by*

$$
J_0(x,y,t_0) := \int_{t_0}^{\infty} e^{-\alpha t} g(t; t_0, x, y)\,\mathrm{d}t, \tag{48}
$$

where $g(t; t_0, x, y)$ is the pdf of $T(x,y)$ when the starting time is equal to t_0. Then, since $J_0(x,y,t_0)$ is actually a deterministic function, we may write that

$$
\Phi(x,y,t_0) := E[J_0(x,y,t_0)] = J_0(x,y,t_0) = M(x,y,t_0;\alpha). \tag{49}
$$

Proceeding as in the above proof, we find that

$$
\begin{aligned}
0 \;=\; & e^{-\alpha t_0} g(t_0; t_0, x, y) + \Phi_{t_0}(x,y,t_0) + \rho(x,y)\,\Phi_x(x,y,t_0) + f(y)\,\Phi_y(x,y,t_0) \\
& + \frac{1}{2} v(y)\,\Phi_{yy}(x,y,t_0) - \lambda \Phi(x,y,t_0) + \lambda \int_{-\infty}^{\infty} \Phi(x,y+z,t_0)\, f_Z(z)\,\mathrm{d}z.
\end{aligned} \tag{50}
$$

We have $g(t_0; t_0, x, y) = 0$ for $(x,y) \in C$. Moreover, using the Leibniz integral rule,

$$
\begin{aligned}
\Phi_{t_0}(x,y,t_0) \;&=\; -e^{-\alpha t_0} g(t_0; t_0, x, y) + \int_{t_0}^{\infty} e^{-\alpha t} g_{t_0}(t; t_0, x, y)\,\mathrm{d}t \\
&=\; 0 - \int_{t_0}^{\infty} e^{-\alpha t} g_t(t; t_0, x, y)\,\mathrm{d}t \\
&=\; -e^{-\alpha t} g(t; t_0, x, y)\Big|_{t_0}^{\infty} - \alpha \int_{t_0}^{\infty} e^{-\alpha t} g(t; t_0, x, y)\,\mathrm{d}t \\
&=\; -\alpha M(x,y,t_0;\alpha),
\end{aligned} \tag{51}
$$

where we used the fact that $g_{t_0}(t;t_0,x,y) = -g_t(t;t_0,x,y)$ *because the two-dimensional process* $(X_u(t), Y_u(t))$ *is time-invariant. Hence, setting* t_0 *equal to zero, we retrieve Equation (10).*

In the case of the killed processes considered in Section 2, the integro-differential Equation (28) reduces to the non-linear PDE

$$\begin{aligned} 0 \;=\;& \theta - \frac{1}{2}\frac{b^2}{q} F_y^2(x,y) + \rho(x,y)\,F_x(x,y) + f(y)\,F_y(x,y) + \frac{1}{2}v(y)\,F_{yy}(x,y) \\ & - \lambda\,F(x,y). \end{aligned} \tag{52}$$

The boundary condition remains the one in Equation (29).

In the next section, a particular problem will be treated. We will find the exact optimal control when the parameter λ is equal to zero, so that there are no jumps, and a numerical solution will be computed in the case when $\lambda > 0$.

4. A Particular Problem

We consider the process $(X(t), Y(t))$, defined by

$$X(t) \;=\; X(0) + \int_0^t [Y(s) - X(s)]\,ds, \tag{53}$$

$$Y(t) \;=\; Y(0) + B(t) + \sum_{n=1}^{N(t)} Z_n. \tag{54}$$

That is, $\{Y(t), t \geq 0\}$ is a standard Brownian motion with jumps. Moreover, we can write that

$$X'(t) = Y(t) - X(t), \tag{55}$$

which implies that

$$X(t) = e^{-t}\,X(0) + \int_0^t Y(s)\,e^{-s}\,ds. \tag{56}$$

Let

$$T(x,y) = \inf\{t > 0 : Y(t) - X(t) = k_1 \text{ or } k_2 \mid (X(0), Y(0)) = (x,y)\}, \tag{57}$$

where $0 \leq k_1 < y - x < k_2$. Notice that $\rho[X(t), Y(t)] = Y(t) - X(t) > 0$ in the continuation region, so that $X(t)$ is strictly increasing with time.

Next, we define

$$Z_1 = \begin{cases} x - y + k_1 & \text{with probability } p_0 \in (0,1), \\ x - y + k_2 & \text{with probability } 1 - p_0. \end{cases} \tag{58}$$

Thus, Z_1 is a discrete random variable such that at time τ_1 the process will leave the continuation region, if it has not already done so. We can write that

$$f_Z(z) = \delta(z - x + y - k_1)\,p_0 + \delta(z - x + y - k_2)\,(1 - p_0), \tag{59}$$

where $\delta(\cdot)$ is the Dirac delta function.

We deduce from Equation (19) that the moment-generating function of $T(x,y)$ satisfies the PDE

$$\alpha\,M(x,y) = \frac{1}{2}M_{yy}(x,y) + (y - x)\,M_x(x,y) + \lambda\,[1 - M(x,y)]. \tag{60}$$

Based on this equation and the boundary conditions $M(x,y) = 1$ if $y - x = k_1$ or k_2, we look for a solution of the form

$$M(x,y) = N(r), \tag{61}$$

where $r := y - x$. This is an application of the method of similarity solutions, and r is called the *similarity variable*. For the method to work, both the equation and the boundary conditions must be expressed in terms of r (after simplification). Here, we find that Equation (60) reduces to the *ordinary* differential equation (ODE)

$$\alpha N(r) = \frac{1}{2} N''(r) - r N'(r) + \lambda [1 - N(r)], \tag{62}$$

subject to the boundary conditions $N(k_i) = 1$, for $i = 1, 2$. With the help of the mathematical software program *Maple*, we find that the general solution of the above equation can be written as

$$N(r) = c_1 r M\left(\frac{1 + \alpha + \lambda}{2}, \frac{3}{2}, r^2\right) + c_2 r U\left(\frac{1 + \alpha + \lambda}{2}, \frac{3}{2}, r^2\right) + \frac{\lambda}{\alpha + \lambda}, \tag{63}$$

where $M(\cdot, \cdot, \cdot)$ and $U(\cdot, \cdot, \cdot)$ are Kummer functions. The constants c_1 and c_2 are uniquely determined from the boundary conditions $N(k_1) = N(k_2) = 1$.

Since, as noted in Section 2 (see Equation (18)), $T(x, y) = \min\{T_0, \tau_1\}$, when λ is large, the function $M(x, y; \alpha)$ should be close to the moment-generating function of an exponential random variable with parameter λ, namely

$$M_0(\alpha) := \int_0^\infty e^{-\alpha t} \lambda e^{-\lambda t} \, dt = \frac{\lambda}{\alpha + \lambda}. \tag{64}$$

In Figure 1, we present the functions $M_0(\alpha)$ and $M(x, y; \alpha)$ for $\alpha \in (0, 10)$, when $\lambda = 1$, $k_1 = 0$, $k_2 = 1$ and $y - x = 0.5$. We observe that the two functions differ significantly. However, the two functions are very similar when $\lambda = 20$, as can be observed in Figure 2. When $\lambda = 100$, $M_0(\alpha)$ and $M(x, y; \alpha)$ practically coincide for $\alpha \in (0, 10)$.

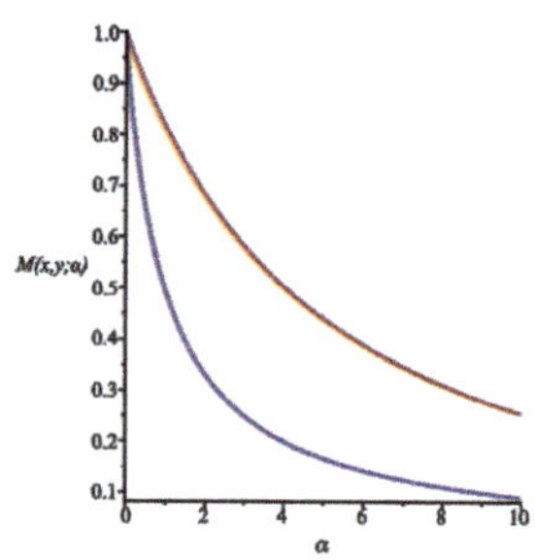

Figure 1. Functions $M_0(\alpha)$ (below) and $M(x, y; \alpha)$ for α in the interval $(0, 10)$, when $\lambda = 1$, $k_1 = 0$, $k_2 = 1$ and $y - x = 0.5$.

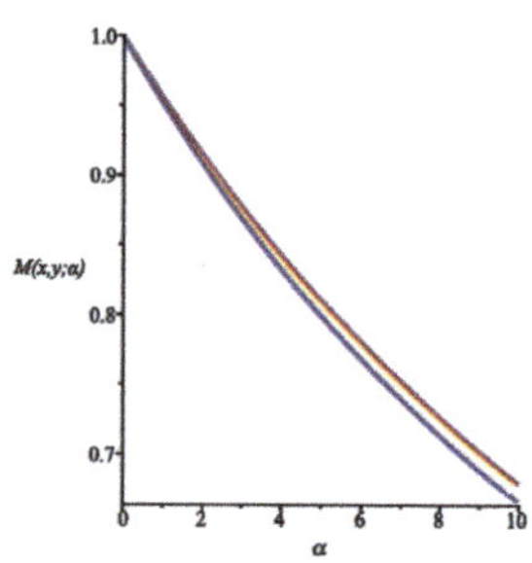

Figure 2. Functions $M_0(\alpha)$ (below) and $M(x, y; \alpha)$ for α in the interval $(0, 10)$, when $\lambda = 20$, $k_1 = 0$, $k_2 = 1$ and $y - x = 0.5$.

Next, the function $m(x,y) = E[T(x,y)]$ satisfies the PDE (see Equation (20))

$$-1 = \frac{1}{2}m_{yy}(x,y) + (y-x)\,m_x(x,y) - \lambda\, m(x,y), \tag{65}$$

subject to $m(x,y) = 0$ if $y - x = k_1$ or k_2. Setting $m(x,y) = n(r)$, we obtain the ODE

$$-1 = \frac{1}{2}n''(r) - r\,N'(r) - \lambda\, N(r), \tag{66}$$

with $n(k_i) = 0$, for $i = 1, 2$. We find that

$$n(r) = c_1\, r\, M\left(\frac{1+\lambda}{2}, \frac{3}{2}, r^2\right) + c_2\, r\, U\left(\frac{1+\lambda}{2}, \frac{3}{2}, r^2\right) + \frac{1}{\alpha}. \tag{67}$$

The particular solution that satisfies the boundary conditions $n(0) = n(1) = 0$ is presented in Figure 3.

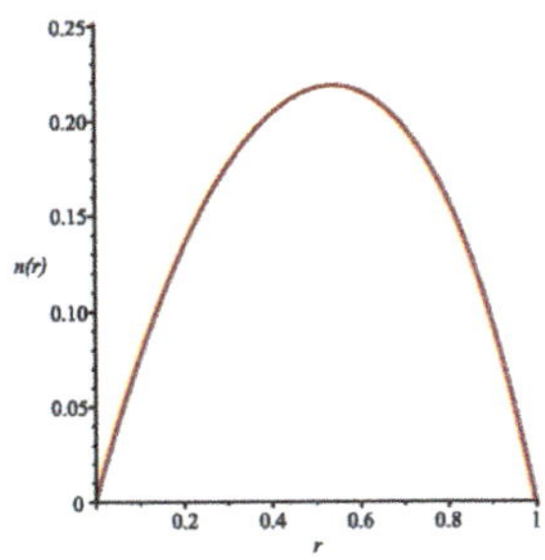

Figure 3. Function $n(r)$ for $0 \leq r \leq 1$, when $\lambda = 1$, $k_1 = 0$ and $k_2 = 1$.

Finally, let

$$p(x,y) = P[Y(T) - X(T) = k_1 \mid X(0) = x, Y(0) = y]. \tag{68}$$

This function is a solution of the PDE

$$0 = \frac{1}{2}p_{yy}(x,y) + (y-x)\,p_x(x,y) + \lambda\,[p_0 - p(x,y)]. \tag{69}$$

Assuming that $p(x,y) = q(r)$, we obtain the ODE

$$0 = \frac{1}{2}q''(r) - r\,q'(r) + \lambda\,[p_0 - q(r)], \tag{70}$$

whose general solution is

$$q(r) = c_1\, r\, M\left(\frac{1+\lambda}{2}, \frac{3}{2}, r^2\right) + c_2\, r\, U\left(\frac{1+\lambda}{2}, \frac{3}{2}, r^2\right) + p_0. \tag{71}$$

The solution that satisfies the boundary conditions $q(0) = 1$ and $q(1) = 0$ is shown in Figure 4, when $\lambda = 1$ and $p_0 = 1/2$.

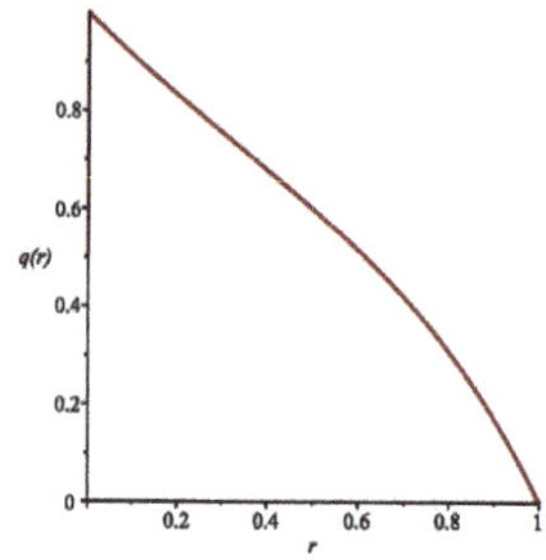

Figure 4. Function $q(r)$ for $0 \leq r \leq 1$, when $\lambda = 1$, $k_1 = 0$, $k_2 = 1$ and $p_0 = 1/2$.

To conclude this section, we will try to find the optimal control of the two-dimensional process $(X_u(t), Y_u(t))$ defined by

$$X_u(t) = X_u(0) + \int_0^t [Y_u(s) - X_u(s)]\,\mathrm{d}s, \tag{72}$$

$$Y_u(t) = Y_u(0) + \int_0^t b u[X_u(s), Y_u(s)]\,\mathrm{d}s + B(t) + \sum_{n=1}^{N(t)} Z_n. \tag{73}$$

To do so, we must solve the non-linear second-order PDE

$$0 = \theta - \frac{1}{2}\frac{b^2}{q} F_y^2(x,y) + (y-x) F_x(x,y) + \frac{1}{2} F_{yy}(x,y) - \lambda F(x,y), \tag{74}$$

subject to $F(x,y) = 0$ if $y - x = k_1$ or k_2.

As above, we make use of the method of similarity solutions. We look for a solution of the form $F(x,y) = G(r = y - x)$. Equation (74) becomes

$$0 = \theta - \frac{1}{2}\frac{b^2}{q} [G'(r)]^2 - r G'(r) + \frac{1}{2} G''(r) - \lambda G(r). \tag{75}$$

If $\lambda = 0$, *Maple* is able to obtain the general solution of the preceding equation:

$$G(r) = -\frac{q}{b^2}\left[w^2 + \ln(\Delta_1/\Delta_2)\right], \tag{76}$$

where

$$\Delta_1 := b^2\left[c_1 M\left(\frac{b^2\theta + q}{2q}, \frac{3}{2}, r^2\right) + c_2 U\left(\frac{b^2\theta + q}{2q}, \frac{3}{2}, r^2\right)\right] \tag{77}$$

and

$$\Delta_2 := (b^2\theta - 2q) U\left(\frac{b^2\theta + q}{2q}, \frac{3}{2}, r^2\right) M\left(\frac{b^2\theta - q}{2q}, \frac{3}{2}, r^2\right) - 2q M\left(\frac{b^2\theta + q}{2q}, \frac{3}{2}, r^2\right) U\left(\frac{b^2\theta - q}{2q}, \frac{3}{2}, r^2\right). \tag{78}$$

The constants c_1 and c_2 are determined by making use of the boundary conditions $G(k_1) = G(k_2) = 0$.

When $\lambda > 0$, *Maple* and *Mathematica* are unable to provide an analytical expression for the solution of Equation (75). It is, however, not difficult to obtain a numerical solution for any choice of the parameters. For instance, if we choose $b = q = \theta = \lambda = 1$, $k_1 = 1$ and $k_2 = 2$, we obtain the function $G(r)$, as shown in Figure 5, together with the function obtained when $\lambda = 0$. Finally, in Figure 6, we present the corresponding optimal controls.

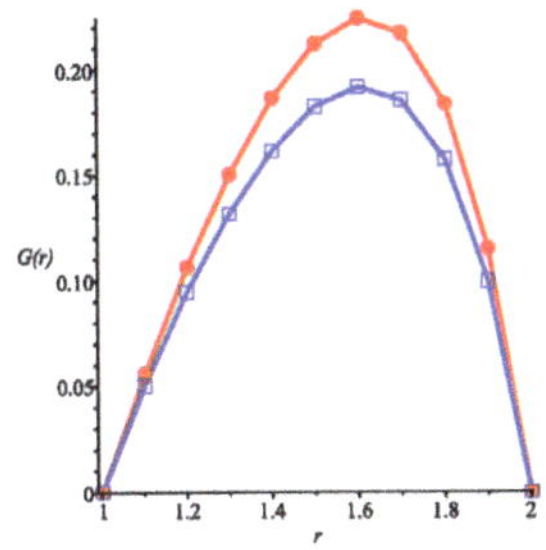

Figure 5. Function $G(r)$ for $1 \leq r \leq 2$, when $b = q = \theta = 1$, $k_1 = 1$, $k_2 = 2$, and $\lambda = 1$ (with the squares) and $\lambda = 0$ (with the circles).

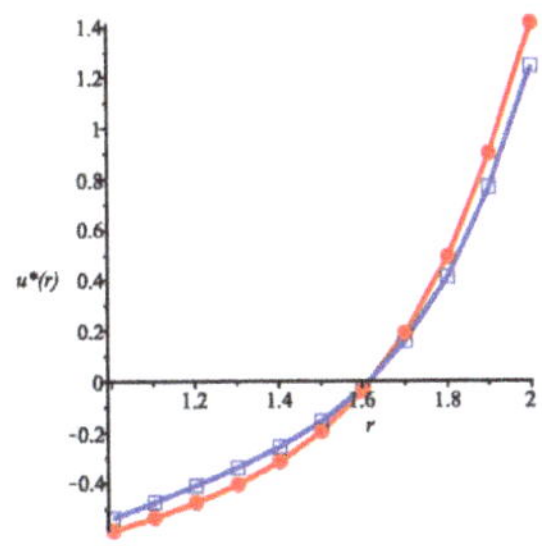

Figure 6. Optimal control in the interval $[1,2]$ when $b = q = \theta = 1$, $k_1 = 1$, $k_2 = 2$, and $\lambda = 1$ (with the squares) and $\lambda = 0$ (with the circles).

5. Conclusions

In this paper, we have considered degenerate two-dimensional jump-diffusion processes, defined in such a way that the first component of the vector $(X(t), Y(t))$ is a strictly increasing or decreasing function with respect to time. This kind of process is more realistic than a one-dimensional diffusion or jump-diffusion process in many applications, especially when $X(t)$ represents the age or wear of a certain device. We could generalize the model by incorporating more than one diffusion process $Y(t)$. The diffusion processes could model the various variables that influence $X(t)$. For example, in the case of wear, important factors to consider are temperature, speed of use, etc.

In Section 2, we obtained equations for functions defined in terms of a first-passage time for the processes $(X(t), Y(t))$. Moreover, we treated an optimal control problem for these processes in Section 3. Finally, a particular problem was solved explicitly in Section 4.

As mentioned in Section 1, there are few first-passage problems for one-dimensional jump-diffusion processes that have been solved exactly and explicitly so far. Here, we were able to find exact analytical expressions for quantities defined in terms of a first-passage time for a (degenerate) two-dimensional jump-diffusion process. Furthermore, in Section 2, we saw that the processes considered in this paper could serve as models in real-life applications, such as the remaining amount of deterioration that a given device can undergo before it needs to be repaired or replaced.

In general, to solve this type of problem, it is necessary to find the solution of an integro-differential equation with partial derivatives. We considered the case when the process leaves the continuation region at the latest when the first event of the Poisson process occurs. In this case, the equation to be solved becomes a partial differential equation. Using the method of similarity solutions, it is sometimes possible to reduce this PDE to an ODE. It should also be possible to find a numerical solution to any particular problem.

As a follow-up to this work, we would like to find exact analytical solutions to problems where the equations to be solved are integro-differential equations; for example, by trying to transform the integro-differential equations into differential equations.

Funding: This research was supported by the Natural Sciences and Engineering Research Council of Canada (NSERC). The author also wishes to thank the anonymous reviewers of this paper for their constructive comments.

Conflicts of Interest: The author declares no conflict of interest.

References

1. Höpfner, R. On a set of data for the membrane potential in a neuron. *Math. Biosci.* **2007**, *207*, 275–301. . [CrossRef] [PubMed]
2. Jahn, P.; Berg, R.W.; Hounsgaard, J.; Ditlevsen, S. Motoneuron membrane potentials follow a time inhomogeneous jump diffusion process. *J. Comput. Neurosci.* **2011**, *31*, 563–579. [CrossRef] [PubMed]
3. Melanson, A.; Longtin, A. Data-driven inference for stationary jump-diffusion processes with application to membrane voltage fluctuations in pyramidal neurons. *J. Math. Neurosci.* **2019**, *9*, 30. [CrossRef] [PubMed]
4. Lefebvre, M. A first-passage problem for exponential integrated diffusion processes. *J. Stoch. Anal.* **2022**, *3*, 2. [CrossRef]
5. Lachal, A. L'intégrale du mouvement brownien. *J. Appl. Probab.* **1993**, *30*, 17–27. [CrossRef]
6. Makasu, C. Exit probability for an integrated geometric Brownian motion. *Stat. Probab. Lett.* **2009**, *79*, 1363–1365. [CrossRef]
7. Metzler, A. The Laplace transform of hitting times of integrated geometric Brownian motion. *J. Appl. Probab.* **2013**, *50*, 295–299. [CrossRef]
8. Caravelli, F.; Mansour, T.; Sindoni, L.; Severini, S. On moments of the integrated exponential Brownian motion. *Eur. Phys. J. Plus* **2008**, *131*, 245. [CrossRef]
9. Levy, E. On the moments of the integrated geometric Brownian motion. *J. Comput. Appl. Math.* **2018**, *342*, 263–273. [CrossRef]
10. Cox, D.R.; Miller, H.D. *The Theory of Stochastic Processes*; Methuen: London, UK, 1965.
11. Kou, S.G.; Wang, H. First passage times of a jump diffusion process. *Adv. Appl. Probab.* **2003**, *35*, 504–531. [CrossRef]
12. Chen, Y.-T.; Sheu, Y.-C.; Chang, M.-C. A note on first passage functionals for hyper-exponential jump-diffusion processes. *Electron. Commun. Probab.* **2013**, *18*, 8. [CrossRef]
13. Yin, C.; Wen, Y.; Zong, Z.; Shen, Y. The first passage time problem for mixed-exponential jump processes with applications in insurance and finance. *Abstr. Appl. Anal.* **2014**, *2014*, 571724 . [CrossRef]
14. Karnaukh, I. Exit problems for Kou's process in a Markovian environment. *Theory Stoch. Process.* **2020**, *25*, 37–60. [CrossRef]
15. Lefebvre, M. Exit problems for jump-diffusion processes with uniform jumps. *J. Stoch. Anal.* **2020**, *1*, 5. [CrossRef]
16. Abundo, M. On first-passage times for one-dimensional jump-diffusion processes. *Probab. Math. Stat.* **2020**, *20*, 399–423.
17. Belkaid, A.; Utzet, F. Efficient computation of first passage times in Kou's jump-diffusion model. *Methodol. Comput. Appl. Probab.* **2017**, *19*, 957–971. [CrossRef]
18. Di Crescenzo, A.; Di Nardo, E.; Ricciardi, L.M. On certain bounds for first-crossing-time probabilities of a jump-diffusion process. *Sci. Math. Jpn.* **2006**, *64*, 449–460.
19. Fernández, L.; Hieber, P.; Scherer, M. Double-barrier first-passage times of jump-diffusion processes. *Monte Carlo Methods Appl.* **2013**, *19*, 107–141. [CrossRef]
20. D'Onofrio, G.; Lanteri, A. Approximating the first passage time density of diffusion processes with state-dependent jumps. *Fractal Fract.* **2023**, *7*, 30. [CrossRef]
21. Lefebvre, M. Exact solutions to optimal control problems for Wiener processes with exponential jumps. *J. Stoch. Anal.* **2021**, *2*, 1. [CrossRef]
22. Rishel, R. Controlled wear process: Modeling optimal control. *IEEE Trans. Autom. Control* **1991**, *36*, 1100–1102. [CrossRef]
23. Whittle, P. *Optimization over Time*; Wiley: Chichester, UK, 1982; Volume 1.
24. Whittle, P. *Risk-Sensitive Optimal Control*; Wiley: Chichester, UK, 1990.

fractal and fractional

MDPI

Article

Numerical Simulation of Nonlinear Stochastic Analysis for Measles Transmission: A Case Study of a Measles Epidemic in Pakistan

Bing Guo [1], Asad Khan [2,*] and Anwarud Din [3,*]

1 School of Mathematic and Statistics, Jishou University, Jishou 416000, China
2 School of Computer Science and Cyber Engineering, Guangzhou University, Guangzhou 510006, China
3 Department of Mathematics, Sun Yat-sen University, Guangzhou 510275, China
* Correspondence: asad@gzhu.edu.cn (A.K.); anwarud@mail.sysu.edu.cn (A.D.)

Abstract: This paper presents a detailed investigation of a stochastic model that rules the spreading behavior of the measles virus while accounting for the white noises and the influence of immunizations. It is hypothesized that the perturbations of the model are nonlinear, and that a person may lose the resistance after vaccination, implying that vaccination might create temporary protection against the disease. Initially, the deterministic model is formulated, and then it has been expanded to a stochastic system, and it is well-founded that the stochastic model is both theoretically and practically viable by demonstrating that the model has a global solution, which is positive and stochastically confined. Next, we infer adequate criteria for the disease's elimination and permanence. Furthermore, the presence of a stationary distribution is examined by developing an appropriate Lyapunov function, wherein we noticed that the disease will persist for $\mathbb{R}_0^{s>1}$ and that the illness will vanish from the community when $\mathbb{R}_0^{s<1}$. We tested the model against the accessible data of measles in Pakistan during the first ten months of 2019, using the conventional curve fitting methods and the values of the parameters were calculated accordingly. The values obtained were employed in running the model, and the conceptual findings of the research were evaluated by simulations and conclusions were made. Simulations imply that, in order to fully understand the dynamic behavior of measles epidemic, time-delay must be included in such analyses, and that advancements in every vaccine campaign are inevitable for the control of the disease.

Keywords: stochastic measles epidemic model; stationary distribution; parameter estimation; real data; Pakistan measles outbreak

Citation: Guo, B.; Khan, A.; Din, A. Numerical Simulation of Nonlinear Stochastic Analysis for Measles Transmission: A Case Study of a Measles Epidemic in Pakistan. *Fractal Fract.* **2023**, *7*, 130. https://doi.org/10.3390/fractalfract7020130

Academic Editor: Mario Abundo

Received: 14 December 2022
Revised: 11 January 2023
Accepted: 25 January 2023
Published: 30 January 2023

1. Introduction

Measles is still a major worldwide health issue, particularly in the less developed countries. Measles (also known as Rubella or morbilli) is an extremely contagious illness caused mostly by the Morbillivirus genus in the Paramyxovirus [1,2]. Although efficient vaccines against this severe illness are commonly accessible, still measles is a leading cause of death among children below five of years of age [3]. The disease infecting hundreds of millions of children each year and resulting in a high mortality and morbidity in the child population, owing primarily to complicated conditions that exist side by side with the disease like malnutrition, diarrhea, and pneumonia [4]. Sneezing and coughing, touching the nasal or aerosol fluids, or close physical touch with an infected person are all ways to spread measles. It can stay extremely contagious for a maximum of two hours in the atmosphere and on the surfaces. Clinical manifestations include soar throat, cough, nasal congestion, blurred vision, and small white patches in the mouth; these signs and symptoms often develop within 10–12 days post-infection. Subsequently, a rash appears, extending downward out of the nose. The period of peak infectivity (virus shed) starts four days before that and four days just after commencement of the rashes. The usual incubation time is 14 days; however, it can range from 7–18 days [5]. In reality, even vaccinated people may still be susceptible if the immunization fails or existing immunity from the vaccine wears off. Despite the fact that vaccination has cut worldwide measles fatalities by 73% between 2000–2018, measles continues to remain a widespread in many underdeveloped nations, particularly in Asia and Africa. Around 140,000 individuals died from the measles in 2018. During 2000–2018, worldwide measles immunization results in an 85 percent drop in measles transmission [6,7]. Despite the abundance of a safe and effective vaccination in 2017, around 110,000 deaths occurred from measles, primarily children under the age of six, as per the report of the World Health Organization (WHO) [8]. Vaccination is amongst the most successful health promotion strategy for reducing death rates and the spreading of epidemics; it has been demonstrated that vaccination saves over 3 million people only in Nigeria every year. The vaccination will work with the immune system of the body to establish the body's natural defenses, reducing the likelihood of relapse [9]. The MMR vaccination can protect against measles, which is a vaccine-preventable illness. The MMR vaccines are highly effective at protecting both adults and children from the epidemic measles. Only one dose of the MMR vaccines is roughly 92% successful in suppressing the measles, whereas two doses are around 95% effective. The MMR vaccine also protects against rubella and mumps [10]. This disease is an endemic one in Nigeria, with epidemics occurring at regular intervals. Measles is present across Nigeria at all seasons; however, it is more widespread in the summer months.

Pakistan is one of the most measles-affected nations in Asia [11]. Every 8–10 years, the nation has a recurring measles epidemic. In fact, 2845 identified measles infections were reported in Pakistan during the year 2016. This figure increased to 6791 in 2017 and, in the year 2018, 33,007 cases were reported. These results represent about 44, 20, and 51 percent of the total number of cases recorded in the East Mediterranean, which includes 22 nations. In 2017, over 130 children lost their lives due to this infection, with the figure rising to nearly 300 in 2018 [12].

It is strongly advised to employ the methods of mathematical modeling to examine the transmission process and prevention of an epidemic disease [7,13–15], modeling with fractional differential equations also have several applications in all fields of science [16–18]. While depicting the natural history of an infectious disease, the tools of modeling can create a balance among the data and its real biology. Thus far, models responsible for describing the dynamics of measles both from population to outbreak levels have demonstrated a wide variety of disease patterns. External noise is usually the main significant feature of physical processes and bio-systems. It has been discovered that environmental variables have a significant impact on the dynamics of measles disease spread [19]. Because of the uncertainty of person-to-person interactions or inherent characteristics of the population, outbreak onset and propagation are fundamentally unpredictable. As a result, the condition of the disease is influenced by the environment's heterogeneity and uncertainty.

Changes in the environment likewise have a significant impact on the parasites' persistence and distribution. Because the stochasticity of parameters and states depicts the exact dynamic behavior of an infectious disease, it is regarded as an essential part in epidemiological studies. Even though the perturbations are varying, these should be autocorrelated in a positive way. Furthermore, the perturbations may be estimated theoretically using the linked problem's probability density function [20–23]. There are two main techniques to epidemic modeling: the deterministic modeling and stochastic modeling. Stochastic differential equation (SDE) models are recommended over deterministic models for mathematical modeling of biological functions because they may provide a higher level of reality than their deterministic equivalents [23–26]. We may choose to use SDEs to generate a distribution of the predicted output(s); for example, the number of infected individuals over time t. Moreover, when tested numerous times, a stochastic model produces more useful outputs than a deterministic one. A deterministic model, on the other hand, will produce only one outcome irrespective of the number of experiments. To explain the viral evolution of COVID-19, numerous deterministic infectious disease models have been suggested; for example, see [27,28].

The rest of the manuscript is organized in the following manner: Section 2 deals with the formulation of the proposed stochastic model for the spreading of the epidemic measles. The uniqueness and existence problem for obtaining a positive global solution is presented in Section 3. In Sections 4 and 5, we characterized sufficient criteria for the stationary distribution and extinction of the disease. We optimize the proposed model using data from Pakistan compiled in the first ten months of 2019 in Section 6. We quantitatively compared the obtained analytical results, and graphical illustrations were presented in Section 7. We concluded the work in Section 8 by presenting the future directions and a comprehensive summary.

2. Model Formulation

Olumuyiwa et al. [29] have recently developed a model of the transmission of rubella disease by using the differential equations. Keeping in view the work of Olumuyiwa et al., here we intend to extend the model to a stochastic model. Furthermore, by considering different stages during the measles epidemic, we stratified the total population into six disjoint classes, namely: susceptible, vaccinated, exposed, infectious, hospitalized and recovered individuals whose sizes in mathematical terms are, respectively, $\mathbb{S}(t), \mathbb{V}(t), \mathbb{E}(t), \mathbb{I}(t), \mathbb{H}(t)$, and $\mathbb{R}(t)$.

The entrance of new persons through this population is captured by the rate ϕ and will be kept in the susceptible compartment. People in the vulnerable group start receiving a vaccination at a rate τ and setback immune function at a vaccine wane rate ω. The contact rate of susceptible persons is α, and thus the term force of infection becomes αSI, with the transition again from exposures to infection stages occurring at a rate of β. People who are infected with the measles seek medical attention at a rate of ρ and recover from the infection after the successful treatment supplemented at a rate of γ. We consider two types of death rates: the natural μ (that is constant for all classes) and the disease-induced mortality rate δ. This study assumes the recovery from measles that is possible due to the treatment only, that is, the study considering no natural recovery. The movements between the compartments are depicted via a flowchart in Figure 1. The above assumptions will lead to the following deterministic system:

$$
\begin{aligned}
\frac{d\mathbb{S}(t)}{t} &= \phi - \alpha\mathbb{S}(t)\mathbb{I}(t) + \omega\mathbb{V}(t) - (\tau+\mu)\mathbb{S}(t),\\
\frac{d\mathbb{V}(t)}{t} &= \tau\mathbb{S}(t) - (\mu+\omega)\mathbb{V}(t),\\
\frac{d\mathbb{E}(t)}{t} &= \alpha\mathbb{S}(t)\mathbb{I}(t) - (\mu+\beta)\mathbb{E}(t),\\
\frac{d\mathbb{I}(t)}{t} &= \beta\mathbb{E}(t) - (\mu+\delta+\rho)\mathbb{I}(t),\\
\frac{d\mathbb{H}(t)}{t} &= \rho\mathbb{I}(t) - (\delta+\gamma+\mu)\mathbb{H}(t),\\
\frac{d\mathbb{R}(t)}{t} &= \gamma\mathbb{H}(t) - \mu\mathbb{R}(t).
\end{aligned} \tag{1}
$$

The threshold parameter has the following expression for model (1) as

$$
\mathbb{R}_0 = \frac{(\mu+\omega)\phi\beta\alpha}{(\mu+\beta)(\mu+\delta+\rho)(\mu+\omega+\tau)\mu}. \tag{2}
$$

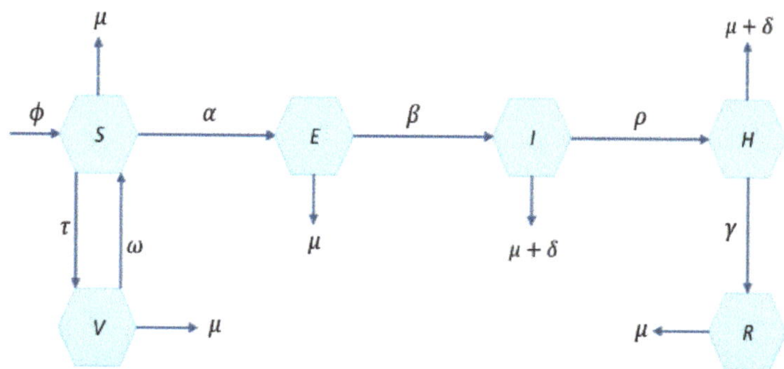

Figure 1. Flow chart of the measles model (1) [29].

In order to consider the noises due the environment in the study (i.e., model (1)), we shall take into account the standard Brownian motions $W_i(t)$ for $i = 1, \cdots, 6$ with $W_i(0) = 0$. Furthermore, system (1) is modified by considering the incidence rate of bi-linear form $\frac{\alpha\mathbb{S}(t)\mathbb{I}(t)}{\mathbb{N}(t)}$. For the noise intensities, we have taken $\xi_1, \xi_2, \xi_3, \xi_4, \xi_5, \xi_6$ as the relative weights. By considering these stochastic terms, the deterministic model (1) takes the form

$$\begin{aligned} d\mathbb{S} &= \left[\phi - \frac{\alpha\mathbb{S}(t)\mathbb{I}(t)}{\mathbb{N}(t)} + \omega\mathbb{V}(t) - (\tau+\mu)\mathbb{S}(t)\right]dt + \xi_1\mathbb{S}(t)d\mathbb{W}_1(t), \\ d\mathbb{V} &= \left[\tau\mathbb{S}(t) - (\mu+\omega)\mathbb{V}(t)\right]dt + \xi_2\mathbb{V}d\mathbb{W}_2(t), \\ d\mathbb{E} &= \left[\frac{\alpha\mathbb{S}(t)\mathbb{I}(t)}{\mathbb{N}(t)} - (\mu+\beta)\mathbb{E}(t)\right]dt + \xi_3\mathbb{E}(t)d\mathbb{W}_3(t), \\ d\mathbb{I} &= \left[\beta\mathbb{E}(t) - (\mu+\delta+\rho)\mathbb{I}(t)\right]dt + \xi_4\mathbb{I}(t)d\mathbb{W}_4(t), \\ d\mathbb{H} &= \left[\rho\mathbb{I}(t) - (\mu+\delta+\gamma)\mathbb{H}(t)\right]dt + \xi_5\mathbb{I}(t)d\mathbb{W}_5(t), \\ d\mathbb{R} &= \left[\gamma\mathbb{H}(t) - \mu\mathbb{R}(t)\right]dt + \xi_6\mathbb{R}(t)d\mathbb{W}_6(t). \end{aligned} \tag{3}$$

Keeping in view system (3), the authors have a keen interest to find the possible answers to the following questions:

Q_1 : What role do the random noises play in the transmission measles?
Q_2 : What role contaminated vaccination in the spreading of measles disease?
Q_3 : Under what condition(s) will the disease tend to go extinct?
Q_4 : Under what condition(s) will the epidemic measles persist in the population?

3. Stochastic Model Analysis

This section investigates the existence/uniqueness of solutions, global asymptotic behavior, derivation of conditions under which the disease tends to go extinct, and the presence of an ergodic stationary distribution for the proposed stochastic model.

Positive Global Solution of the Model

The very first crucial question in studying the dynamic behavior is whether there is a possibility of the existence of a global solution to the model. Furthermore, for a system modeling the population dynamics, the nature of the solution's value is of major relevance. In addition, we demonstrate in this section that the solution of randomized system (3) is global in nature and positive. It is well established that, for every given initial amount, the coefficients of a stochastic equation must fulfill the normal growth constraint and the local Lipschitz criterion in order to have a unique global solution (i.e., without any explosion in a limited period).

Theorem 1. *There exists a unique solution* $(\mathbb{S}(t), \mathbb{V}(t), \mathbb{E}(t), \mathbb{I}(t), \mathbb{H}(t), \mathbb{R}(t))$ *of system (3) for* $t \geq 0$ *under the initial conditions from the state* $\mathbb{R}^6_+$. *Moreover, the solution remains in the same space (i.e.,* $\mathbb{R}^6_+$*) surely for* $\forall t \geq 0$.

Proof. Keeping in view the Lipschitzness of the coefficients used in the model and from the fact $((\mathbb{S}(0), \mathbb{V}(0), \mathbb{E}(0), \mathbb{I}(0), \mathbb{H}(0), \mathbb{R}(0))) \in \mathbb{R}^6_+$, we can say that the proposed system has a unique local solution in $[0, \tau_e)$ and $t \geq 0$. The term τ_e stands for the explosion time, and readers are referred to [30,31] for a detailed analysis. By showing that, actually $\tau_e = \infty$, we reach the conclusion that such a solution is global in nature. To do so, it is necessary that we assume a large $k_0 > 0$ in such a way that $[\frac{1}{k_0}, k_0]$ contains all parts of the solution. Assume $k_0 \leq k$ and let us define

$$\tau_k = inf\{t \in [0, \tau_e) : \frac{1}{k} \geq \min\{\mathbb{S}(t), \mathbb{V}(t), \mathbb{E}(t), \mathbb{I}(t), \mathbb{H}(t), \mathbb{R}(t)\} \text{ or } k \leq \max\{\mathbb{S}(t), \mathbb{V}(t), \mathbb{E}(t), \mathbb{I}(t), \mathbb{H}(t), \mathbb{R}(t)\}. \quad (4)$$

Whenever ϕ represents the empty set, then we shall write $\inf \phi = \infty$. By increasing the value of k, one can notice that it increases τ_k. We apply the limit $k \to \infty$ and assume that the $\tau_k \to \tau_\infty$ and a.s. $\tau_e \geq \tau_\infty$. Thus, if we show $\tau_\infty = \infty$ a.s., it ensures $\tau_e = \infty$. Proving all these guarantees that $(\mathbb{S}(t), \mathbb{V}(t), \mathbb{E}(t), \mathbb{I}(t), \mathbb{H}(t), \mathbb{R}(t)) \in \mathbb{R}^6_+$ for any time $t \geq 0$. Let us assume the contrary case that $\tau_e \neq \infty$; then, there must exist positive real numbers T and $\epsilon \in (0,1)$ in such a way

$$\epsilon < P\{\tau_\infty \leq T\}. \quad (5)$$

Thus, for an integer $k_0 \leq k_1$, we have

$$P\{T \geq \tau_k\} \geq \epsilon, \; \forall \, k_1 \leq k.$$

To begin, first let us establish a Lyapunov function of the type

$$G = (\mathbb{S} - 1 - \log\mathbb{S}) + (\mathbb{E} - \log\mathbb{E} - 1) + (I - \log\mathbb{I} - 1) + (\mathbb{H} - \log\mathbb{H} - 1) + (\mathbb{R} - \log\mathbb{R} - 1), \quad (6)$$

By utilizing the formula due to *Itô*, letting $k_0 \leq k$ and assuming a very large non-negative real T, Equation (6) can be written in the form of

$$dG(\mathbb{S}, \mathbb{V}, \mathbb{E}, \mathbb{I}, \mathbb{H}, \mathbb{R}) = LG(\mathbb{S}, \mathbb{V}, \mathbb{E}, \mathbb{I}, \mathbb{H}, \mathbb{R})dt + \xi_1(\mathbb{S} - 1)d\mathbb{W}_1(t) + \xi_2(\mathbb{V} - 1)d\mathbb{W}_2(t) + \xi_3(\mathbb{E} - 1)d\mathbb{W}_3(t) + \xi_4(\mathbb{I} - 1)d\mathbb{W}_4(t) + \xi_5(\mathbb{H} - 1)d\mathbb{W}_5(t) + \xi_6(\mathbb{R} - 1)d\mathbb{W}_6(t). \quad (7)$$

In Equation (23), the $L\mathbb{V}$ operator is from the space $\mathbb{R}^6_+$ to $\mathbb{R}_+$. □

The remaining parts of the proof are merely similar to Theorem 2.1 in [26,30]. Thus, it is very simple for the reader to follow the result and and hence, is not completely proved here.

4. Extinction

While modeling the dynamical aspects of any epidemic diseases, it is important to investigate the situations under which the disease will become exterminated or tend to become extinct in the community. Within this section, we demonstrate that, when the white noises are large enough, the solution of the associated stochastic model (3) surely vanishes. Let us define

$$\langle \mathbb{B}(t) \rangle = \frac{1}{t} \int_0^t \mathbb{B}(s) ds.$$

Lemma 1. *(Strong Law)* [32,33] *Let $\mathbb{Z} = \{\mathbb{Z}\}_{0 \leq t}$ be continuous and real valued along with a local martingale, which vanishes as $t \to 0$, then*

$$\begin{aligned} &\lim_{t\to\infty} \langle \mathbb{Z}, \mathbb{Z} \rangle_t = \infty, \text{ a.s.}, \Rightarrow \lim_{t\to\infty} \frac{\mathbb{Z}_t}{\langle \mathbb{Z}, \mathbb{Z} \rangle_t} = 0, \text{ a.s.} \\ &\limsup_{t\to\infty} \frac{\langle \mathbb{Z}, \mathbb{Z} \rangle_t}{t} < 0, \text{ a.s.}, \Rightarrow \lim_{t\to\infty} \frac{\mathbb{Z}_t}{t} = 0, \text{ a.s.} \end{aligned} \tag{8}$$

Lemma 2. *Assume that $(\mathbb{S}, \mathbb{V}, \mathbb{E}, \mathbb{I}, \mathbb{H}, \mathbb{R})$ corresponds to initial data $\mathbb{S}(0), \mathbb{V}(0), \mathbb{E}(0), \mathbb{I}(0), \mathbb{H}(0), \mathbb{R}(0))$ in the space $\mathbb{R}^6_+$ and is a solution of model (3). Then,*

$$\begin{aligned} &\limsup_{t\to\infty} \frac{\ln \mathbb{S}(t)}{t} = 0, \ \limsup_{t\to\infty} \frac{\mathbb{V}(t)}{t} = 0, \ \limsup_{t\to\infty} \frac{\ln \mathbb{E}(t)}{t} = 0, \\ &\limsup_{t\to\infty} \frac{\ln \mathbb{I}(t)}{t} = 0, \ \limsup_{t\to\infty} \frac{\ln \mathbb{H}(t)}{t} = 0, \ \limsup_{t\to\infty} \frac{\ln \mathbb{R}(t)}{t} = 0, \text{ a.s.} \end{aligned} \tag{9}$$

Furthermore, if $\mu > \frac{\xi_1^2 \vee \xi_2^2 \vee \xi_3^2 \vee \xi_4^2}{2}$, and $d > \frac{\xi_5^2}{2}$, then

$$\begin{aligned} &\lim_{t\to\infty} \frac{\int_0^t \mathbb{S}(s) d\mathbb{W}_1(s)}{t} = 0, \ \lim_{t\to\infty} \frac{\int_0^t \mathbb{V}(u) d\mathbb{W}_2(u)}{t} = 0, \ \lim_{t\to\infty} \frac{\int_0^t \mathbb{E}(u) d\mathbb{W}_3(u)}{t} = 0, \\ &\lim_{t\to\infty} \frac{\int_0^t \mathbb{I}(s) d\mathbb{W}_4(s)}{t} = 0, \ \lim_{t\to\infty} \frac{\int_0^t \mathbb{H}(s) d\mathbb{W}_5(s)}{t} = 0, \ \lim_{t\to\infty} \frac{\int_0^t \mathbb{R}(s) d\mathbb{W}_6(s)}{t} = 0, \text{ a.s.} \end{aligned} \tag{10}$$

Then, the solution of (3)

$$\begin{aligned} \limsup_{t\to\infty} \mathbb{S}(t) &= \frac{(\mu + \omega)\phi}{(\mu + \omega + \tau)\mu}, \\ \limsup_{t\to\infty} \mathbb{V}(t) &= \frac{\phi\tau}{(\mu + \omega + \tau)\mu}, \\ \limsup_{t\to\infty} \mathbb{E}(t) &= 0, \\ \limsup_{t\to\infty} \mathbb{I}(t) &= 0, \\ \limsup_{t\to\infty} \mathbb{H}(t) &= 0, \\ \limsup_{t\to\infty} \mathbb{R}(t) &= 0, \text{ a.s.} \end{aligned} \tag{11}$$

To prove Lemma 2, we follow almost the same techniques as performed in proving Lemmas 2.1 and 2.2 carried out in the work of Zhao [32], and thus the readers can simply prove the results.

Next, to develop the extinction theory related to model (3), we first define the threshold quantity $\mathbb{R}_s$ for the proposed stochastic model which can be written in the form of

$$\mathbb{R}_s = \frac{\alpha}{\left[(\mu + \beta)(\mu + \delta + \rho) + \frac{\xi_3^2}{2} + \frac{\xi_4^2}{2} \right]}. \tag{12}$$

Theorem 2. *The exposed population $\mathbb{E}(t)$ and infected population $\mathbb{I}(t)$ of system ((3) tends to zero exponentially almost surely if $\mathbb{R}_s < 1$, where $\mathbb{R}_s$ is given by Equation (12).*

Proof. Let $(\mathbb{S}(t), \mathbb{V}(t), \mathbb{E}(t), \mathbb{I}(t), \mathbb{H}(t), \mathbb{R}(t))$ correspond to initial data $\mathbb{S}(0), \mathbb{V}(0), \mathbb{E}(0), \mathbb{I}(0), \mathbb{H}(0), \mathbb{R}(0))$ in the space $\mathbb{R}^6_+$, being a solution of model (3).

Define

$$G_1(t) = \beta\mathbb{E}(t) + (\mu+\beta)\mathbb{I}(t). \tag{13}$$

Differentiating Equation (13) following Ito's formula, one can obtain

$$\begin{aligned}
d(lnG_1(t)) &= \frac{1}{G_1}\left[\frac{\alpha\beta\mathbb{S}\mathbb{I}}{\mathbb{N}} - (\mu+\beta)(\mu+\delta+\rho)\mathbb{I}\right] - \frac{\beta^2\mathbb{E}^2\xi_3^2 + (\mu+\beta)^2\xi_4^2\mathbb{I}^2}{2(G_1)^2} \\
&+ \frac{\beta\xi_3}{[\beta\mathbb{E}(t) + (\mu+\beta)\mathbb{I}]}\mathbb{E}d\mathbb{W}_3(t) + \frac{(\mu+\beta)\xi_4}{[\mathbb{E} + (\mu+\beta)\mathbb{I}]}\mathbb{I}d\mathbb{W}_4(t), \\
&\leq \frac{1}{G_1}\left[\alpha\beta\mathbb{I} - (\mu+\beta)(\mu+\delta+\rho)\mathbb{I}\right] - \frac{\beta^2\mathbb{E}^2\xi_3^2}{2(G_1)^2} - \frac{(\mu+\beta)^2\xi_4^2\mathbb{I}^2}{2(G_1)^2} \\
&+ \frac{\beta\xi_3}{[\beta\mathbb{E}(t) + (\mu+\beta)\mathbb{I}]}\mathbb{E}d\mathbb{W}_3(t) + \frac{(\mu+\beta)\xi_4}{[\mathbb{E} + (\mu+\beta)\mathbb{I}]}\mathbb{I}d\mathbb{W}_4(t), \quad [\because \mathbb{S} \leq \mathbb{N}] \\
&\leq \frac{1}{(\mu+\beta)}\left[\alpha - (\mu+\beta)(\mu+\delta+\rho)\right] - \frac{\xi_3^2}{2} - \frac{\xi_4^2}{2} \\
&+ \frac{\beta\xi_3}{[\beta\mathbb{E}(t) + (\mu+\beta)\mathbb{I}]}\mathbb{E}d\mathbb{W}_3(t) + \frac{(\mu+\beta)\xi_4}{[\mathbb{E} + (\mu+\beta)\mathbb{I}]}\mathbb{I}d\mathbb{W}_4(t). \quad [\because \mathbb{I} \leq \mathbb{I} + \frac{\beta\mathbb{E}}{(\mu+\beta)}]
\end{aligned} \tag{14}$$

By taking the integral of either sides of the above inequality over the interval $[0, t]$, we have

$$\begin{aligned}
lnG_1(t) &\leq \frac{1}{(\mu+\beta)}\left\{\alpha - \left[(\mu+\beta)(\mu+\delta+\rho) + \frac{\xi_3^2}{2} + \frac{\xi_4^2}{2}\right]\right\} \\
&+ \int_0^t \frac{\beta\xi_3\mathbb{E}d\mathbb{W}_3(t)}{[\beta\mathbb{E}(t) + (\mu+\beta)\mathbb{I}]} + \int_0^t \frac{(\mu+\beta)\xi_4\mathbb{I}d\mathbb{W}_4(t)}{[\mathbb{E} + (\mu+\beta)\mathbb{I}]}, \\
&\leq \frac{1}{(\mu+\beta)}\left\{\alpha - \left[(\mu+\beta)(\mu+\delta+\rho) + \frac{\xi_3^2}{2} + \frac{\xi_4^2}{2}\right]\right\} \\
&+ \int_0^t \frac{\beta\xi_3\mathbb{E}d\mathbb{W}_3(t)}{[\beta\mathbb{E}(t) + (\mu+\beta)\mathbb{I}]} + \int_0^t \frac{(\mu+\beta)\xi_4\mathbb{I}d\mathbb{W}_4(t)}{[\mathbb{E} + (\mu+\beta)\mathbb{I}]}, \\
&\leq \frac{\left[(\mu+\beta)(\mu+\delta+\rho) + \frac{\xi_3^2}{2} + \frac{\xi_4^2}{2}\right]}{(\mu+\beta)}\left[\mathbb{R}_s - 1\right] + \int_0^t \frac{\beta\xi_3\mathbb{E}d\mathbb{W}_3(t)}{[\beta\mathbb{E}(t) + (\mu+\beta)\mathbb{I}]} + \int_0^t \frac{(\mu+\beta)\xi_4\mathbb{I}d\mathbb{W}_4(t)}{[\mathbb{E} + (\mu+\beta)\mathbb{I}]}.
\end{aligned} \tag{15}$$

By assuming the lim sup as $t \to \infty$ and multiplying both sides of relation (15) by $\frac{1}{t}$ while considering Lemma 2, we obtain

$$\limsup_{t\to\infty}(lnG_1(t)) \leq \frac{\left[(\mu+\beta)(\mu+\delta+\rho) + \frac{\xi_3^2}{2} + \frac{\xi_4^2}{2}\right]}{(\mu+\beta)}\left[\mathbb{R}_s - 1\right]. \tag{16}$$

If $\mathbb{R}_s < 1$, then $\lim_{t\to\infty} G_1 = \lim_{t\to\infty}[\beta\mathbb{E}(t) + (\mu+\beta)\mathbb{I}(t)] = 0,$ a.s if $\mathbb{R}_s < 1$. Again, $\beta > 0, (\mu+\beta) > 0$, and we assert that $\lim_{t\to\infty}[\beta\mathbb{E}(t) + (\mu+\beta)\mathbb{I}(t)] = 0 \Longrightarrow \lim_{t\to\infty}\mathbb{E} = \lim_{t\to\infty}\mathbb{I} = 0$—hence the result. □

5. The Stationary Distribution of the Disease

We understand that there are no endemic states in stochastic models. As a result, the stability of the system cannot be used to predict how long an illness would endure. As a result, one must concentrate on the existence/uniqueness assumption of the stationary distribution. In certain aspects, this assists the illness with survival. For this purpose, we employ a well-known method due to Hasminskii [34].

Assume a homogeneously time-dependent Markov process $X(t)$ in the space R^n_+ that satisfies the below stochastic model

$$dX(t) = b(X)dt + \sum_{r}^{k} \sigma_r dB_r(t).$$

The diffusion matrix can be demonstrated as:

$$A(X) = [a_{ij}(\varkappa)], \ a_{ij}(\varkappa) = \sum_{r=1}^{k} \sigma_r^i(\varkappa)\sigma_j^r(\varkappa).$$

Lemma 3. [34]. *The process $X(t)$ has a one and only one stationary distribution $m(.)$ whenever there is a bounded domain having a regular boundary in such a way that $\bar{U} \in R^d \setminus \bar{U}$ closure $\bar{U} \in R^d$, and having the below characteristics*

1. *The smallest eigenvalue of the matrix $A(t)$ is bounded below from the origin in the open domain U and in its neighborhood.*
2. *For $\varkappa \in R^d \setminus U$, the average time period τ (at which a path starts from $\varkappa$ reaching the set U) is bounded, and for every compact subset $K \subset R^n$, we have $Sup_{\varkappa \in k} E^{\varkappa}\tau < \infty$. When $f(.)$ is an integrable function with the measure pi, then we have*

$$P\left\{ \lim_{T\to\infty} \frac{1}{T} \int_0^T f(X_{\varkappa}(t))dt = \int_{R^d} f(\varkappa)\pi(dx) \right\} = 1,$$

for all $\varkappa \in R^d$.

Let us define the following parameter for our future use:

$$\mathbb{R}_0^s = \frac{\mu\beta\alpha}{\left(\xi + \mu + \frac{\sigma_2^2}{2}\right)\left(\alpha + \mu + \frac{\sigma_3^2}{2}\right)\left(\delta + \mu + \frac{\sigma_4^2}{2}\right)}. \tag{17}$$

Theorem 3. *For $\mathbb{R}_0^s > 1$, then a solution $(\mathbb{S}(t), \mathbb{V}(t), \mathbb{E}(t), \mathbb{I}(t), \mathbb{H}(t), \mathbb{R}(t))$ of system (3) is ergodic and has one and only one stationary distribution $\pi(.)$.*

Proof. For verifying condition (2) of Lemma (3), we must develop a positive C^2-function $G_2 : R^6_+ \to R_+$. To do so, we first formulate

$$G_2 = \mathbb{S} + \mathbb{V} + \mathbb{E} + \mathbb{I} + \mathbb{H} + \mathbb{R} - a_1 \ln \mathbb{S} - a_2 \ln \mathbb{E} - a_3 \ln \mathbb{I},$$

where a_1, a_2 and a_3 are all real and positive constants, and must be calculated later on. By assuming the formula due to Itô's and keeping in view system (3), we obtain

$$
\begin{aligned}
\mathcal{L}(\mathbb{V}+\mathbb{S}+\mathbb{I}+\mathbb{E}+\mathbb{H}+\mathbb{R}) =& \phi-\mu(\mathbb{V}+\mathbb{S}+\mathbb{I}+\mathbb{E}+\mathbb{H}+\mathbb{R})-\delta(\mathbb{I}+\mathbb{H}),\\
\mathcal{L}(-\ln\mathbb{S}) =& -\frac{\Pi}{\mathbb{S}}+\frac{\alpha\mathbb{I}}{\mathbb{N}}-\frac{\omega\mathbb{V}}{\mathbb{S}}+(\tau+\mu)+\frac{\xi_1^2}{2},\\
\mathcal{L}(-\ln\mathbb{V}) =& -\frac{\tau\mathbb{S}}{\mathbb{V}}+(\mu+\omega)+\mu+\frac{\xi_2^2}{2},\\
\mathcal{L}(-\ln\mathbb{E}) =& -\frac{\alpha\mathbb{S}\mathbb{I}}{\mathbb{E}\mathbb{N}}+(\mu+\beta)+\frac{\xi_3^2}{2},\\
\mathcal{L}(-\ln\mathbb{I}) =& -\frac{\beta\mathbb{E}}{\mathbb{I}}+(\mu+\delta+\rho)+\frac{\xi_4^2}{2},\\
\mathcal{L}(-\ln\mathbb{H}) =& -\frac{\rho\mathbb{I}}{\mathbb{H}}+(\gamma+\delta+\mu)+\frac{\xi_5^2}{2},\\
\mathcal{L}(-\ln\mathbb{R}) =& -\frac{\gamma\mathbb{H}}{R}+\mu+\frac{\xi_6^2}{2}.
\end{aligned}
\tag{18}
$$

□

Hence,

$$
\mathcal{L}G_2=\phi-\mu(\mathbb{V}+\mathbb{S}+\mathbb{I}+\mathbb{E}+\mathbb{H}+\mathbb{R})-\delta(\mathbb{I}+\mathbb{H})-\frac{a_1\Pi}{\mathbb{S}}+\frac{a_1\alpha\mathbb{I}}{\mathbb{N}}-\frac{a_1\omega\mathbb{V}}{\mathbb{S}}+a_1(\tau+\mu)+\frac{a_1\xi_1^2}{2}
$$
$$
-\frac{a_2\alpha\mathbb{S}\mathbb{I}}{\mathbb{E}\mathbb{N}}+a_2(\mu+\beta)+\frac{a_2\xi_3^2}{2}-\frac{a_3\beta\mathbb{E}}{\mathbb{I}}+a_3(\mu+\delta+\rho)+\frac{a_3\xi_4^2}{2}.
$$

The preceding calculation indicates that

$$
\mathcal{L}G_2\leq-4\left[\mu\mathbb{N}\times\frac{a_1\phi}{\mathbb{S}}\times\frac{a_2\alpha\mathbb{S}\mathbb{I}}{\mathbb{E}\mathbb{N}}\times\frac{a_3\beta\mathbb{E}}{\mathbb{I}}\right]^{\frac{1}{4}}+a_1(\tau+\mu+\frac{\xi_1^2}{2})+a_2(\mu+\beta+\frac{\xi_3^2}{2})+a_3(\mu+\delta+\rho+\frac{\xi_4^2}{2})
$$
$$
+\phi+\frac{a_1\alpha\mathbb{I}}{\mathbb{N}}-\frac{a_1\omega\mathbb{V}}{\mathbb{S}}-\delta(\mathbb{I}+\mathbb{H}).
$$

Suppose

$$
a_1(\tau+\mu+\frac{\xi_1^2}{2})=a_2(\mu+\beta+\frac{\xi_3^2}{2})=a_3(\mu+\delta+\rho+\frac{\xi_4^2}{2})=\phi.
$$

Namely,

$$
\begin{aligned}
a_1 &= \frac{\phi}{(\tau+\mu+\frac{\xi_1^2}{2})},\\
a_2 &= \frac{\phi}{(\mu+\beta+\frac{\xi_3^2}{2})},\\
a_3 &= \frac{\phi}{(\mu+\delta+\rho+\frac{\xi_4^2}{2})}.
\end{aligned}
\tag{19}
$$

As a result,

$$
\mathcal{L}G_2\leq-4\left[\left(\frac{\phi^4\mu\alpha\beta}{(\tau+\mu+\frac{\xi_1^2}{2})(\mu+\beta+\frac{\xi_3^2}{2})(\mu+\delta+\rho+\frac{\xi_4^2}{2})}\right)^{\frac{1}{4}}-\phi\right]+\frac{a_1\alpha\mathbb{I}}{\mathbb{N}}-\frac{a_1\omega\mathbb{V}}{\mathbb{S}}-\delta(\mathbb{I}+\mathbb{H}),
$$

$$
\mathcal{L}G_2\leq-4\Pi\left[(\mathbb{R}_0^s)^{1/4}-1\right]+\frac{a_1\alpha\mathbb{I}}{\mathbb{N}}-\frac{a_1\omega\mathbb{V}}{\mathbb{S}}-\delta(\mathbb{I}+\mathbb{H}).
$$

Furthermore, we obtain

$$G_2 = a_4(\mathbb{E}+\mathbb{S}+\mathbb{V}+\mathbb{H}+\mathbb{I}+\mathbb{R}-a_1\ln\mathbb{S}-a_2\ln\mathbb{E}-a_3\ln\mathbb{I})-\ln\mathbb{S}-\ln\mathbb{V}-\ln\mathbb{H}-\ln\mathbb{R}$$
$$\mathbb{E}+\mathbb{S}+\mathbb{V}+\mathbb{H}+\mathbb{I}+\mathbb{R},$$
$$=(a_4+1)(\mathbb{E}+\mathbb{S}+\mathbb{V}+\mathbb{H}+\mathbb{I}+\mathbb{R})-(a_1a_4+1)\ln\mathbb{S}-a_2a_4\ln\mathbb{E}-a_3a_4\ln\mathbb{I}-\ln\mathbb{V}-\ln\mathbb{H}-\ln\mathbb{R},$$

where $a_4 > 0$ is an unknown real number and must be determined later. It is very helpful to state

$$\liminf_{(\mathbb{S},\mathbb{V},\mathbb{E},\mathbb{I},\mathbb{H},\mathbb{R})\in\mathbb{R}^6_+\setminus U_k} G_3(\mathbb{S},\mathbb{V},\mathbb{E},\mathbb{I},\mathbb{H},\mathbb{R}) = +\infty, \quad \text{as} \quad k\to\infty, \tag{20}$$

where $U_k = (\frac{1}{k},k)\times(\frac{1}{k},k)\times(\frac{1}{k},k)$.

In the below steps, we show that actually the function $G_3(\mathbb{S},\mathbb{V},\mathbb{E},\mathbb{I},\mathbb{H},\mathbb{R})$ has the minimum value, $G_3(\mathbb{S}(0),\mathbb{V}(0),\mathbb{E}(0),\mathbb{I}(0),\mathbb{H}(0),\mathbb{R}(0))$. The partial derivatives of the function G_3 with respect to the state variables are given by:

$$\begin{aligned}
\frac{\partial G_3(\mathbb{S},\mathbb{V},\mathbb{E},\mathbb{I},\mathbb{H},\mathbb{R})}{\partial\mathbb{S}} &= 1+a_4-\frac{1+a_1a_4}{\mathbb{S}},\\
\frac{\partial G_3(\mathbb{S},\mathbb{V},\mathbb{E},\mathbb{I},\mathbb{H},\mathbb{R})}{\partial\mathbb{V}} &= 1+a_4-\frac{1}{\mathbb{V}},\\
\frac{\partial G_3(\mathbb{S},\mathbb{V},\mathbb{E},\mathbb{I},\mathbb{H},\mathbb{R})}{\partial\mathbb{E}} &= 1+a_4-\frac{a_2a_4}{\mathbb{E}},\\
\frac{\partial G_3(\mathbb{S},\mathbb{V},\mathbb{E},\mathbb{I},\mathbb{H},\mathbb{R})}{\partial\mathbb{I}} &= 1+a_4-\frac{a_3a_4}{\mathbb{I}},\\
\frac{\partial G_3(\mathbb{S},\mathbb{V},\mathbb{E},\mathbb{I},\mathbb{H},\mathbb{R})}{\partial\mathbb{H}} &= 1+a_4-\frac{c_3c_4}{\mathbb{H}},\\
\frac{\partial G_3(\mathbb{S},\mathbb{V},\mathbb{E},\mathbb{I},\mathbb{H},\mathbb{R})}{\partial\mathbb{R}} &= 1+a_4-\frac{1}{\mathbb{R}}.
\end{aligned}$$

One can verify very easily that the function G_3 has only one stagnation point.

$$(\mathbb{S}(0),\mathbb{V}(0),\mathbb{E}(0),\mathbb{I}(0),\mathbb{H}(0),\mathbb{R}(0)) = \left(\frac{1+a_1a_4}{1+a_4},\frac{1}{1+a_4},\frac{a_2a_4}{1+a_4},\frac{a_3a_4}{1+a_4},\frac{1}{1+a_4},\frac{1}{1+a_4}\right). \tag{21}$$

Furthermore, at $(\mathbb{S}(0),\mathbb{V}(0),\mathbb{E}(0),\mathbb{I}(0),\mathbb{H}(0),\mathbb{R}(0))$ for $V_2(\mathbb{S},\mathbb{V},\mathbb{E},\mathbb{I},\mathbb{H},\mathbb{R})$, the Hessian matrix is as follows:

$$B = \begin{bmatrix}
\frac{1+a_1a_4}{\mathbb{S}^2(0)} & 0 & 0 & 0 & 0 & 0\\
0 & \frac{1}{\mathbb{V}^2(0)} & 0 & 0 & 0 & 0\\
0 & 0 & \frac{a_2a_4}{\mathbb{E}^2(0)} & 0 & 0 & 0\\
0 & 0 & 0 & \frac{a_3a_4}{\mathbb{I}^2(0)} & 0 & 0\\
0 & 0 & 0 & 0 & \frac{1}{\mathbb{H}^2(0)} & 0\\
0 & 0 & 0 & 0 & 0 & \frac{1}{\mathbb{R}^2(0)}
\end{bmatrix}. \tag{22}$$

The Hessian matrix is obviously positive definite. As a result, the minimum value of $G_3(\mathbb{S},\mathbb{V},\mathbb{E},\mathbb{I},\mathbb{H},\mathbb{R})$ is $G_3(\mathbb{S}(0),\mathbb{V}(0),\mathbb{E}(0),\mathbb{I}(0),\mathbb{H}(0),\mathbb{R}(0))$. We may assert that $G_3(\mathbb{S},\mathbb{V},\mathbb{E},\mathbb{I},\mathbb{H},\mathbb{R})$ has one and only one minimum value $G_3(\mathbb{S}(0),\mathbb{V}(0),\mathbb{E}(0),\mathbb{I}(0),\mathbb{H}(0),\mathbb{R}(0))$ inside $\mathbb{R}^6_+$ based on Equation (20) and the continuity of $G_3(\mathbb{S},\mathbb{V},\mathbb{E},\mathbb{I},\mathbb{H},\mathbb{R})$. Then, as follows, we delineate a non-negative C^2-function $G:\mathbb{R}^6_+\to\mathbb{R}_+$:

$$G(\mathbb{S},\mathbb{V},\mathbb{E},\mathbb{I},\mathbb{H},\mathbb{R}) = G_3(\mathbb{S},\mathbb{V},\mathbb{E},\mathbb{I},\mathbb{H},\mathbb{R}) - G_3(\mathbb{S}(0),\mathbb{V}(0),\mathbb{E}(0),\mathbb{I}(0),\mathbb{H}(0),\mathbb{R}(0)).$$

Using *Ito's* formula and the suggested system, we arrive at

$$
\begin{aligned}
\mathcal{L}(G) \leq a_4 \Big\{ & -4\Pi\Big[(\mathbb{R}_0^s)^{1/4}-1\Big]+\frac{a_1\alpha\mathbb{I}}{\mathbb{N}}-\frac{a_1\omega\mathbb{V}}{\mathbb{S}}-\delta(\mathbb{I}+\mathbb{H})\Big\} \\
& +\phi-\mu(\mathbb{S}+\mathbb{V}+\mathbb{E}+\mathbb{I}+\mathbb{H}+\mathbb{R})-\delta(\mathbb{I}+\mathbb{H})-\frac{\Pi}{\mathbb{S}}+\frac{\alpha\mathbb{I}}{\mathbb{N}}-\frac{\omega\mathbb{V}}{\mathbb{S}}+(\tau+\mu)+\frac{\zeta_1^2}{2} \\
& -\frac{\tau\mathbb{S}}{\mathbb{V}}+(\mu+\omega)+\mu+\frac{\zeta_2^2}{2}-\frac{\rho\mathbb{I}}{\mathbb{H}}+(\gamma+\delta+\mu)+\frac{\zeta_5^2}{2}-\frac{\gamma\mathbb{H}}{R}+\mu+\frac{\zeta_6^2}{2}.
\end{aligned} \tag{23}
$$

Based on the above result, we have the following assertion:

$$
\begin{aligned}
\mathcal{L}V \leq & -a_4a_5+(a_1a_4+1)\frac{\alpha\mathbb{I}}{\mathbb{N}}-(a_1a_4+1)\frac{\omega\mathbb{V}}{\mathbb{S}}-\delta(a_4+1)(\mathbb{I}+\mathbb{H})\Big\}+\phi-\mu\mathbb{N}-\frac{\Pi}{\mathbb{S}} \\
& +(\tau+\mu)+\frac{\zeta_1^2}{2}-\frac{\tau\mathbb{S}}{\mathbb{V}}+(\mu+\omega)+\mu+\frac{\zeta_2^2}{2}-\frac{\rho\mathbb{I}}{\mathbb{H}}+(\gamma+\delta+\mu)+\frac{\zeta_5^2}{2}-\frac{\gamma\mathbb{H}}{R}+\mu+\frac{\zeta_6^2}{2},
\end{aligned} \tag{24}
$$

where

$$
a_5 = 4\phi\Big[(\mathbb{R}_0^s)^{1/4}-1\Big] > 0.
$$

Overall, for the solution to the model, we have the following set

$$
D = \{\epsilon_1 < \mathbb{S} < \frac{1}{\epsilon_2},\ \epsilon_1 < \mathbb{V} < \frac{1}{\epsilon_2},\ \epsilon_1 < \mathbb{E} < \frac{1}{\epsilon_2},\ \epsilon_1 < \mathbb{I} < \frac{1}{\epsilon_2},\ \epsilon_1 < \mathbb{H} < \frac{1}{\epsilon_2},\ \epsilon_1 < \mathbb{R} < \frac{1}{\epsilon_2}\},
$$

where $\epsilon_i > 0$ are extremely small positive real values to be calculated later on. For the sake of simplicity, the whole set was partitioned into the following subsets:

$$
\begin{aligned}
D_1 &= \Big\{(\mathbb{S},\mathbb{V},\mathbb{E},\mathbb{I},\mathbb{H},\mathbb{R}) \in \mathbb{R}_+^6, 0 < \mathbb{S} \leq \epsilon_1\Big\}, \\
D_2 &= \Big\{(\mathbb{S},\mathbb{V},\mathbb{E},\mathbb{I},\mathbb{H},\mathbb{R}) \in \mathbb{R}_+^6, 0 < \mathbb{V} \leq \epsilon_1, \mathbb{S} > \epsilon_2\Big\}, \\
D_3 &= \Big\{(\mathbb{S},\mathbb{V},\mathbb{E},\mathbb{I},\mathbb{H},\mathbb{R}) \in \mathbb{R}_+^6, 0 < \mathbb{E} \leq \epsilon_1, \mathbb{V} > \epsilon_2\Big\}, \\
D_4 &= \Big\{(\mathbb{S},\mathbb{V},\mathbb{E},\mathbb{I},\mathbb{H},\mathbb{R}) \in \mathbb{R}_+^6, 0 < \mathbb{I} \leq \epsilon_1, \mathbb{E} > \epsilon_2\Big\}, \\
D_5 &= \Big\{(\mathbb{S},\mathbb{V},\mathbb{E},\mathbb{I},\mathbb{H},\mathbb{R}) \in \mathbb{R}_+^6, 0 < \mathbb{H} \leq \epsilon_1, \mathbb{I} > \epsilon_2\Big\}, \\
D_6 &= \Big\{(\mathbb{S},\mathbb{V},\mathbb{E},\mathbb{I},\mathbb{H},\mathbb{R}) \in \mathbb{R}_+^6, 0 < \mathbb{R} \leq \epsilon_1, \mathbb{I} > \epsilon_2\Big\}, \\
D_7 &= \Big\{(\mathbb{S},\mathbb{V},\mathbb{E},\mathbb{I},\mathbb{H},\mathbb{R}) \in \mathbb{R}_+^6, \mathbb{S} \geq \frac{1}{\epsilon_2}\Big\}, \\
D_8 &= \Big\{(\mathbb{S},\mathbb{V},\mathbb{E},\mathbb{I},\mathbb{H},\mathbb{R}) \in \mathbb{R}_+^6, \mathbb{V} \geq \frac{1}{\epsilon_2}\Big\}, \\
D_9 &= \Big\{(\mathbb{S},\mathbb{V},\mathbb{E},\mathbb{I},\mathbb{H},\mathbb{R}) \in \mathbb{R}_+^6, \mathbb{E} \geq \frac{1}{\epsilon_2}\Big\}, \\
D_{10} &= \Big\{(\mathbb{S},\mathbb{V},\mathbb{E},\mathbb{I},\mathbb{H},\mathbb{R}) \in \mathbb{R}_+^6, \mathbb{I} \geq \frac{1}{\epsilon_2}\Big\}, \\
D_{11} &= \Big\{(\mathbb{S},\mathbb{V},\mathbb{E},\mathbb{I},\mathbb{H},\mathbb{R}) \in \mathbb{R}_+^6, \mathbb{H} \geq \frac{1}{\epsilon_2}\Big\}, \\
D_{12} &= \Big\{(\mathbb{S},\mathbb{V},\mathbb{E},\mathbb{I},\mathbb{H},\mathbb{R}) \in \mathbb{R}_+^6, \mathbb{R} \geq \frac{1}{\epsilon_2}\Big\}.
\end{aligned}
$$

Then, we show that $\mathcal{L}G(\mathbb{S},\mathbb{V},\mathbb{E},\mathbb{I},\mathbb{H},\mathbb{R}) < 0$ on $\mathbb{R}^6_+ \backslash D$, which is the same is displaying the solution on the eight sub-regions.

Case 1. If $(S,V,E,I,R) \in D_1$, then, by Equation (24), we obtain

$$\begin{aligned}
\mathcal{L}G \leq & -a_4a_5 + (a_1a_4+1)\frac{\alpha\mathbb{I}}{\mathbb{N}} - (a_1a_4+1)\frac{\omega\mathbb{V}}{\mathbb{S}} - \delta(a_4+1)(\mathbb{I}+\mathbb{H})\Big\} + \phi - \mu\mathbb{N} - \frac{\Pi}{\mathbb{S}} \\
& + (\tau+\mu) + \frac{\xi_1^2}{2} - \frac{\tau\mathbb{S}}{\mathbb{V}} + (\mu+\omega) + \mu + \frac{\xi_2^2}{2} - \frac{\rho\mathbb{I}}{\mathbb{H}} + (\gamma+\delta+\mu) + \frac{\xi_5^2}{2} - \frac{\gamma\mathbb{H}}{R} + \mu + \frac{\xi_6^2}{2}, \\
\leq & +(a_1a_4+1)\frac{\alpha\mathbb{I}}{\mathbb{N}}\Big\} + \phi - \frac{\Pi}{\epsilon_1} + (\tau+\mu) + \frac{\xi_1^2}{2} + (\mu+\omega) + \mu + \frac{\xi_2^2}{2} + (\gamma+\delta+\mu) + \frac{\xi_5^2}{2} + \mu + \frac{\xi_6^2}{2},
\end{aligned}$$

For every $(\mathbb{S},\mathbb{V},\mathbb{E},\mathbb{I},\mathbb{H},\mathbb{R}) \in D_1$., picking $\epsilon_1 > 0$, returns $\mathcal{L}G < 0$.

Just as in the proof above, we conclude that $\mathcal{L}G < 0$ for $(\mathbb{S},\mathbb{V},\mathbb{E},\mathbb{I},\mathbb{H},\mathbb{R}) \in D_i$, $i = 2,3...12$..

As a result, we arrived to the point that there must be positive constant $W > 0$ in such a way

$$\mathcal{L}G(\mathbb{S},\mathbb{V},\mathbb{E},\mathbb{I},\mathbb{H},\mathbb{R}) < -W < 0 \text{ for all } (\mathbb{S},\mathbb{V},\mathbb{E},\mathbb{I},\mathbb{H},\mathbb{R}) \in \mathbb{R}^6_+ \backslash D.$$

Thus,

$$\begin{aligned}
dG(\mathbb{S},\mathbb{V},\mathbb{E},\mathbb{I},\mathbb{H},\mathbb{R}) < & -Wdt + [(a_4+1)\mathbb{S} - (a_1a_4+1)\xi_1]d\mathbb{W}_1(t) + [(a_4+1)\mathbb{V} - \xi_2]d\mathbb{W}_2(t) \\
& + [(a_4+1)\mathbb{E} - a_2a_4\xi_3]d\mathbb{W}_3(t) + [(a_4+1)I - a_3a_4\xi_4]d\mathbb{W}_4(t) \\
& + [(a_4+1)\mathbb{H} - \xi_5]d\mathbb{W}_5(t) + [(a_4+1)\mathbb{R} - \xi_5]d\mathbb{W}_6(t).
\end{aligned} \tag{25}$$

Consider $(\mathbb{S}(0),\mathbb{V}(0),\mathbb{E}(0),\mathbb{I}(0),\mathbb{H}(0),\mathbb{R}(0)) = (\varkappa_1,\varkappa_2,\varkappa_3,\varkappa_4,\varkappa_5) = \varkappa \in \mathbb{R}^6_+ \backslash D$, and $\tau^{\varkappa}$ is the amount of time it takes for a path to start from $\varkappa$ to achieve set D,

$$\tau_n = inf\{t : |X(t)| = n\} \ \&\ \tau^{(n)}(t) = \min\{\tau^{\varkappa}, t, \tau_n\}.$$

The next relation could be obtained if one integrates the above inequality from 0 to $\tau^{(n)}(t)$, considering the expectation and using Dynkin's formula:

$$\begin{aligned}
& \mathbb{E}G(\mathbb{S}(\tau^{(n)}(t)),\mathbb{V}(\tau^{(n)}(t)),\mathbb{E}(\tau^{(n)}(t)),\mathbb{I}(\tau^{(n)}(t)),\mathbb{H}(\tau^{(n)}(t)),\mathbb{R}(\tau^{(n)}(t))) - G(\varkappa) \\
& = \mathbb{E}\int_0^{\tau(n)(t)} \mathcal{L}G(\mathbb{S}(u),\mathbb{V}(u),\mathbb{E}(u),\mathbb{I}(u),H(u),\mathbb{R}(u))du, \\
& \leq \mathbb{E}\int_0^{\tau(n)(t)} -Wdu = -W\mathbb{E}\tau^{(n)}(t).
\end{aligned}$$

Hence, $G(\varkappa)$ is non-negative; then,

$$\mathbb{E}\tau^{(n)}(t) \leq \frac{V(\varkappa)}{W}.$$

We have $P\{\tau_e = \infty\} = 1$ as a result of the proof of Theorem (3). This also shows that model (3) is regular and, consequently, by letting $n,t \to \infty$, almost surely we have $\tau(n)(t) \to \tau^{\varkappa}$.

Moreover, by utilizing the Fatou's lemma, we have

$$\mathbb{E}\tau^{(n)}(t) \leq \frac{G(\varkappa)}{W} < \infty,$$

$sup_{\varkappa \in K}\mathbb{E}\tau^{\varkappa} < \infty$, where K is the compact set, i.e., a subset of $\mathbb{R}^6_+$. This proves the condition (2) of Lemma (3) in an alternative approach.

Additionally, the diffusion matrix for the system (3) is

$$B = \begin{bmatrix} \xi_1^2\mathbb{S}^2 & 0 & 0 & 0 & 0 & 0 \\ 0 & \xi_2^2\mathbb{V}^2 & 0 & 0 & 0 & 0 \\ 0 & 0 & \xi_3^2\mathbb{E}^2 & 0 & 0 & 0 \\ 0 & 0 & 0 & \xi_4^2\mathbb{I}^2 & 0 & 0 \\ 0 & 0 & 0 & 0 & \xi_5^2\mathbb{H}^2 & 0 \\ 0 & 0 & 0 & 0 & 0 & \xi_6^2\mathbb{R}^2 \end{bmatrix}.$$

Taking $M = \min_{(\mathbb{S},\mathbb{V},\mathbb{E},\mathbb{I},\mathbb{H},\mathbb{R})\in\overline{D}\in R_+^6}\{\xi_1^2\mathbb{S}^2, \xi_2^2\mathbb{V}^2, \xi_3^2\mathbb{E}^2, \xi_4^2\mathbb{I}^2, \xi_3^2\mathbb{E}^2, \xi_5^2\mathbb{H}^2, , \xi_6^2 R^2\}$, we obtain

$$\sum_{i,j=1}^{6} a_{ij}(\mathbb{S},\mathbb{V},\mathbb{E},\mathbb{I},\mathbb{H},\mathbb{R})\xi_i\xi_j = \xi_1^2\mathbb{S}^2\sigma^2 + \xi_2^2\mathbb{V}^2\sigma_2^2 + \xi_3^2\mathbb{E}^2\sigma^2 + \xi_4^2\xi_4^2\mathbb{I}^2 + \xi_5^2\sigma_5^2 + \xi_6^2\sigma_6^2\mathbb{R}^2 \geq M|\sigma|^2,$$
$$(\mathbb{S},\mathbb{V},\mathbb{E},\mathbb{I},\mathbb{H},\mathbb{R}) \in \overline{D},$$

where $\xi = (\sigma_1, \sigma_2, \sigma_3, \sigma_4, \sigma_5, \sigma_6) \in \mathbb{R}_+^6$. Similarly, this proves condition (1) of Lemma (3). Keeping in view the previous statements, the ergodicity of model (3) is insured by Lemma (3) and hence proving that the model has stationary distribution.

6. Parameter Estimation

Among the most important processes to carry through out the testing process is the utilization of real data (if available) to acquire findings for certain unidentified biological factors used in the epidemiology system under study. Real-world measles cases, as shown in Table 1, are used to test the proposed rubella model and to choose the best fitted parameters for numerous unknown biological parameters that emerge in the system. Considering the 2018 statistics of WHO, the natural death rate of a Pakistani individual is 1/66.5 since the life expectancy of a Pakistani is 66.5. In addition, the entire size of the country is 207, 862, 518, whereas the recruitment rate is calculated to be $\Pi = 207,862,518 \times \mu \approx 260,479$. Additionally, Memon et al. [13] indicates that the rate of measles vaccination is generally 97 percent effective, implying that the vaccines' effectiveness τ equals 0.97. Some values of the parameters are estimated and others were fitted against the data, and these were presented in Table 2. In Figure 2, the results via simulations for measles resurgence cases obtained by adapting the proposed model (3) with real data from the first ten months of the year 2019 are shown. As shown in Figure 2, it gives a rather good match, lending veracity to the predictions generated from the proposed measles model (3). We employed the relation $\frac{1}{10}\sum_{k=1}^{10}\left|\frac{\varkappa_k^{real} - \varkappa_k^{approximate}}{\varkappa_k^{real}}\right| \approx 1.4685 \times 10^{-1}$ to measure the associated relative error for fitting the model against the data.

Table 1. Real cases of the measles epidemic in Pakistan during January and October 2019.

Jan (01)	Feb (02)	Mar (03)	Apr (04)	May (05)	June (06)	July (07)	Aug (08)	Sep (09)	Oct (10)
237	252	397	399	276	168	70	28	23	19

Table 2. Values of the parameters estimated and fitted against the real measles cases.

Parameter	Description	Source
ϕ	260,479	Estimated
α	1.253133×10^{-3}	Estimated
ω	0.97	Estimated
τ	1.60056×10^{-7}	Fitted
μ	9.3408	Fitted
β	9.2373×10^{-1}	Fitted
δ	5.8306×10^{-1}	Fitted
ρ	0	Estimated
γ	5.8306×10^{-1}	Fitted

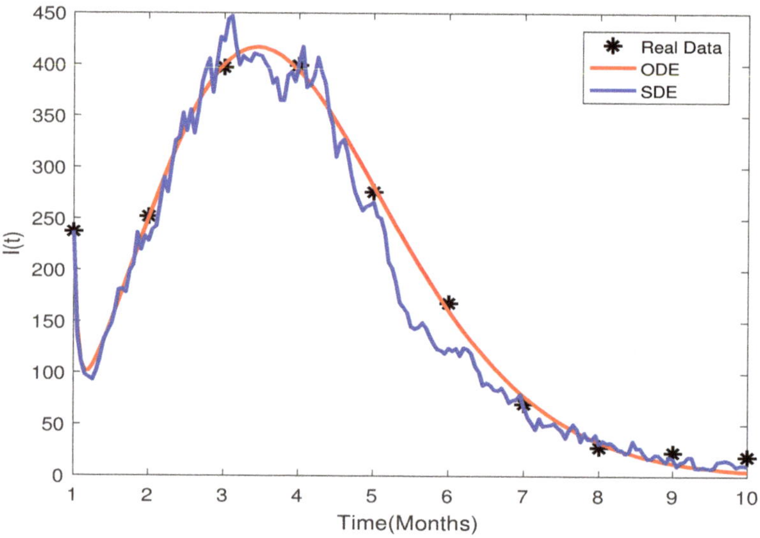

Figure 2. The figure shows the fitting of both ODE and SDE models against the real data by using values of the parameters shown in Table 2.

7. Numerical Simulations and Discussion

It is of great concern to simulate each mathematical model against the real data and to verify theoretical results, and this step is very important when dealing with modeling biological problems. The researchers seek to simulate problem (3) employing classic numerical algorithms that rapidly converged. Almost all of the theoretical conclusions are quantitatively validated by using the well-known RK-4 approach.

$$
\begin{aligned}
\mathbb{S}_{i+1} &= \mathbb{S}_i + \Big[\phi - \alpha\mathbb{S}_i\mathbb{I}_i + \omega\mathbb{V}_i - (\tau+\mu)\mathbb{S}_i\Big] \triangle t + \xi_1\mathbb{S}_i\sqrt{\triangle t}\varsigma_{1,i} + \frac{\xi_1^2}{2}\mathbb{S}_i(\varsigma_{1,i}^2 - 1) \triangle t,\\
\mathbb{V}_{i+1} &= \mathbb{V}_i + \Big[\tau\mathbb{S}_i - (\mu+\omega)\mathbb{V}_i\Big] \triangle t + \xi_2\mathbb{V}_i\sqrt{\triangle t}\varsigma_{2,i} + \frac{\xi_2^2}{2}\mathbb{V}_i(\varsigma_{2,i}^2 - 1) \triangle t,\\
\mathbb{E}_{i+1} &= \mathbb{E}_i + \Big[\alpha\mathbb{S}_i\mathbb{I}_i - (\mu+\beta)\mathbb{E}_i\Big] \triangle t + \xi_3\mathbb{E}_i\sqrt{\triangle t}\varsigma_{3,i} + \frac{\xi_3^2}{2}\mathbb{E}_i(\varsigma_{3,i}^2 - 1) \triangle t,\\
\mathbb{I}_{i+1} &= \mathbb{I}_i + \Big[\beta\mathbb{E}_i - (\mu+\delta+\rho)\mathbb{I}_i\Big] \triangle t + \xi_4\mathbb{I}_i\sqrt{\triangle t}\varsigma_{4,i} + \frac{\xi_4^2}{2}\mathbb{I}_i(\varsigma_{4,i}^2 - 1) \triangle t,\\
\mathbb{H}_{i+1} &= \mathbb{H}_i + \Big[\rho\mathbb{I}_i - (\gamma+\delta+\mu)\mathbb{H}_i\Big] \triangle t + \xi_5\mathbb{H}_i\sqrt{\triangle t}\varsigma_{5,i} + \frac{\xi_5^2}{2}\mathbb{H}_i(\varsigma_{5,i}^2 - 1) \triangle t,\\
\mathbb{R}_{i+1} &= \mathbb{R}_i + \Big[\gamma\mathbb{H}_i - \mu\mathbb{R}_i\Big] \triangle t + \xi_6\mathbb{R}_i\sqrt{\triangle t}\varsigma_{5,i} + \frac{\xi_6^2}{2}\mathbb{R}_i(\varsigma_{6,i}^2 - 1) \triangle t.
\end{aligned}
\tag{26}
$$

where $\varsigma_{i,j}(i = 1,2,3,4,5,6)$ stands for the standard Gaussian stochastic variables, having distribution $N(0,1)$, and Δt is the constant time-step. The terms $\xi_i > 0, (i = 1,2,3,4,5,6)$ reflect the intensities of the white noises.

To quantitatively validate the analytical conclusions, we require the parameters' value being used in model (3). In Example 1 and 2, we established two sets of parameters' values for this reason, and the population levels of each compartment at $t = 0$ were displayed. For every scenario, we simulated the model over the interval $[0, 80]$.

We formulated Theorem 2 based on the stochastic theory of stability, which indicates that, under the condition of $\mathbb{R}_s < 1$, the infection would continue to perish from the community regardless of the levels of the variables at $t = 0$. Furthermore, the theory demonstrates that the infection will be eradicated from the community with unit probability. Figure 3 shows that the random curves reach the infection-free state after a limited period of time, confirming the analytical results.

Example 1. *The values of the parameter are assumed as:* $\phi = 0.12, \tau = 0.002, \beta = 0.008, \omega = 0.09, \mu = 0.005, \delta = 0.005, \alpha = 0.02, \gamma = 0.004$ *and* $\rho = 0.05$, *where the initial values of the state vector are* $\mathbb{S}(0) = 50, \mathbb{V}(0) = 30, \mathbb{E}(0) = 10, \mathbb{I}(0) = 15, \mathbb{H}(0) = 25, \mathbb{R}(0) = 20$. *Similarly, the intensities of the white noises are:* $\xi_1 = 0.55, \xi_2 = 0.25, \xi_3 = 0.25, \xi_4 = 0.33, \xi_5 = 0.55, \xi_6 = 0.50$. *Using those same model parameters, we estimated* $\mathbb{R}_s$, *which was found to be lower than 1. As a result, the assumption of Theorem 2 is met, and hence the components of the solution of the model adhere to the following relations:*

$$\limsup_{t \to \infty} \frac{log\mathbb{E}(t)}{t} \leq 0, \quad a.s.$$

$$\limsup_{t \to \infty} \frac{log\mathbb{I}(t)}{t} \leq 0, \quad a.s.$$

Essentially, these relations explain the elimination of the virus from the community, as seen by Figure 3. As a consequence, the derived research findings on elimination are valid and may be relied on.

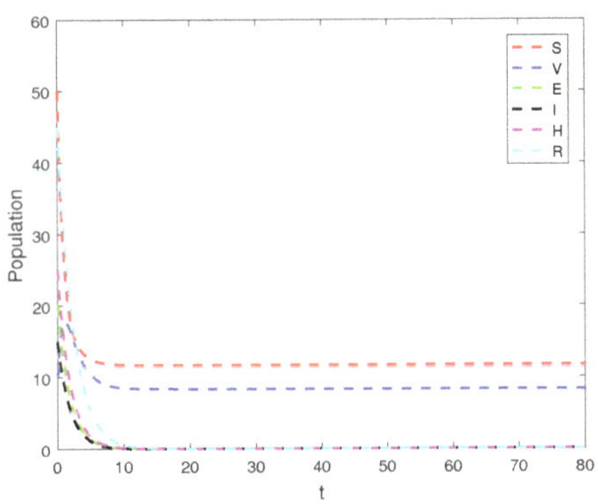

(**a**) Derministic System (1)

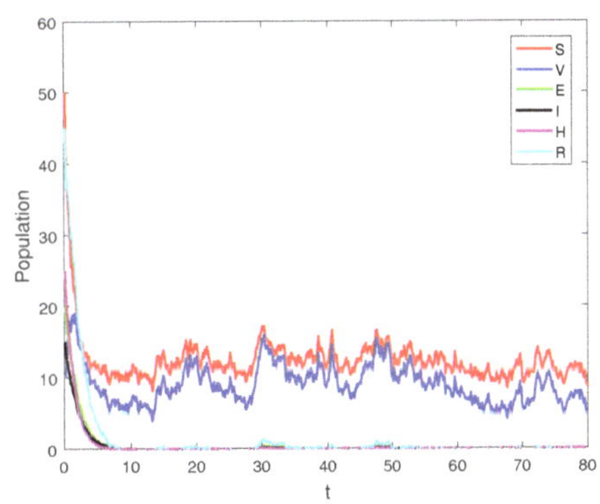

(**b**) Stochastic System (3)

Figure 3. Sample solutions of stochastic systems (3) with its associated deterministic counterpart.

Likewise, Theorem 3 guarantees the disease's prevalence in the population given permissible limits. By considering data from Example 2, we calculated the value of $\mathbb{R}_0^s$, and it was found that it is greater than unity. Figure 4 depicts the numerical results based on the theorem's assumptions. The figure implies that the infection persists inside the population, and that there is persistence of the solution of the proposed model (3). This verifies the conclusion of Theorem 3 in the case of deterministic model (1). These results further elaborate that, when the related threshold parameter of the perturbed system exceeds unity, the solution of the model (3) lies within the neighborhood of endemic equilibrium. Thus, policymakers must provide strong preventative measures against various variations in order to restrict the spreading of multiple strains throughout the community. Moreover, Theorem 3 ensures the existence of an ergodic stationary distribution for model (3), and it is confirmed by Figure 5.

Example 2. *In this case, the chosen parameter values are of the form:* $\phi = 2.12, \beta = 0.08, \omega = 0.07, \mu = 0.001, \delta = 0.005, \alpha = 0.2, \rho = 0.004$ *and* $\gamma = 0.08$. *Similarly, the initial population sizes of the state variables are* $\mathbb{S}(0) = 50, \mathbb{V}(0) = 30, \mathbb{E}(0) = 10, \mathbb{I}(0) = 15, \mathbb{H}(0) = 25,$

$\mathbb{R}(0) = 20$, whereas the intensities are given by $\xi_1 = 0.50, \xi_2 = 0.35, \xi_3 = 0.70, \xi_4 = 0.50$, $\xi_5 = 0.41, \xi_5 = 0.45$. Considering the measles data and hence the estimated and fitted parameters, we find that $\mathbb{R}_0^s$ exceeds unity. It is also explored that the model parameters taken in this example satisfy the premise of Theorem 3. We tested the model using this input, and the outcomes are displayed visually in Figure 4. The figure shows that the disease tends to stay inside the community, and the model exhibits a homogeneous stationary distribution in this situation.

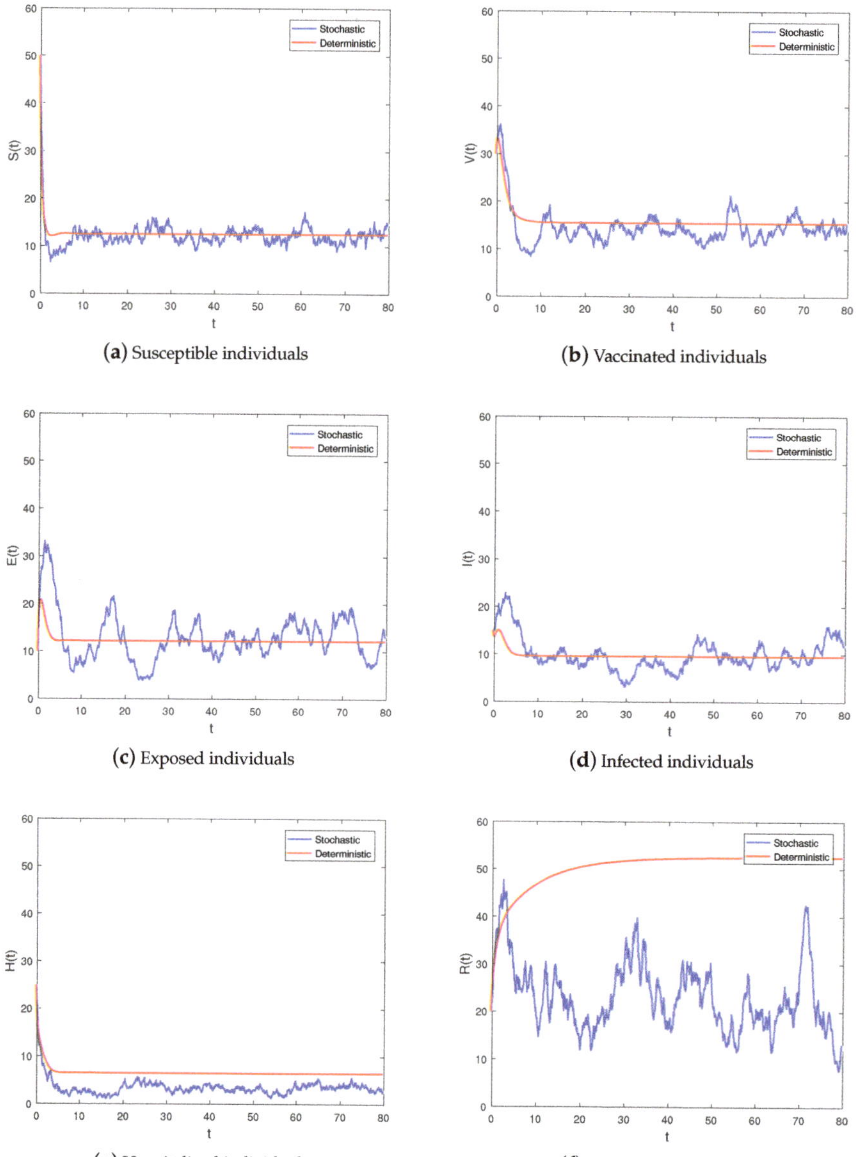

(a) Susceptible individuals **(b)** Vaccinated individuals **(c)** Exposed individuals **(d)** Infected individuals **(e)** Hospitalized individuals **(f)** Recovered individuals

Figure 4. Sample solution profiles of the system (3) correspond to its deterministic counterpart.

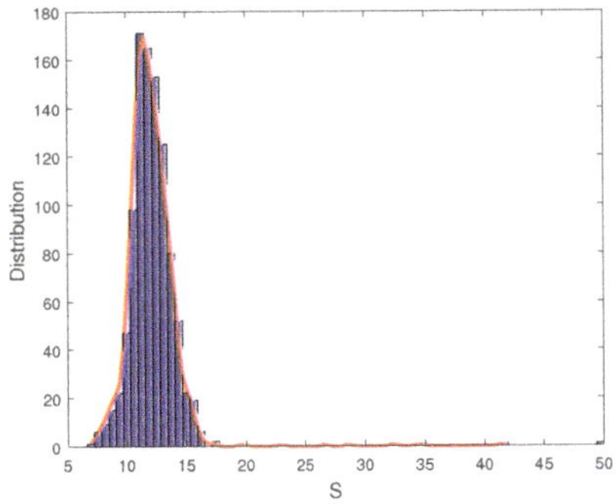

(a) Histogram of S

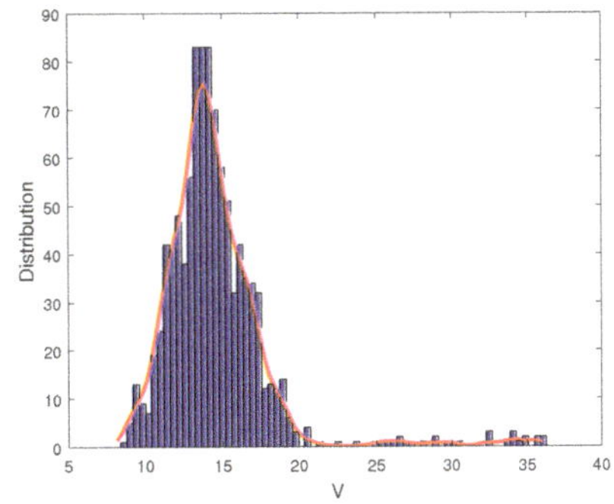

(b) Histogram of V

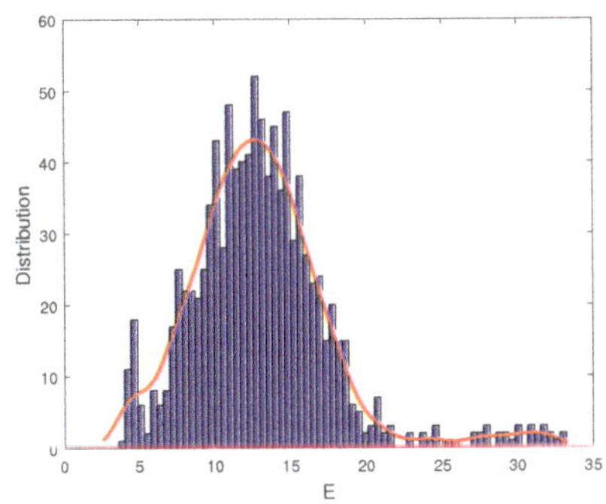

(c) Histogram of E

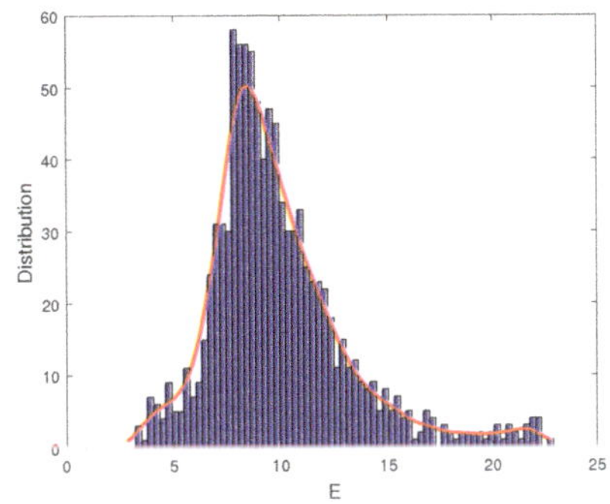

(d) Histogram of I

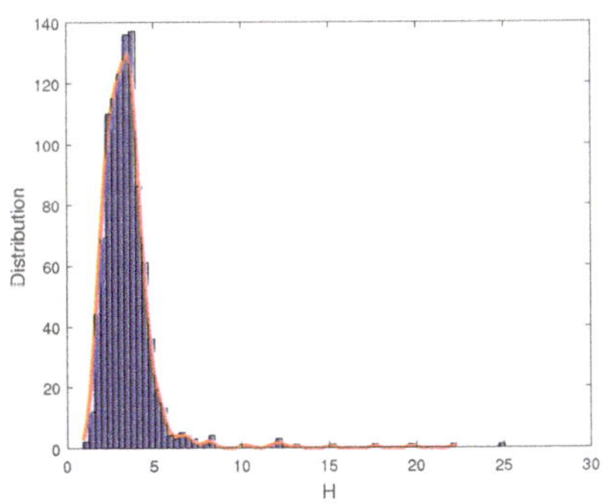

(e) Histogram of H

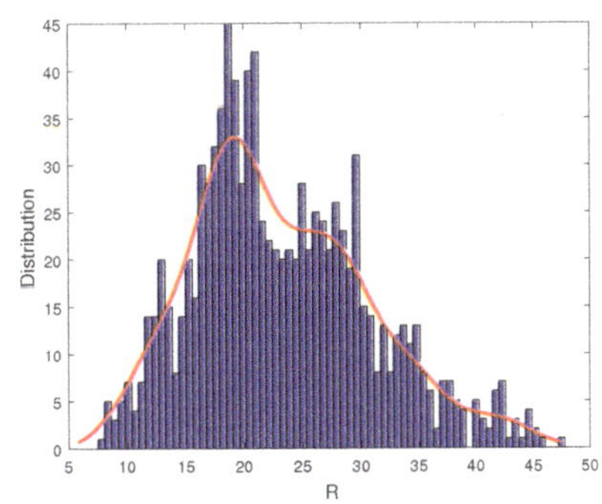

(f) Histogram of R

Figure 5. Ergodic stationary distribution of (3).

From Theorem 3, there is a single stationary distribution of system (3). To numerically illustrate this statistical property, we present in Figure 4 and 5, the trajectories and the associated joint density function for each class of the population. For a good visibility, we offer the 3D and upper view of the aforementioned joint densities in Figures 6–10. This indicates that the infection is still present in the population over time. We talk here about persistence in the mean of the epidemic. In Figure 4, we illustrate the continuation of all groups of the studied population. We remark that the stochastic trajectories fluctuate around the deterministic solution with reasonable distances according to magnitude of the noises.

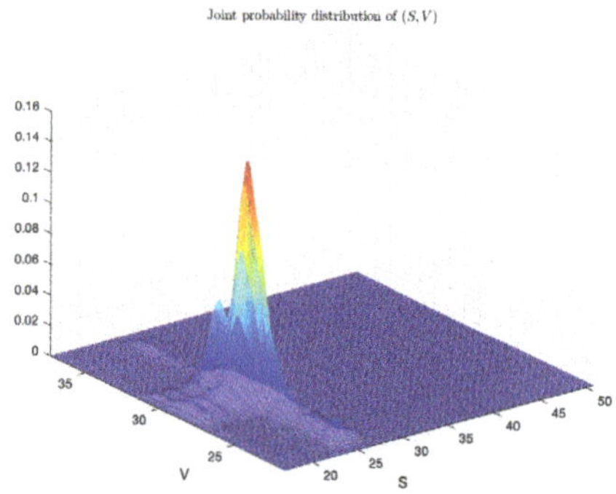

(a) $S(t)$ individuals and $V(t)$ individuals

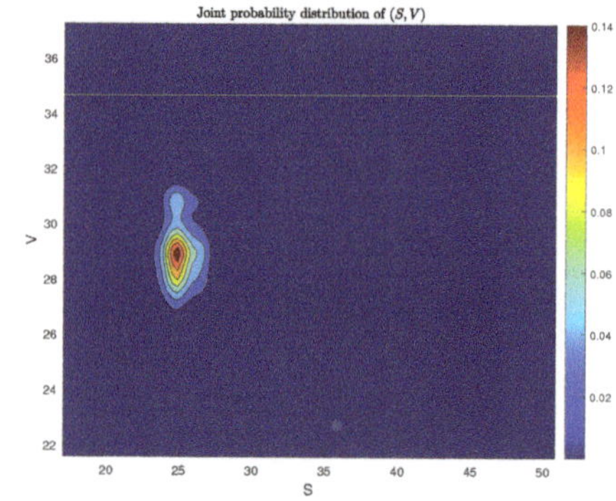

(b) $S(t)$ individuals and $V(t)$ individuals

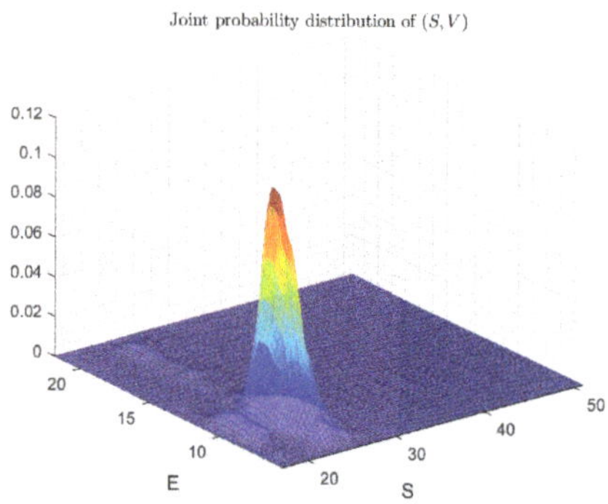

(c) $S(t)$ individuals and $E(t)$ individuals

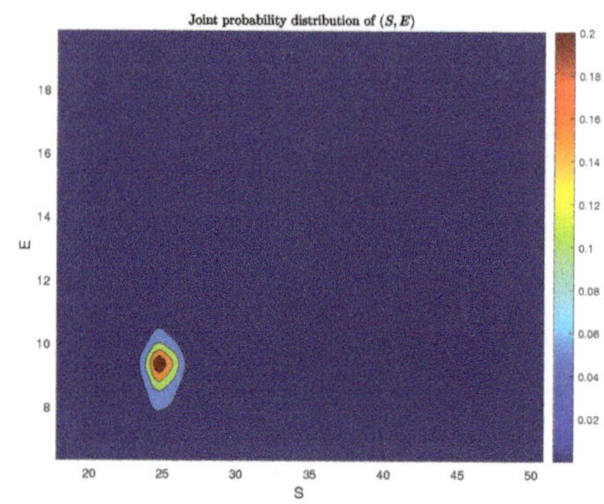

(d) $S(t)$ individuals and $E(t)$ individuals

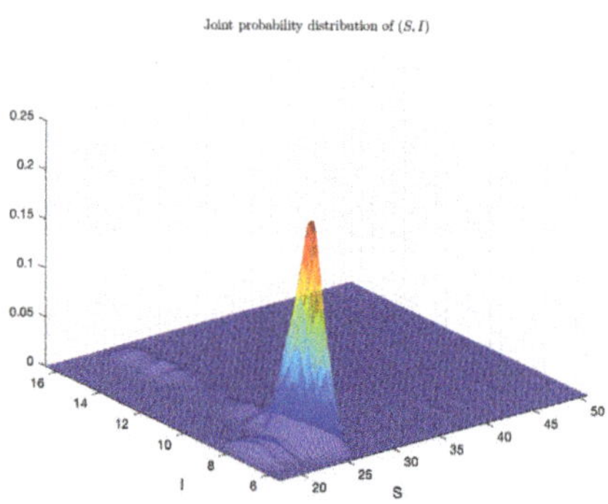

(e) $S(t)$ individuals and $I(t)$ individuals

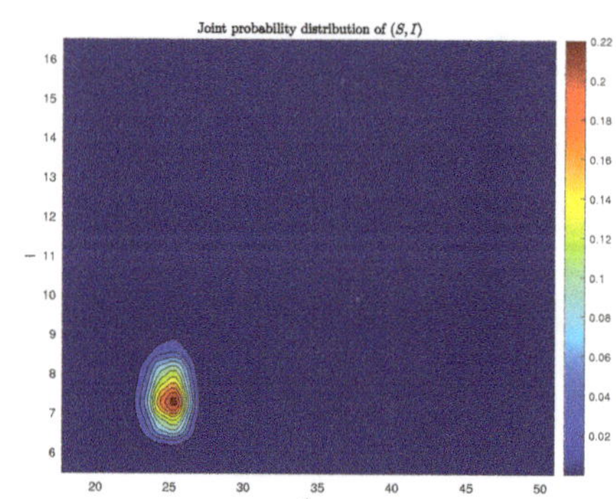

(f) $S(t)$ individuals and $I(t)$ individuals

Figure 6. The right panel of the figure shows the joint two-dimensional densities at $t = 3000$ of individuals $\mathbb{S}$, $\mathbb{V}$, $\mathbb{E}$ and $\mathbb{I}$ of system (3) correspond to data from Example 2—Test 2 (2nd column), where different colors depict the density sizes. The panel on the left demonstrates the 3D graph of the two-dimensional densities of $\mathbb{S}$, $\mathbb{V}$, $\mathbb{E}$ and $\mathbb{I}$ collectively (in this case, $\alpha = 1.99$).

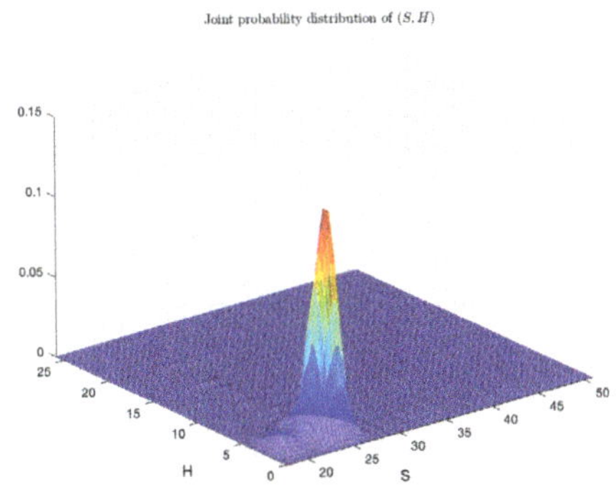

(a) $S(t)$ individuals and $H(t)$ individuals

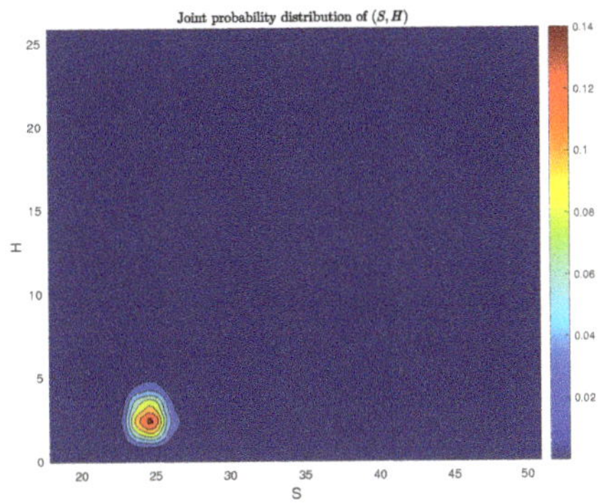

(b) $S(t)$ individuals and $H(t)$ individuals

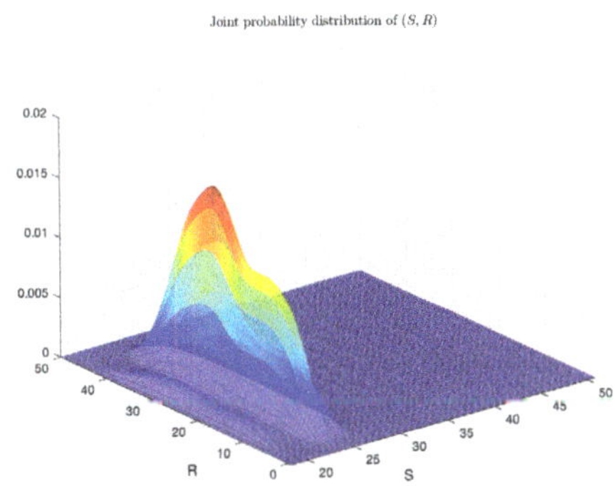

(c) $S(t)$ individuals and $R(t)$ individuals

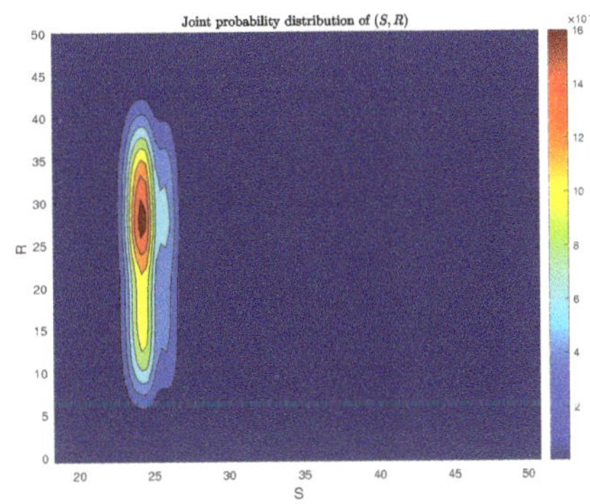

(d) $S(t)$ individuals and $R(t)$ individuals

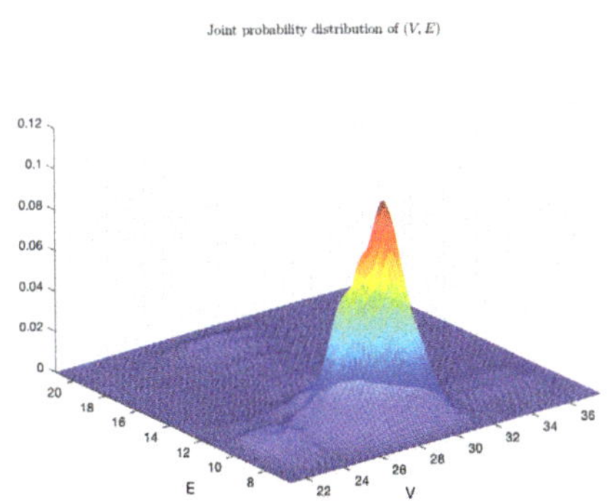

(e) $V(t)$ individuals and $E(t)$ individuals

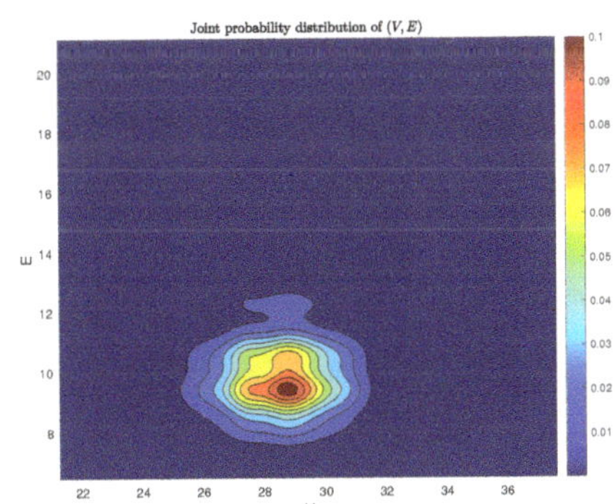

(f) $V(t)$ individuals and $E(t)$ individuals

Figure 7. The right panel of the figure shows the joint two-dimensional densities at $t = 3000$ of individuals $\mathbb{S}$, $\mathbb{H}$, $\mathbb{R}$, $\mathbb{V}$ and $\mathbb{E}$ of system (3) correspond to data from Example 2—Test 2 (2nd column), where different colors depict the density sizes. The panel on the left demonstrates the 3D graph of the two-dimensional densities of $\mathbb{S}$, $\mathbb{H}$, $\mathbb{R}$, $\mathbb{V}$ and $\mathbb{E}$ collectively (in this case, $\alpha = 1.99$).

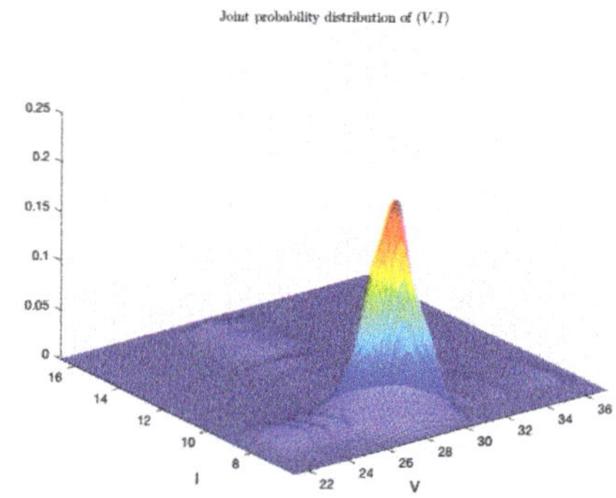

(a) $V(t)$ individuals and $I(t)$ individuals

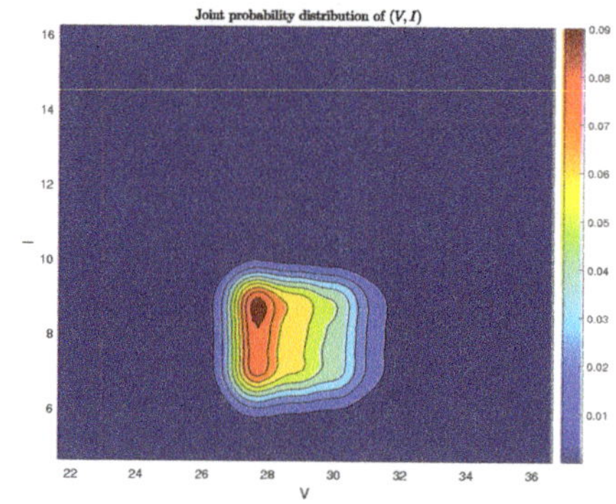

(b) $V(t)$ individuals and $I(t)$ individuals

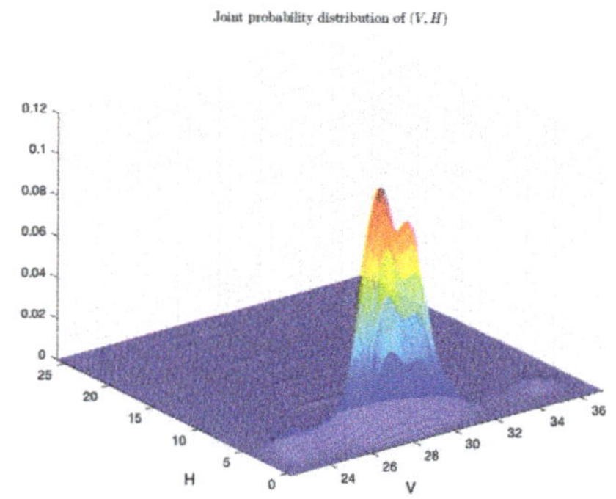

(c) $V(t)$ individuals and $H(t)$ individuals

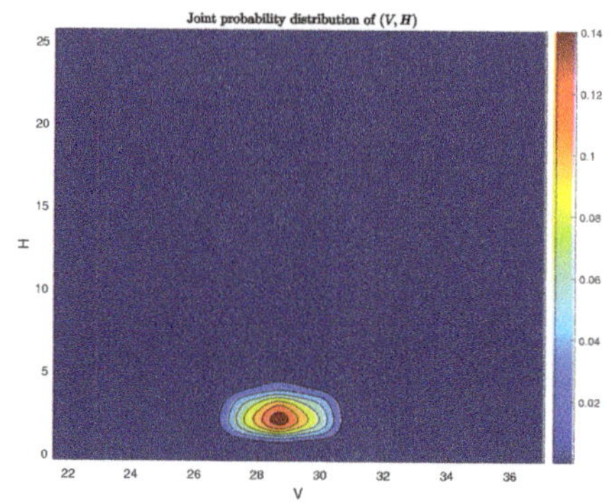

(d) $V(t)$ individuals and $H(t)$ individuals

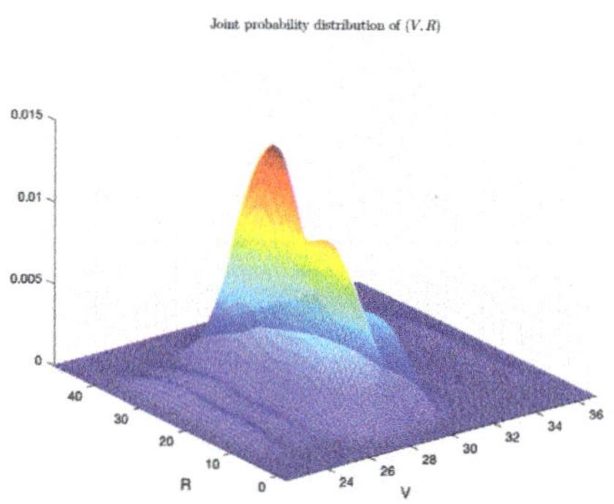

(e) $V(t)$ individuals and $R(t)$ individuals

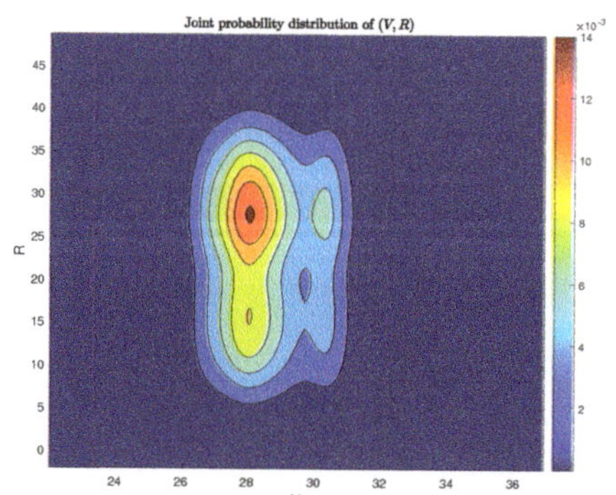

(f) $V(t)$ individuals and $R(t)$ individuals

Figure 8. The right panel of the figure shows the joint two-dimensional densities at $t = 3000$ of individuals $\mathbb{V}$, $\mathbb{I}$, $\mathbb{H}$, and $\mathbb{R}$ of system (3) correspond to data from Example 2—Test 2 (2nd column), where different colors depict the density sizes. The panel on the left demonstrates the 3D graph of the two-dimensional densities of $\mathbb{V}$, $\mathbb{I}$, $\mathbb{H}$, and $\mathbb{R}$ collectively (in this case, $\alpha = 1.99$).

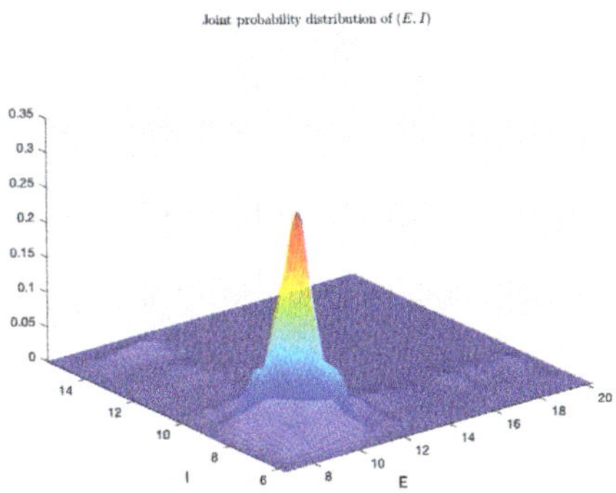

(a) $E(t)$ individuals and $I(t)$ individuals

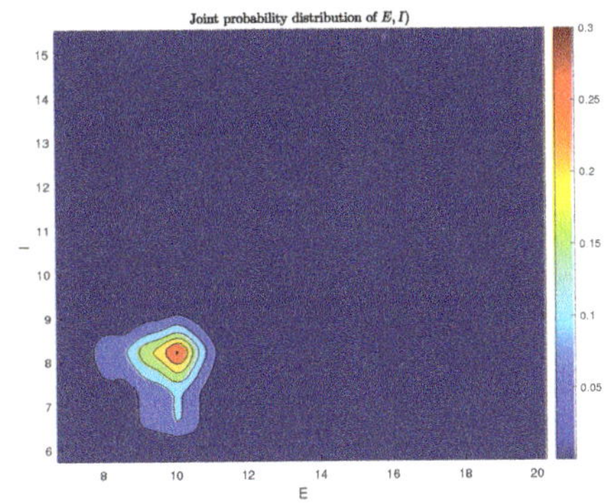

(b) $E(t)$ individuals and $I(t)$ individuals

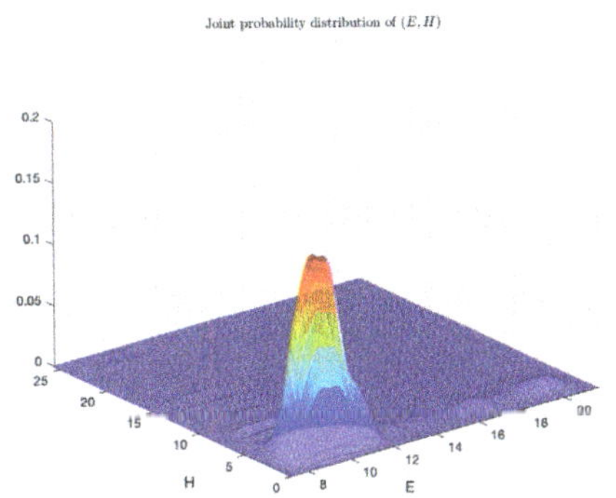

(c) $E(t)$ individuals and $H(t)$ individuals

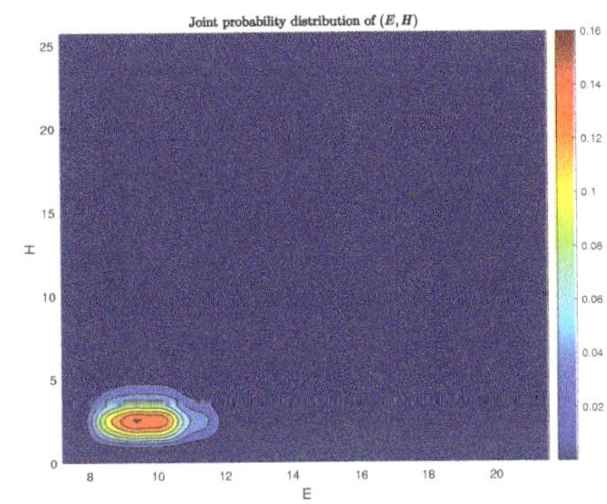

(d) $E(t)$ individuals and $H(t)$ individuals

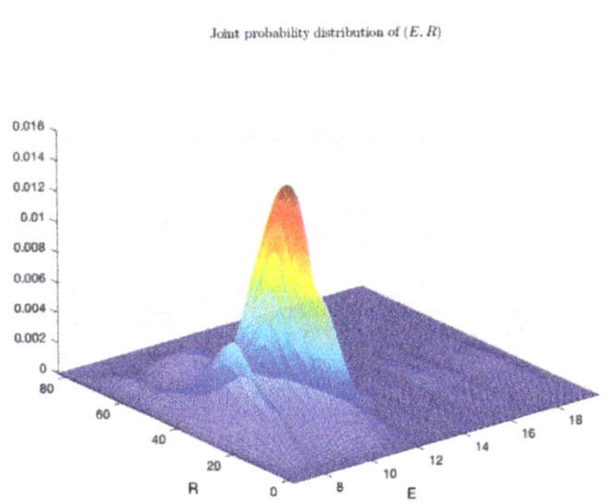

(e) $E(t)$ individuals and $R(t)$ individuals

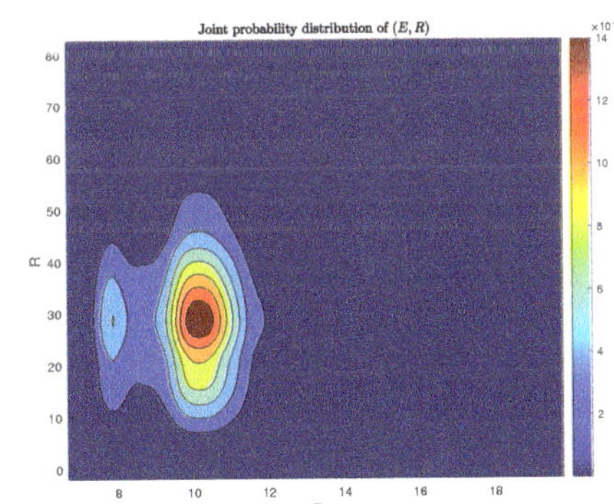

(f) $E(t)$ individuals and $R(t)$ individuals

Figure 9. The right panel of the figure shows the joint two-dimensional densities at $t = 3000$ of individuals $\mathbb{E}$, $\mathbb{I}$, $\mathbb{H}$, and $\mathbb{R}$ of system (3) correspond to data from Example 2—Test 2 (2nd column), where different colors depict the density sizes. The panel on the left demonstrates the 3D graph of the two-dimensional densities of $\mathbb{E}$, $\mathbb{I}$, $\mathbb{H}$, and $\mathbb{R}$, collectively (in this case, $\alpha = 1.99$).

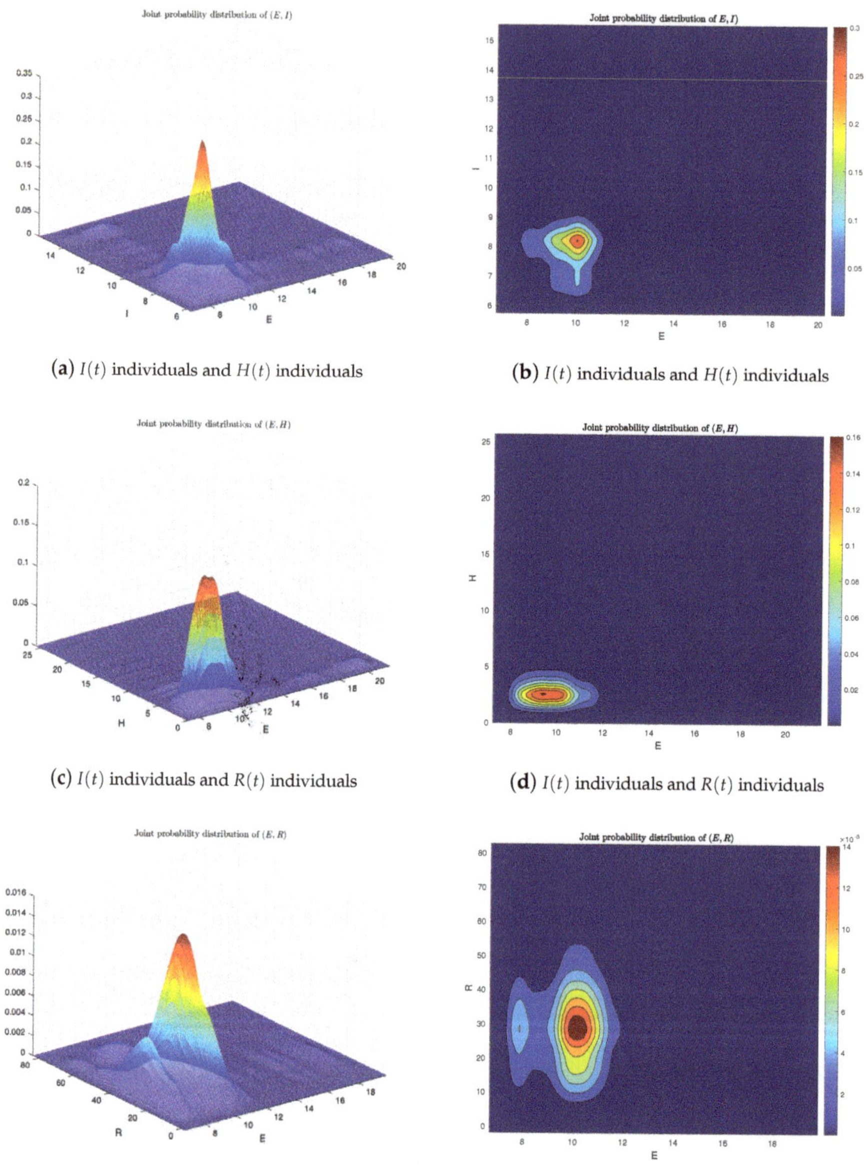

(**a**) $I(t)$ individuals and $H(t)$ individuals

(**b**) $I(t)$ individuals and $H(t)$ individuals

(**c**) $I(t)$ individuals and $R(t)$ individuals

(**d**) $I(t)$ individuals and $R(t)$ individuals

(**e**) $H(t)$ individuals and $R(t)$ individuals

(**f**) $H(t)$ individuals and $R(t)$ individuals

Figure 10. (First part) The right panel of the figure shows that the joint two-dimensional densities at $t = 3000$ of individuals $\mathbb{I}$, $\mathbb{H}$, and $\mathbb{R}$ of system (3) correspond to data from Example 2–Test 2 (2nd column), where different colors depict the density sizes. The panel on the left demonstrates the 3D graph of the two-dimensional densities of $\mathbb{I}$, $\mathbb{H}$, and $\mathbb{R}$ collectively (in this case, $\alpha = 1.99$).

8. Conclusions and Future Research

In this study, we presented a detail analysis of a stochastic model that describes the spreading behavior of the measles virus while accounting for the white noises and the influence of immunizations. It is assumed that the randomness being used in the model is nonlinear, and that a person may lose the resistance after vaccination, implying that vaccination might create temporary protection against the disease. First of all, we formulated a deterministic model and then it was expanded to a stochastic model. It is shown that the stochastic model is both theoretically and practically viable by demonstrating that the model has a global solution which is positive and stochastically bounded. Next, we developed sufficient criteria for the disease's elimination and permanence. Furthermore, the presence of a stationary distribution is examined by constructing a suitable Lyapunov function, wherein we noticed that the disease will persist for $\mathbb{R}_0^{s>1}$ and that the illness will vanish from the community for $\mathbb{R}_0^{s<1}$. We simulated the model against the available data of measles in Pakistan during the first ten months of 2019, by using the conventional curve fitting methods, and the values of the parameters were calculated therein. These values of the parameters were used in simulating the model, and the theoretical findings of the research were evaluated and conclusions were made. Simulations of the study suggest that, in order to fully understand the dynamic behavior of the measles epidemic, time-delay must be included in such analyses, and that advancements in every vaccine campaign are unavoidable in order to stop or minimize the spreading of the disease.

We fit both stochastic and determinism models to known data on rubella in Pakistan during the first ten months of 2019, and the values of parameters were obtained using *lsqcurvefit* methods. These model parameters are used, and the threshold number, which is around 13, is determined. This demonstrates that measles is extremely harmful and might have a negative impact on this community if adequate control tactics are not implemented in time. It also suggests that, if no appropriate measures are implemented, the cases of the measles may increase in the coming years. To further reduce the transmission rate, health authorities and lawmakers must launch awareness campaigns, mass immunizations, particularly among youngsters, and, most crucially, care and treatment for people having chronic conditions. This was also discovered that α serves as the most critical indicator to the threshold parameter; thus, lowering the disease spreading co-efficient (by lowering the household and sexual interactions of chronic patients with vulnerable) for acutely infected individuals who become chronic is also an effective control to reduce the spread of measles.

In the next research studies, the authors intend to use fractional calculus and modify this and the related model while utilizing different definitions of fractional derivatives such as Riesz, Caputo, Atangana–Baleanu, Caputo–Fabrizio, and many others for capturing the real dynamics of such and related diseases.

Author Contributions: Methodology, B.G. Formal analysis, B.G. Investigation, B.G. Data curation, A.K. and A.D.; Writing—original draft, B.G. and A.D. Writing—review and editing, A.K. Supervision, A.D. Project administration, A.K. All authors have read and agreed to the published version of the manuscript.

Funding: This research was sponsored by the Guangzhou Government Project under Grant No. 62216235, and the National Natural Science Foundation of China (Grant No. 12250410247).

Institutional Review Board Statement: Not applicable.

Informed Consent Statement: Not applicable.

Data Availability Statement: Not applicable.

Acknowledgments: Authors are grateful to the handling editor and reviewers for their constructive comments and remarks which helped to improve the quality of the manuscript.

Conflicts of Interest: The authors declare that they have no competing interests.

References

1. Mose, O.F.; Sigey, J.K.; Okello, J.A.; Okwoyo, J.M.; Kang'ethe, G.J. Mathematical modeling on the control of measles by vaccination: Case study of KISII County, Kenya. *SIJ Trans. Comput. Sci. Eng. Appl. CSEA* **2014**, *2*, 61–69.
2. Garba, S.M.; Safi, M.A.; Usaini, S. Mathematical model for assessing the impact of vaccination and treatment on measles transmission dynamics. *Math. Meth. Appl. Sci.* **2017**, *40*, 6371–6388. [CrossRef]
3. Roberts, M.G.; Tobias, M.I. Predicting and preventing measles epidemic in New Zealand: Application of mathematical model. *Epidem. Infect.* **2000**, *124*, 279–287. [CrossRef] [PubMed]

4. World Health Organization. Measles, Preprint. 2018. Available online: https://www.who.int/news-room/fact-sheets/detail/measles (accessed on 27 December 2019).
5. Perry, R.T.; Halsey, N.A. The clinical significance of measles: A review. *J. Infect. Dis.* **2004**, *189*, S4–S16.
6. Ejima, K.; Omori, R.; Aihara, K.; Nishiura, H. Real-time investigation of measles epidemics with estimate of vaccine efficacy. *Int. J. Biol. Sci.* **2012**, *8*, 620–629. [CrossRef]
7. Mossong, J.; Muller, C.P. Modelling measles re-emergence as a result of waning of immunity in vaccinated populations. *Vaccine* **2003**, *21*, 4597–4603. [CrossRef]
8. Bolarin, G. On the dynamical analysis of a new model for measles infection. *Int. J. Math. Trends Technol.* **2014**, *7*, 144–154. [CrossRef]
9. Taiwo, L.; Idris, S.; Abubakar, A.; Nguku, P.; Nsubuga, P.; Gidado, S. Factors affecting access to information on routine immunization among mothers of under 5 children in Kaduna state Nigeria, 2015. *Pan Afr. Med. J.* **2017**, *27*, 186. [CrossRef]
10. Center for Disease Control. Available online: https://www.cdc.gov/vaccines/vpd/measles/index.html (accessed on 26 January 2021).
11. World Health Organization. *Eastern Mediterranean Vaccine Action Plan 2016–2020: A Framework for Implementation of the Global Vaccine Action Plan (No. WHO-EM/EPI/353/E)*; World Health Organization, Regional Office for the Eastern Mediterranean: Cairo, Egypt, 2019.
12. Dawn. Curbing Measles. Available online: https://www.dawn.com/news/1520931 (accessed on 7 December 2019).
13. Memon, Z.; Qureshi, S.; Memon, B.R. Mathematical analysis for a new nonlinear measles epidemiological system using real incidence data from Pakistan. *Eur. Phys. J. Plus* **2020**, *135*, 1–21. [CrossRef]
14. Asamoah, J.K.K.; Nyabadza, F.; Jin, Z.; Bonyah, E.; Khan, M.A.; Li, M.Y.; Hayat, T. Backward bifurcation and sensitivity analysis for bacterial meningitis transmission dynamics with a nonlinear recovery rate. *Chaos Solitons Fractals* **2020**, *140*, 110237. [CrossRef]
15. Khan, F.M.; Khan, Z.U.; Lv, Y.-P.; Yusuf, A.; Din, A. Investigating of fractional order dengue epidemic model with ABC operator. *Results Phys.* **2021**, *24*, 104075. [CrossRef]
16. Li, X.-P.; Bayatti, H.A.; Din, A.; Zeb, A. A vigorous study of fractional order COVID-19 model via ABC derivatives. *Results Phys.* **2021**, *29*, 104737. [CrossRef]
17. Din, A.; Li, Y. Lévy noise impact on a stochastic hepatitis B epidemic model under real statistical data and its fractal–fractional Atangana–Baleanu order model. *Phys. Scr.* **2021**, *96*, 124008. [CrossRef]
18. Srivastava, H.M.; Saad, K.M. Numerical simulation of the fractal-fractional Ebola virus. *Fractal Fract.* **2020**, *4*, 49. [CrossRef]
19. Liu, P.; Ikram, R.; Khan, A. The measles epidemic model assessment under real statistics: An application of stochastic optimal control theory. *Comput. Methods Biomech. Biomed. Eng.* **2022**, *26*, 138–159. [CrossRef]
20. Liu, G.; Qi, H.; Chang, Z.; Meng, X. Asymptotic stability of a stochastic May mutualism system. *Comput. Math. Appl.* **2020**, *79*, 735–745. [CrossRef]
21. Tran, K.Q.; Yin, G. Optimal harvesting strategies for stochastic ecosystems. *IET Control. Theory Appl.* **2017**, *11*, 2521–2530. [CrossRef]
22. Sabbar, Y.; Din, A. Probabilistic analysis of a marine ecological system with intense variability. *Mathematics* **2022**, *10*, 2262. [CrossRef]
23. Din, A. The stochastic bifurcation analysis and stochastic delayed optimal control for epidemic model with general incidence function. *Chaos Interdiscip. J. Nonlinear Sci.* **2021**, *31*, 123101. [CrossRef]
24. Khan, A.; Sabbar, Y. Stochastic modeling of the Monkeypox 2022 epidemic with cross-infection hypothesis in a highly disturbed environment. *Math. Biosci. Eng.* **2022**, *19*, 13560–13581. [CrossRef]
25. Din, A.; Khan, A.; Sabbar, Y. Long-Term Bifurcation and Stochastic Optimal Control of a Triple-Delayed Ebola Virus Model with Vaccination and Quarantine Strategies. *Fractal Fract.* **2022**, *6*, 578. [CrossRef]
26. Din, A.; Ain, Q.T. Stochastic Optimal Control Analysis of a Mathematical Model: Theory and Application to Non-Singular Kernels. *Fractal Fract.* **2022**, *6*, 279. [CrossRef]
27. Omame, A.; Abbas, M.; Din, A. Global asymptotic stability, extinction and ergodic stationary distribution in a stochastic model for dual variants of SARS-CoV-2. *Math. Comput. Simul.* **2022**, *204*, 302–336. [CrossRef] [PubMed]
28. Liu, P.; Rahman, M.u.; Din, A. Fractal fractional based transmission dynamics of COVID-19 epidemic model. *Comput. Methods Biomech. Biomed. Eng.* **2022**, *25*, 1852–1869. [CrossRef]
29. Olumuyiwa, J.P.; Ojo, M.M.; Viriyapong, R.; Oguntolu, F.A. Mathematical model of measles transmission dynamics using real data from Nigeria. *J. Differ. Equations Appl.* **2022**, *28*, 753–770.
30. Jin, X.; Jia, J. Qualitative study of a stochastic SIRS epidemic model with information intervention. *Phys. A Stat. Mech. Appl.* **2020**, *547*, 123866. [CrossRef]
31. Rajasekar, S.P.; Pitchaimani, M. Qualitative analysis of stochastically perturbed SIRS epidemic model with two viruses. *Chaos Solitons Fractals* **2019**, *118*, 207–221. [CrossRef]
32. Bao, K.; Zhang, Q. Stationary distribution and extinction of a stochastic SIRS epidemic model with information intervention. *Adv. Differ. Equations* **2017**, *2017*, 352. [CrossRef]
33. Zhao, Y.N.; Jiang, D.Q. The threshold of a stochastic SIS epidemic model with vaccination. *Appl. Math. Comput.* **2014**, *243*, 718–727. [CrossRef]
34. Khasminskii R. *Stochastic Stability of Differential Equations*; Springer Science and Business Media: Berlin/Heidelberg, Germany, 2011.

fractal and fractional

MDPI

Article

On the Estimation of the Persistence Exponent for a Fractionally Integrated Brownian Motion by Numerical Simulations

Mario Abundo [1,†] and Enrica Pirozzi [2,*,†]

1 Dipartimento di Matematica, Università "Tor Vergata", Via della Ricerca Scientifica, I-00133 Roma, Italy
2 Dipartimento di Matematica e Applicazioni, Università di Napoli "Federico II", Via Cintia, Complesso Monte S. Angelo, I-80126 Napoli, Italy
* Correspondence: enrica.pirozzi@unina.it
† These authors contributed equally to this work.

Abstract: For a fractionally integrated Brownian motion (FIBM) of order $\alpha \in (0,1]$, $X_\alpha(t)$, we investigate the decaying rate of $P(\tau_S^\alpha > t)$ as $t \to +\infty$, where $\tau_S^\alpha = \inf\{t > 0 : X_\alpha(t) \geq S\}$ is the first-passage time (FPT) of $X_\alpha(t)$ through the barrier $S > 0$. Precisely, we study the so-called persistent exponent $\theta = \theta(\alpha)$ of the FPT tail, such that $P(\tau_S^\alpha > t) = t^{-\theta+o(1)}$, as $t \to +\infty$, and by means of numerical simulation of long enough trajectories of the process $X_\alpha(t)$, we are able to estimate $\theta(\alpha)$ and to show that it is a non-increasing function of $\alpha \in (0,1]$, with $1/4 \leq \theta(\alpha) \leq 1/2$. In particular, we are able to validate numerically a new conjecture about the analytical expression of the function $\theta = \theta(\alpha)$, for $\alpha \in (0,1]$. Such a numerical validation is carried out in two ways: in the first one, we estimate $\theta(\alpha)$, by using the simulated FPT density, obtained for any $\alpha \in (0,1]$; in the second one, we estimate the persistent exponent by directly calculating $P(\max_{0 \leq s \leq t} X_\alpha(s) < 1)$. Both ways confirm our conclusions within the limit of numerical approximation. Finally, we investigate the self-similarity property of $X_\alpha(t)$ and we find the upper bound of its covariance function.

Keywords: fractional integrals; first-passage time; decaying rate; tail distribution

Citation: Abundo, M.; Pirozzi, E. On the Estimation of the Persistence Exponent for a Fractionally Integrated Brownian Motion by Numerical Simulations. *Fractal Fract.* **2023**, 7, 107. https://doi.org/10.3390/fractalfract7020107

Academic Editor: Vassili Kolokoltsov

Received: 19 December 2022
Revised: 11 January 2023
Accepted: 16 January 2023
Published: 20 January 2023

1. Introduction

The study of integrals with respect to the time-parameter of assigned stochastic processes constitutes one of the main chapters of stochastic calculus and one of the main tools for designing phenomenological models (see, for instance, [1–3] and references therein). Fractionally integrated stochastic processes are the natural extensions of the above processes in the context of the fractional calculus applied to the stochastic one (see, for instance, [4,5]). They are a rather new topic which appears to be of interest both from a theoretical point of view and for application (see [1] and references therein). Here, in particular, we focus on the (Liouville) fractionally integrated Brownian motion (FIBM) of order $\alpha \in (0,1]$, denoted by $X_\alpha(t)$ rigorously defined below. Our aim is to study the distribution of $\tau_S^\alpha = \inf\{t > 0 : X_\alpha(t) \geq S\}$, also denoted only by τ if there is no ambiguity about the specified boundary S. This is the first-passage time (FPT) of $X_\alpha(t)$ through a boundary (otherwise called a barrier or threshold) $S > 0$. We specifically address the problem of studying the decaying rate of $P(\tau_S^\alpha > t)$ as $t \to +\infty$.

The study of the distribution of the FPT τ of a stochastic process through a boundary is a classic problem in probability theory; generally, it is difficult to obtain an explicit expression of this law. However, it has been observed that in many interesting cases, the survival function has a polynomial decay which does not depend on the boundary:

$$P(\tau > t) = t^{-\theta+o(1)}, \text{ as } t \to +\infty, \tag{1}$$

or, equivalently:

$$\lim_{t\to+\infty} \frac{\log P(\tau > t)}{\log t} = -\theta, \tag{2}$$

where θ is a positive constant called *persistence exponent* and characterizes the profile of the tail of the distribution of τ for large t values. The computation of this exponent turns out to have connections with various problems in probability and mathematical physics. In general, for self-similar processes, the persistent exponent θ does not depend on the boundary, so this fact is natural in our case, since $X_\alpha(t)$ is actually self-similar (see Section 4.1).

A well-known result concerns Brownian motion: in this case, the persistence exponent turns out to be equal to $1/2$; another important result is that of Goldman-Sinai, regarding the case of integrated Brownian motion (see [6,7]), for which the persistence exponent is $\theta = 1/4$. A generalization of this result regards the study of the persistence exponent for twice integrated, or more generally n-th time integrated, Brownian motion (see [3,8] and the references therein). Another example is the study of the persistence exponent for integrated fractional Brownian motion with Hurst parameter H (it was conjectured in [9] that θ should be $H(1-H)$). Moreover, the persistence exponent for the integrated stable Lévy process was studied in [8].

Furthermore, the persistence exponent was studied for an $\alpha-$fractionally integrated centered Lévy process; in [10], it was proved that the corresponding persistence exponent is a non-increasing function of the fractional order α; the class of processes considered includes FIBM.

1.1. The Motivation

To our knowledge, none of the known results in the literature regard the theoretical computation of the persistence exponent for FIBM $X_\alpha(t)$, nor have numerical estimates of $\theta(\alpha)$ been previously obtained. Thus, the aim of the present article is to numerically estimate $\theta(\alpha)$ as a function of $\alpha \in (0,1)$ by using simulated trajectories of $X_\alpha(t)$. In particular, we are able to validate the following conjecture numerically:

$$\text{for } \alpha \in [0,1] \text{ the persistence exponent of FIBM } X_\alpha(t) \text{ is given by } \theta(\alpha) = \frac{1}{2(\alpha+1)}. \quad (3)$$

Note that this formula agrees with the known results in the cases $\alpha = 0$ and $\alpha = 1$, and it is also a non-increasing function of α, according to results in [10]. The idea and some motivations of the conjecture are given in Section 2; the numerical validation of the conjecture will be illustrated in detail in Section 3. Note that our validation is strongly based on numerical simulations of long enough trajectories of the process $X_\alpha(t)$, which require a lot of computation time; so, our analysis could be improved by using a more powerful computer dedicated to the purpose. Unfortunately, we cannot compare our results with those of other authors, since to our knowledge no numerical result of this kind is actually present in the literature.

We emphasize that our study about the persistent exponent of $X_\alpha(t)$ with regard only to the values of $\alpha \in [0,1]$; actually, (3) is only a local conjecture, holding for α belonging to the unit interval. We have not considered extensions of the process $X_\alpha(t)$ to negative values of α or to $\alpha > 1$, nor have we studied the persistent exponent for these values of α. In fact, the conjecture (3) cannot hold for negative α (see [3]) and for $\alpha > 1$ (see [11]). The reason we have limited ourselves to study the process $X_\alpha(t)$ and its persistent exponent for $\alpha \in [0,1]$ is due to the fact that we are mainly interested in stochastic processes, such as the FIBM, that model neuronal activities, for which the appropriate range of the fractionally integration parameter α is the unit interval.

The mathematical interest of a such study relies essentially on the need to further investigate the probability laws of $X_\alpha(t)$ and its FPT in order to refine and complete the mathematical setting of the FIBM. Furthermore, this study presents the possibility of shedding light upon a wider class of fractionally integrated stochastic processes and their applications. Indeed, the FIBM has interesting applications in the description of the time evolution of stochastic systems: it appears, e.g., in the framework of certain diffusion models for neuronal activity (see [1], but also [12] for similar models with different processes), where one expects that the inter-spike instants will have a heavy tail distribution, i.e., a power-law decaying rate. The specific choice of fractionally integrated stochastic processes (or, specifically, diffusions) in neuronal modeling allows us to devise models that are more adherent to phenomenological evidence, such those affecting the neuronal spike activity "with memory"for which, after a sequence of short inter-spike times, sequences of long inter-spike times are detected, due to a sort of "adaptation "([12]).

We can essentially remark that this article is motivated by the aim to study the behavior of the persistent exponent for stochastic processes "with memory" such as the FIBM, by varying the order $\alpha \in [0,1]$ of fractional integration.

1.2. The Results

Our study of the decaying rate of the tail distribution of τ_S^α (namely its persistence exponent) is essentially based on numerical simulations. Indeed, by using an ad hoc algorithm implemented in an R-script, we perform simulations of long enough trajectories of the process $X_\alpha(t)$, and the results confirm that for $\alpha = 0$ (i.e., case of BM), one has $\theta(\alpha) = \theta = 1/2$, while for $\alpha = 1$ (case of integrated BM), one has $\theta = \theta(\alpha) = 1/4$ (see [6,7]).

For $0 < \alpha < 1$, our numerical investigation shows that the exponent $\theta(\alpha)$ decreases as α increases (cf., for instance [3], in which the persistence exponent is revealed to be as a non-increasing function of α). We provide numerical estimates and comparisons by means of graphs and tables (see Section 3 for details and discussion of results).

1.3. In Summary

The paper is organized as follows. In the next section, we give the essential ingredients of our study and the main known results from which it starts, and we explain our conjecture, successively supported by simulation results. In Section 3, we describe the specialized algorithm for the simulation paths of the process $X_\alpha(t)$ and of its FPT. We provide graphic results in some figures in order to show and compare the profiles of the FPT density approximations for different values of fractional order α. We explain our method to obtain the estimation of the persistence exponent for the simulated cases. Our results are in agreement with the well-known result for the case of $\alpha = 1$, and they provide quantitative approximations for cases of $\alpha = 0.75, 0.5, 0.25$, suitably justified under our conjecture. Additionally, we also provide numerical estimates of the probability in (1) in order to show the agreement between the study of the density and of the distribution of the FPT under the conjecture assumption. In Section 4, we highlight some properties of the fractionally integrated processes such as self-similarity that can be useful for implementation purposes of numerical simulations, and long-range dependence that, together with the magnitude of persistence exponent, makes such processes suitable tools for modeling biological dynamics with "memory". In Section 5, we discuss the possible strategy to be adopted for special Gaussian processes, including Gauss–Markov (GM) processes such as Ornstein–Uhlenbeck (OU). Indeed, thanks to the fact that a GM can be transformed into an OU process (which in turn can be written in terms of BM), one finds that the FPT of a GM process is finite with probability one, and information about the tail behavior of the FPT may be analytically obtained.

2. The Persistence Exponent for Integrated BM: Known Results and a Conjecture

Now, we recall the definition of the fractional integral of order $\alpha \in (0,1)$ of Brownian motion (FIBM):

$$X_\alpha(t) = \frac{1}{\Gamma(\alpha)} \int_0^t (t-s)^{\alpha-1} B(s) ds, \tag{4}$$

where $B_t = B(t)$ is a standard Brownian motion (BM), and Γ denotes the Gamma Euler function, i.e., $\Gamma(z) = \int_0^{+\infty} t^{z-1} e^{-t} dt$, $z > 0$.

Taking the limit for $\alpha \to 0^+$, one finds that $X_0(t)$ is BM itself, while for $\alpha = 1$, one obtains the ordinary integral of BM. The process $X_\alpha(t)$ starts from zero at $t = 0$ with probability 1 (w.p.1) and turns out to be Gaussian with mean zero (cf. [4]); its covariance function as well its variance were studied at length in [1]. Actually, in [1] we have also performed numerical simulations of trajectories of $X_\alpha(t)$, and the probability distribution of τ_S^α was numerically studied.

As the case of BM is concerned, it is well-known that its FPT through the barrier S, say τ_S^B, is finite with probability one, though the expectation $E(\tau_S^B) = +\infty$ and the exact formula holds (see, e.g., [13]):

$$P(\tau_S^B \leq t) = 2(1 - \Phi(S/\sqrt{t})) \tag{5}$$

where Φ denotes the cumulative distribution function of a standard Gaussian variable. Then (see also [14]):

$$P(\tau_S^B > t) \sim \sqrt{\frac{2}{\pi}}\frac{S}{\sqrt{t}}, \text{ as } t \to \infty, \tag{6}$$

i.e., the persistence exponent is $\theta = 1/2$ (see also Example 2.2.2. in [12]). Instead, for non-Markov Gaussian processes such as $X_\alpha(t)$ is, very few results are known about the FPT through a barrier S.

Actually, there is an objective difficulty in numerically estimating the FPT distribution using simulated trajectories of the process, since detecting the instant of the first passage through the barrier S can be an arduous task, because the trajectory considered could hit the barrier, but only after a number of simulation time steps which possibly exceed the maximum allowed by the computer algorithm. Therefore, this kind of trajectory is disregarded in the computation of those crossing the boundary within that maximum time of simulation.

Now, we recall the behavior of the tail of the FPT τ^α of $X_\alpha(t)$ through the boundary S for well-known cases.

<u>CASE $\alpha = 0$</u>

This is the case when $X(t) := X_0(t) = B(t)$. From Formula (2), it follows that:

$$P(\tau^B > t) = 2\Phi\left(\frac{S}{\sqrt{t}}\right) - 1;$$

then, as easily seen by using the Hospital rule, one gets:

$$\lim_{t \to +\infty} P(\tau^B > t)\sqrt{t} = \lim_{t \to +\infty} \sqrt{t}\left(2\Phi\left(\frac{S}{\sqrt{t}}\right) - 1\right) = S\sqrt{\frac{2}{\pi}},$$

that is $P(\tau^B > t) \sim c/\sqrt{t}$, as $t \to +\infty$, and the persistence exponent is $\theta = 1/2$.

<u>CASE $\alpha = 1$</u>

Now, we have $X(t) := X_1(t) = \int_0^t B(s)ds$ i.e., the ordinary integral of BM. The exact result (see, specifically, [6,15]) is that $\theta = 1/4$. Actually, the estimation of the FPT of integrated BM through the barrier S, numerically obtained by computer simulation, indicates that its probability density behaves as $c \cdot t^{-5/4}$, as $t \to +\infty$, that is, the persistence exponent is $\theta = 1/4$. According to this, we find that the tail of the FPT distribution of integrated BM is heavier than that of τ^B, which behaves as $t^{-1/2}$.

The constant c in front of $t^{-5/4}$ was exactly calculated in [15] (see the last formula at pg. 1292), and it is:

$$c = \frac{S^{1/6}3^{25/12}\Gamma(5/12)}{2^{65/12}\pi\sqrt{\pi}}\,{}_2F_1(5/12; 7/4; 3/2; 3/4). \tag{7}$$

This is the product of the result of Goldman ([6]) and $1/6$. We also evaluated it by R functions; for the case of $S = 1$, we obtained the value $c = 0.1795595$.

Remark 1. *One can observe that* $X(t) \sim \mathcal{N}(0, t^3/3)$,; *so, for fixed* t, $X(t)$ *has the same distribution as* $\tilde{X}(t) = B(t^3/3)$. *Note that* $X(t)$ *is different by* $\tilde{X}(t)$, *which is a time-changed BM. Indeed, if* $\tilde{\tau}$ *denotes the FPT of* $\tilde{X}(t)$, *then* $\tilde{\tau}^3/3 = \inf\{r > 0 : B_r > S\} = \tau^B$, *and so:*

$$P(\tilde{\tau} > t) = P(\tilde{\tau}^3/3 > t^3/3) = P(\tau^B > t^3/3) \sim \frac{c}{t^{3/2}}, \text{ as } t \to +\infty. \tag{8}$$

This is expected, since the process $\tilde{X}(t)$ *reaches the barrier* S *more quickly than in the case of BM (being* $t^3/3$ *much greater than* t, *for* t *large).*

CASE $\alpha \in (0,1)$: a conjecture

For the other values of $\alpha \in (0,1)$, our numerical estimations show that the tail of the FPT τ^α of the fractionally integrated BM X_α through the barrier S is heavier than that of τ^B (corresponding to the case of $\alpha = 0$). Precisely, we find that, as α increases in $(0,1]$, the tail becomes heavier and heavier; that is, the persistence exponent $\theta(\alpha)$ does not increase. Finally, we confirm that the persistence exponent $\theta(\alpha)$ is a non-increasing function of the fractional order α. Indeed, we are confident that our following conjecture holds:

The persistence exponent for $X_\alpha(t)$ *is:*

$$\theta = \theta(\alpha) = \frac{1}{2(\alpha+1)}, \ \alpha \in (0,1). \tag{9}$$

Our conjecture is born from the following reasoning. First, we recall the results for the Brownian motion and its FPT τ^B through, e.g., the boundary $S = 1$ (recall that the persistent exponent is independent of the boundary), i.e.,

$$P(\tau^B > t) = P(\max_{0\le s\le t} B(s) < 1) \sim t^{-1/2}, \quad \text{for large t} \tag{10}$$

and for the integrated Brownian motion and its FPT τ, i.e.,

$$P(\tau > t) = P\left(\max_{0\le s\le t} \int_0^s B(z)dz < 1\right) \sim t^{-1/4}, \quad \text{for large t.} \tag{11}$$

By recalling the following distribution equality (see, for instance, [3]):

$$P\left(\max_{0\le s\le t} \int_0^s B(z)dz < 1\right) = P\left(\max_{0\le s\le t} \int_0^s (s-z)dB(z) < 1\right) \tag{12}$$

we also have that

$$P(\tau > t) = P\left(\max_{0\le s\le t} \int_0^s (s-z)dB(z) < 1\right) \sim t^{-1/4} \quad \text{for large t.} \tag{13}$$

Note that the last approximation can be interpreted in the two following ways:

$$P\left(\max_{0\le s\le t} \int_0^s (s-z)dB(z) < 1\right) \sim (t^{1/2})^{-1/2} \tag{14}$$

or

$$P\left(\max_{0\le s\le t} \int_0^s (s-z)dB(z) < 1\right) \sim (t^{-1/2})^{1/2}. \tag{15}$$

From this, we do our conjecture for $X_\alpha(t)$. We consider that

$$P(\tau > t) = P\left(\max_{0\le s\le t} X_\alpha(s) < 1\right) = P\left(\max_{0\le s\le t} \int_0^s (s-z)^{\alpha-1} B(z)dz < 1\right) \tag{16}$$

and, by using the following distribution equality

$$P\left(\max_{0\le s\le t}\int_0^s (s-z)^{\alpha-1}B(z)dz < 1\right) = P\left(\max_{0\le s\le t}\frac{1}{\alpha}\int_0^s (s-z)^{\alpha}dB(z) < 1\right)$$

we conjecture that, for large t,

$$P(\tau > t) = P\left(\max_{0\le s\le t}\frac{1}{\alpha}\int_0^s (s-z)^{\alpha}dB(z) < 1\right) \sim \left(t^{\frac{1}{\alpha+1}}\right)^{-1/2} = t^{-\frac{1}{2(\alpha+1)}}. \quad (17)$$

The conjecture (17) is equivalent to:

$$P(\tau > t) = P\left(\max_{0\le s\le t}\frac{1}{\alpha}\int_0^s (s-z)^{\alpha}dB(z) < 1\right) \sim \left(t^{-\frac{1}{\alpha+1}}\right)^{1/2} = t^{-\frac{1}{2(\alpha+1)}}. \quad (18)$$

In particular, the conjecture, expressed as in (17), can also be explained by means of (2) and by interpreting the persistence exponent θ for large t as a function of the time t and α; i.e., $\theta(t,\alpha)$, such that:

$$\theta(t,0) = \frac{1}{2} = \frac{1}{2\cdot 1} \quad \text{in case of } (t^{1/(0+1)})^{-1/2} \sim P\left(\max_{0\le s\le t}\int_0^s (z-s)^0 dB(z) < 1\right)$$

$$\theta(t,1) = \frac{1}{4} = \frac{1}{2\cdot 2} \quad \text{in case of } (t^{1/(1+1)})^{-1/2} \sim P\left(\max_{0\le s\le t}\int_0^s (z-s)^1 dB(z) < 1\right)$$

and consequently,

$$\theta(t,\alpha) = \frac{1}{2(\alpha+1)} \quad \text{in case of } (t^{1/(\alpha+1)})^{-1/2} \sim P\left(\max_{0\le s\le t}\frac{1}{\alpha}\int_0^s (z-s)^{\alpha} dB(z) < 1\right).$$

Note that in the conjecture we include the case of $\alpha = 0$ (i.e., that of the BM) with persistence exponent $\theta = \frac{1}{2(\alpha+1)} = \frac{1}{2}$, and the case of $\alpha = 1$ (i.e., that of the integrated BM) with persistence exponent $\theta = \frac{1}{2(\alpha+1)} = \frac{1}{2(1+1)} = \frac{1}{4}$; all other cases for $\alpha \in (0,1)$ have a persistence exponent θ, such that $\frac{1}{4} \le \theta \le \frac{1}{2}$ with θ non increasing function of α.

Unfortunately, we are not able to show an analytical reason for the conjecture (17); our heuristic motivation comes by comparing the above equations, and it is confirmed by our numerical computations.

About the FPT density:

Actually, by taking the derivative in the expression $P(\tau^{\alpha} > t) \sim t^{-\theta(\alpha)}$, the conjecture (9) implies that the FPT density of τ^{α} behaves as $const \cdot t^{-(\theta(\alpha)+1)}$, as $t \to \infty$, where $\theta(\alpha)$ is the persistent exponent of FIBM $X^{\alpha}(t)$. Then, inspired by Goldman (see [6,15]), we will suppose that the density of the FPT τ^{α} through the boundary S behaves as

$$c_{\alpha} \cdot S^{1/6} t^{-(3+2\alpha)/(2\alpha+2)}, \quad (19)$$

as $t \to \infty$, where the multiplicative constant c_{α} is also estimated as a suitable constant multiplied by α. We will validate our conjecture, that is, (19), by means of long trajectories of the process obtained by computer simulation, samples of their FPT and the approximation of the respective densities.

About the FPT distribution:

In addition, we will also work with the purpose

(i) to obtain a numerical estimate of the following probability for t "large enough"

$$P\left(\max_{0\le s\le t}\int_0^s (s-z)^{\alpha-1}B(z)dz < 1\right)$$

(ii) to compare it with the function $t^{-\frac{1}{2(\alpha+1)}}$ and from their ratio to derive an estimate of the multiplicative c_α such that the asymptotic (in time) tail behavior of the FPT distribution for $S = 1$ can be finally characterized as (see Section 3.2 for details)

$$c_\alpha \cdot t^{-\frac{1}{2(\alpha+1)}}.$$

All numerical validations of such a conjecture and approximation results are described in detail in the next section.

3. The Simulation Algorithm and Numerical Results

In the following, we show and compare graphically the behaviour of the tail of the density of FPT τ^α for some samples of simulated paths. The simulation algorithm partially follows the **R** script adopted in [1]. Here, the script is specialized in order to investigate the decaying profile of the FPT densities as the value of the fractional order α varies. The algorithm requires knowledge of the covariance matrix of $X_\alpha(t)$, whose mathematical expression can be found here in Section 4.2. After some statistical comparisons taking into account the execution time of the simulation algorithm, the accuracy of the results and the theoretical expected number of the sample paths crossing the specified boundary, the time discretization step h is chosen as $h = 0.05$ here. Moreover, by means of the same sequence of pseudo-random Gaussian numbers, we perform several cases corresponding to different α values.

Specifically, and referring to the methods for generation of pseudo-random numbers (see also [16]), we give the scheme of our R-algorithm:

STEP1. IN INPUT: we specify the size of the sample of paths we want to simulate, the boundary, the fractional order, the time step, the seed for the random generation routine, and the maximum time length of each path, i.e., the maximum number N of time steps for each path;

STEP2. we calculate the covariance matrix $C(t_i, t_j)$ with dimension $N \times N$ in $h-$equi-spaced times $t_i, t_j,\ i, j = 1, \ldots, N$;

STEP3. we construct the lower triangular matrix $L(i, j)$ such that $C = LL^T$ by applying the Cholesky decomposition algorithm to the covariance matrix C;

STEP4. by the ad hoc R-function (rnorm), for each path, we generate an N-dimensional array $\mathbf{z}$ of standard pseudo-Gaussian numbers;

STEP5. we construct the simulated path with the specified correlation matrix C as the array $\mathbf{x} = L\mathbf{z}$;

STEP6. at each step of the path, we check if crossing of the boundary occurs, and if this is the case, the corresponding FPT is recorded;

STEP7. we repeat STEPS 4-6 for each path in order to simulate the specified number of paths;

STEP8. we provide the array of simulated FPTs of dimension equal to the size of the total number of simulated paths (note that if a simulated path does not cross the boundary in the specified (in input) maximum number of steps, a zero value is recorded as its FPT).

STEP9. IN OUTPUT: the array of FPT is cleaned from zero values and it is used for histograms and kernel density approximations for visualizations and comparisons in figures.

3.1. Graphical and Approximation Results

On the left of Figure 1, we plotted the empirical FPT densities of $X_\alpha(t)$ (on the horizontal axes time t) through the threshold $S = 0.1$ for values of α specified in the figure. The empirical density has been obtained by means of 5000 simulated trajectories with a time step equal to 0.05; to obtain it, we have counted only the trajectories which have crossed the threshold within the time bound of 300 time steps $h = 0.05$, i.e., until time 15. The simulated paths are obtained by applying the algorithm described at the beginning of this section, and their first passage times have been recorded. The plotted curves in the figure are the results of the *plot* R-function of the *density* R-function applied to the array of the FPT values of the simulated paths. We underline that the function *density* computes (Gaussian) kernel density estimates with values for bandwidth "*adjust* "parameter between 4 and 12, in such a way that greater values determine smoother curves. The plots of such approximations are provided in the same figure to show how the fractional order α affects the profiles of the FPT densities, and to allow comparisons.

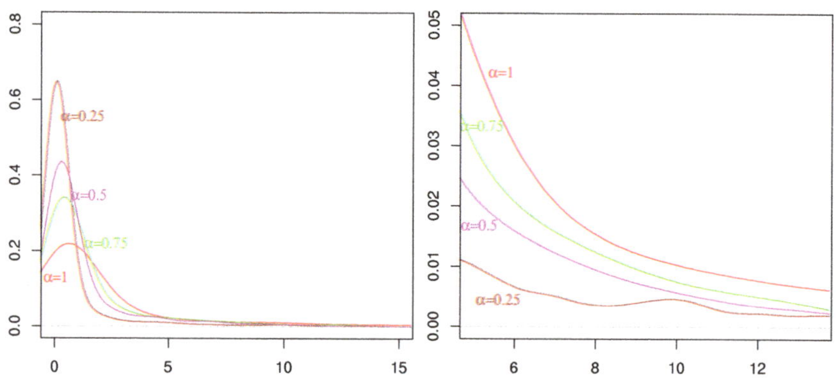

Figure 1. Left: Comparisons of FPT densities of $X_\alpha(t)$ through the boundary $S = 0.1$ for several values of α. The samples are until 5000 simulation paths and the time discretization step is 0.05. **Right**: Zoom on tails of densities (shown on the left) for several values of α.

Finally, on the right of Figure 1, we give an enlarged visualization of the tail of the densities on the left of the same figure. We note that as α increases, the tails become heavier, determining smaller values for the corresponding persistence exponents.

First, we show our simulation results for the well-known case of $\alpha = 1$ in order to validate our investigation approach. In Figure 2, (left) we show the simulated FPT density (in red) of $X_1(t)$ through the threshold $S = 0.1$. The black curve is the plot of the function we denote as

$$a(t) = S^{1/6} \cdot t^{-(3+2\alpha)/(2\alpha+2)} = S^{1/6} t^{-5/4}$$

with $S = 0.1$. Then, we estimated the value of the constant c_α as the ratio between the simulated density and the black curve $a(t)$ at times ≥ 12 obtaining $c_\alpha = 0.177$. Actually, if we denote the simulated density by $d(t)$, we estimate:

$$c_\alpha = \frac{d(t)}{a(t)}$$

in values of $t \geq \bar{t}$, $\bar{t}$ being the starting time at which we observe a quite constant difference between $d(t)$ and $a(t)$ values. The choice of times "large enough"depends on the value of α. In this case, we apply this strategy in order to validate it, because we already know the value of the constant c_1, is, as previously specified, 0.1795595.

Moreover, the blue curve is obtained as the product of the black curve and the constant c_α, i.e., by denoting as $b(t)$ the blue curve, we plot, in blue color, the function

$$b(t) = c_\alpha \cdot a(t) = c_\alpha \cdot S^{1/6} \cdot t^{-(3+2\alpha)/(2\alpha+2)}.$$

that, for $\alpha = 1$, $S = 0.1$ and $c_\alpha = 0.17$, specifically is

$$b(t) = 0.17 \cdot (0.1)^{1/6} \cdot t^{-5/4}.$$

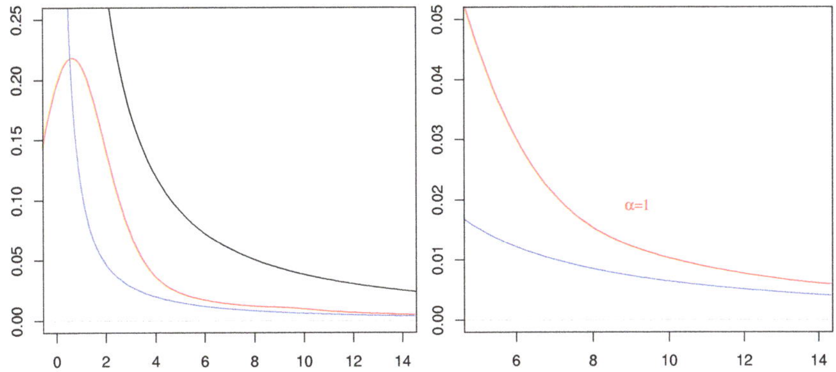

Figure 2. Case $\alpha = 1$. Simulation details as in Figure 1. On the **left**: the red curve is simulated. The black curve is the plot of $S^{1/6} \cdot t^{-(3+2\alpha)/(2\alpha+2)} = S^{1/6}t^{-5/4}$ with $S = 0.1$. The blue curve is then obtained as the product of the black curve and the constant c_α, whose estimated value is $c_\alpha = 0.177$. On the **right** is a zoom of the simulated density (red) and the approximation curve (blue) corresponding to those on the left.

Ultimately, the matching between the two curves (visible in the enlarged visualization on the right of Figure 2) is very good for large enough t. This means that our simulation results are in according to well-known behavior of the FPT tail distribution, i.e., $const \cdot t^{-1/4}$, corresponding to the persistence exponent $\theta(1) = 1/4$.

In Figures 3–5, we show the results of the corresponding investigations, as performed to provide the results shown in Figure 2 for the specified cases of $\alpha = 0.75, 0.5, 0.25$, respectively.

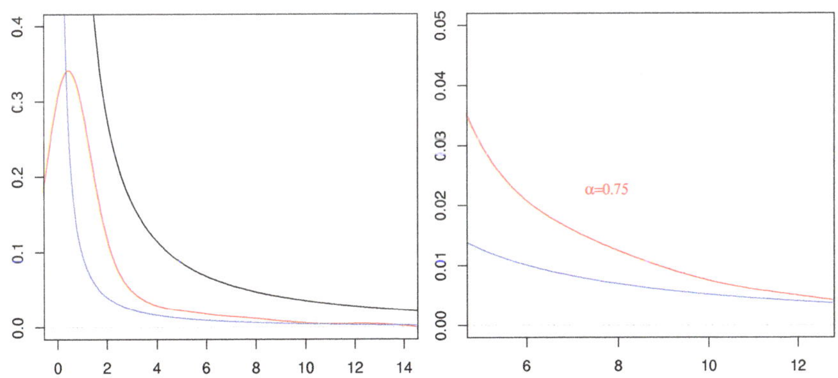

Figure 3. Case $\alpha = 0.75$. Simulation details as in Figure 1. On the **left**, the red curve is simulated. The black curve is the plot of $a(t) = S^{1/6} \cdot t^{-(3+2\alpha)/(2\alpha+2)}$ with $S = 0.1$. The blue curve is then obtained as $b(t) = c_\alpha \cdot a(t)$, with an estimated value of $c_\alpha = 0.15$. On the **right**, a zoom of the simulated density (red) and the approximation curve (blue), corresponding to those on the left.

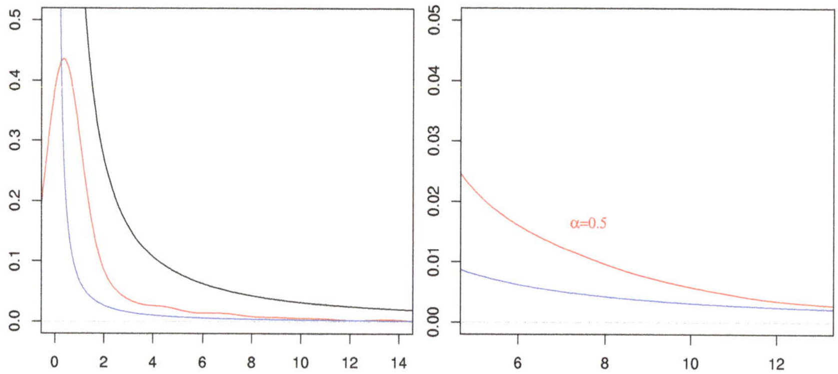

Figure 4. Case $\alpha = 0.5$. Simulation details as in Figure 1. On the **left**, the red curve is simulated. The black curve is the plot of $a(t) = S^{1/6} \cdot t^{-(3+2\alpha)/(2\alpha+2)}$ with $S = 0.1$. Blue curve is then obtained as $b(t) = c_\alpha \cdot a(t)$, with estimated value of $c_\alpha = 0.1$. On the **right**, zoom of the simulated density (red) and the approximation curve (blue), corresponding to those on the left.

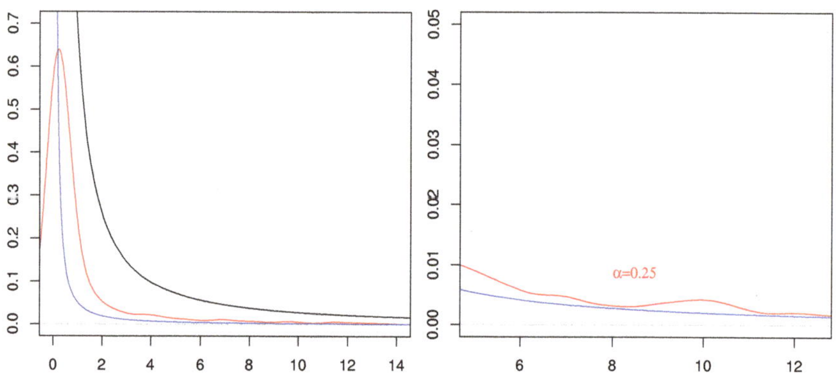

Figure 5. Case $\alpha = 0.25$. Simulation details as in Figure 1. On the **left**, the red curve is simulated. The black curve is the plot of $a(t) = S^{1/6} \cdot t^{-(3+2\alpha)/(2\alpha+2)}$ with $S = 0.1$. The blue curve is then obtained as $b(t) = c_\alpha \cdot a(t)$, with estimated value of $c_\alpha = 0.075$. On the **right**, a zoom of the simulated density (red) and the approximation curve (blue), corresponding to those on the left.

In these figures, the simulated FPT densities $d(t)$ are plotted in red for processes $X_\alpha(t)$ through the same threshold $S = 0.1$. The black curves are the plot of the function

$$a(t) = S^{1/6} \cdot t^{-(3+2\alpha)/(2\alpha+2)} \tag{20}$$

with the specific value of $\alpha = 0.75, 0.5, 0.25$, respectively. The estimates of c_α are obtained as $c_\alpha = \frac{d(t)}{a(t)}$ for $t \geq \bar{t}_\alpha$, whose values are specified in Table 1. From the presented study, we hypothesize that c_α is approximately equal to the product $\alpha \cdot c_1 \approx \alpha \cdot 0.177$, even if we believe further investigations needed to validate this. Finally, our approximation curves, i.e., the blue curves, are obtained as

$$b(t) = c_\alpha \cdot a(t) = c_\alpha \cdot S^{1/6} \cdot t^{-(3+2\alpha)/(2\alpha+2)} \tag{21}$$

for the values of $\alpha = 0.75, 0.5, 0.25$, respectively.

Furthermore, the values of the persistence exponents for the considered cases are reported in Table 1.

Table 1. For the specified values of fractional order α, the persistence exponent θ, the estimates of c_α, the power decaying function $a(t) \cdot S^{-1/6}$, the value $\bar{t}_\alpha$ and the value of the *adjust* parameter are reported. Note that here, $S = 0.1$, and the value of c_α is evaluated as $d(\bar{t}_\alpha)/a(\bar{t}_\alpha)$, with $d(t)$ the simulated density and $a(t)$ is as in (20). The values of *adjust* are used in the evaluation of the *density* R-function to obtain the smoothed simulated density $d(t)$.

Numerical Approximations for FPT Density					
α	θ	c_α	$a(t) \cdot S^{-1/6}$	$\bar{t}_\alpha$	*adjust*
1.0	1/4	0.17	$t^{-5/4}$	14	4
0.75	2/7	0.15	$t^{-9/7}$	13.5	4
0.5	1/3	0.10	$t^{-4/3}$	13.9	8
0.25	2/5	0.075	$t^{-7/5}$	14.5	4

We also recall that our approximation function (the blue curve in Figures 3–5) for the specified values of $\alpha = 0.75, 0.5, 0.25$ is finally obtained as $b(t)$ in (21) by means of values of Table 1, respectively.

Ultimately, for the considered cases, the obtained numerical results, the visible good agreement between the simulated densities (red) and the proposed approximation curves (blue) present in the figures support the proposed conjecture.

3.2. A Further Numerical Approximation

Now, specifically for the tail of the FPT distribution, we aim to give an additional validation of our conjecture, but also a verification tool for previous results inherent to the FPT density. We have already explained the idea of such numerical verification at the end of Section 2. To this end, we realized a specific algorithm in an ad hoc R-script that is implemented with the following steps:

- For a fixed value of $\hat{t}$ and for a given integer n, we consider the equispaced times $0 < t_1 < \cdots < t_k < \cdots < t_n = \hat{t}$, and we generate the pseudo-random values:

$$E_{t_k} = \left\{ \int_0^{t_k} (t_k - z)^{\alpha-1} B(z) dz \right\} \qquad \text{for } k = 1, \ldots, n$$

  by means of the following approximation procedure (with m is specified in input):

$$\hat{E}_{t_k} = \frac{t_k}{m} \sum_{i=1}^{m-1} (t_k - \vartheta_i)^{\alpha-1} z(\sqrt{\vartheta_i}) \tag{22}$$

  where $0 < \theta_1 < \cdots < \theta_m = t_k$ are equispaced times and $z(\sqrt{\vartheta_i})$, with $i = 1, \ldots, m-1$, are pseudo-random numbers generated with Gaussian distribution, having zero mean and $\sqrt{\vartheta_i}$ standard deviation.
- We record the value

$$M_n(\hat{t}) = \max_{k=1,\ldots,n} \hat{E}_{t_k}. \tag{23}$$

- We repeat the previous steps for N times in order to obtain a sample of N pseudo-random values

$$M_n^{(j)}(\hat{t}) = \max_{k=1,\ldots,n} \hat{E}_{t_k}^{(j)}, \; j = 1, \ldots, N;$$

  we count how many $M_n^{(j)}(\hat{t})$ are such that

$$M_n^{(j)}(\hat{t}) < 1, \qquad j = 1, \ldots, N,$$

  and we denote such number as $\mathbb{M}_n(\hat{t})$.

- Hence, we approximate the following probability as

$$P\left(\max_{0\le s\le \hat{t}} \int_0^s (s-z)^{\alpha-1} B(z) dz < 1\right) \approx \frac{\mathbb{M}_n(\hat{t})}{N}$$

and we denote such estimated probability as $p(\hat{t}) = \frac{\mathbb{M}_n(\hat{t})}{N}$.
- Under the conjecture $\theta = -\frac{1}{2(\alpha+1)}$, we compute the value $\hat{c}_\alpha$ as follows

$$\hat{c}_\alpha = \frac{p(\hat{t})}{\hat{t}^{-\frac{1}{2(\alpha+1)}}} = \frac{p(\hat{t})}{\hat{t}^{-\theta}}. \tag{24}$$

We did many executions of such algorithm with different $\hat{t}$ in order to find the value of $\hat{t}$ for which we can obtain the agreement between the estimated probability $p(\hat{t})$ and the expected value,i.e.,

$$p(\hat{t}) \approx \alpha c_1 \cdot \hat{t}^{-\theta} \qquad \text{with} \qquad c_1 = 0.179.$$

Indeed, from the previous investigation by means of simulations, we already knew that the approximation $c_\alpha \approx \alpha c_1$ can be adopted for the considered cases.

Note that the value of $\hat{c}_\alpha$ in (24) depends on the value of $\hat{t}$. This numerical strategy also provides the value of the time $\hat{t}$ at which a quite satisfactory agreement of the above approximations is obtained.

The above algorithm has been implemented many times in order to also find the optimal parameters n, m, N to guarantee a sufficient accuracy of the results (we limit ourselves to the second digital digit) and not too long execution times. The results are reported in Table 2 for specified values of α.

We remark that the above probability approximation $p(\hat{t})$ is a punctual estimation; indeed, it is computed in the time instant $\hat{t}$ for which we found some results in agreement. We also note that the values of $\hat{t}$ in Table 2 are different from those of $\bar{t}$ in Table 1, even if these times are those for which we obtain the validation of our conjecture. This depends on the two different numerical strategies used; however, we can say that both times, $\bar{t}$ and $\hat{t}$, provide information about the starting time of the power decaying of the FPT probability laws (density and distribution, respectively) whose order has been conjectured here.

Table 2. For the specified values of fractional order α, we report the following: the conjectured persistence exponent θ, the number N of trials, the number n for the evaluation of the maximum as in (23), the number m of nodes of the quadrature as in (22), the product αc_1, the estimate $\hat{c}_\alpha$ as in (24), the time value $\hat{t}$ in which the probability is computed, the approximate probability $p(\hat{t})$ and the expected (conjectured) value $\alpha c_1 \hat{t}^{-\theta}$.

Numerical Approximations for FPT Distribution									
α	θ	N	n	m	αc_1	$\hat{c}_\alpha$	$\hat{t}$	$p(\hat{t})$	$\alpha c_1 \hat{t}^{-\theta}$
1.0	1/4	100	75	30	0.179	0.177	15	0.09	0.0909
0.75	2/7	100	70	20	0.13425	0.1275	14	0.06	0.0631
0.5	1/3	500	60	20	0.0895	0.0732	12	0.032	0.0390
0.25	2/5	2000	20	20	0.0425	0.0425	8	0.0185	0.01849

4. Some Further Properties of the Process $X_\alpha(t)$

4.1. On the Self-Similarity Property of $X_\alpha(t)$

Let c be a positive constant; from the definition of the FIBM $X_\alpha(t)$, it follows:

$$X_\alpha(ct) = \frac{1}{\Gamma(\alpha)} \int_0^{ct} (ct-s)^{\alpha-1} B(s) ds$$

$$= \frac{1}{\Gamma(\alpha)} \int_0^{ct} \left(ct - c\frac{s}{c}\right)^{\alpha-1} B(s)ds;$$

by the variable change $u = s/c$, and by using the self-similarity property of Brownian motion ($B(ct) = \sqrt{c}B(t)$), one finds the following equality in distribution:

$$X_\alpha(ct) = \frac{c^\alpha}{\Gamma(\alpha)} \int_0^t (t-u)^{\alpha-1} \sqrt{c}\, B(u)du = \frac{c^{\alpha+1/2}}{\Gamma(\alpha)} \int_0^t (t-u)^{\alpha-1} B(u)du = c^{\alpha+1/2} X_\alpha(t).$$

Thus, for $c > 0$ we obtain the self-similarity property of the FIBM:

$$X_\alpha(ct) = c^{\alpha+1/2} X_\alpha(t),$$

or

$$X_\alpha(t) = c^{-(\alpha+1/2)} X_\alpha(ct), \tag{25}$$

where equality is meant in distribution.

From the self-similarity property (25), we obtain:

$$X_\alpha(\tau_S) = S \Rightarrow c^{-(\alpha+1/2)} X_\alpha(c\tau_S) = S,$$

and so $X_\alpha(c\tau_S) = c^{\alpha+1/2} S$.

Therefore, if $S' = c^{\alpha+1/2} S$, it holds $\tau_{S'} = c\tau_S$, namely

$$\tau_S = \frac{1}{c}\tau_{S'}, \tag{26}$$

where equality is meant in distribution.

Equation (26) provides a relation between the FPT-distribution of the FIBM $X_\alpha(t)$ through the boundary S and the FPT-distribution through the boundary $S' = c^{\alpha+1/2} S$; note that, if $c > 1$, one finds that $S' > S$.

Now, setting $a = S' = c^{\alpha+1/2} S$, namely

$$c = \left(\frac{a}{S}\right)^{\frac{1}{\alpha+1/2}},$$

we obtain:

$$\tau_a = c\tau_S = \left(\frac{a}{S}\right)^{\frac{1}{\alpha+1/2}} \tau_S. \tag{27}$$

Thus, without loss of generality, one can study, e.g., the distribution of τ_1, that is, the FPT of the FIBM through the barrier $S = 1$. In fact, from (27), it follows:

$$\tau_a = a^{\frac{1}{\alpha+1/2}} \tau_1. \tag{28}$$

This fact can be useful during the numerical procedure to obtain the FPT of $X_\alpha(t)$ through the barrier a.

4.2. On the Covariance Function of $X_\alpha(t)$

Let $C_\alpha(u,t) = cov(X_\alpha(u), X_\alpha(t))$ be the covariance function of $X_\alpha(t)$, $u, t \geq 0$; we aim to obtain an upper bound for $C_\alpha(u,t)$ for $0 \leq u \leq t$.

We recall the following explicit expressions of $C_\alpha(u,t)$ and of $Var(X_\alpha(t))$ (see Equations (10) and (13) of [1]):

$$C_\alpha(u,t) = \frac{1}{\Gamma^2(\alpha)} \left[\frac{t^{\alpha+1} u^\alpha}{\alpha^2(\alpha+1)} - \frac{tH_\alpha(u,t)}{\alpha(\alpha+1)} + \frac{J_\alpha(u,t)}{\alpha(\alpha+1)} \right], \tag{29}$$

$$Var(X_\alpha(t)) = \frac{t^{2\alpha+1}}{(2\alpha+1)\Gamma^2(\alpha+1)}, \tag{30}$$

where

$$J_\alpha(u,t) = \int_0^u s(u-s)^{\alpha-1}(t-s)^\alpha ds \text{ and } H_\alpha(u,t) = \int_0^u (u-s)^{\alpha-1}(t-s)^\alpha ds. \quad (31)$$

Let us suppose that $0 \leq u \leq t$; then, we obtain:

$$0 \leq H_\alpha(u,t) \leq \int_0^u (t-s)^{2\alpha-1} ds; \quad (32)$$

moreover:

$$0 \leq J_\alpha(u,t) \leq \int_0^u s(t-s)^{2\alpha-1} ds \leq u \int_0^u (t-s)^{2\alpha-1} ds$$
$$= \frac{u}{2\alpha}\left[t^{2\alpha} - (t-u)^{2\alpha}\right] \leq \frac{ut^{2\alpha}}{2\alpha} \leq \frac{t^{2\alpha+1}}{2\alpha}. \quad (33)$$

From (29), by using (33) and the fact that $H_\alpha(u,t) \geq 0$, we get:

$$C_\alpha(u,t) \leq \frac{1}{\Gamma^2(\alpha)}\left[\frac{t^{2\alpha+1}}{\alpha^2(\alpha+1)} + \frac{t^{2\alpha+1}}{2\alpha^2(\alpha+1)}\right],$$

namely:

$$C_\alpha(u,t) \leq \frac{3t^{2\alpha+1}}{2\Gamma^2(\alpha)\alpha^2(\alpha+1)}, \; 0 \leq u \leq t. \quad (34)$$

Thus, for fixed u and $t \geq u$, the covariance function of $X_\alpha(t)$ increases as t increases, but for $t \to +\infty$ it grows at most as a constant $\times$ the power $t^{2\alpha+1}$, as the variance does.

5. Some Details on Gauss–Markov Process Fractionally Integrated

Remark 2. *In principle, one could use the previous arguments to study the tail behavior of the FPT of a fractionally integrated Gauss–Markov process $Y(t)$. We recall that a continuous GM process (see [1]) is in the form:*

$$Y(t) = m(t) + h_2(t)B(\rho(t)), \quad (35)$$

where $m(t)$, $h_1(t)$, $h_2(t)$ are continuous functions of $t \geq 0$, which are C^1 in $(0, +\infty)$, such that $h_2(t) \neq 0$, and $\rho(t) = h_1(t)/h_2(t)$ is differentiable non-negative, with $\rho(0) = 0$ and $\rho'(t) > 0$ for $t > 0$. The process $Y(t)$ has mean $m(t)$ and covariance $c(s,t) = cov(Y(s), Y(t)) = h_1(s)h_2(t)$, for $0 \leq s \leq t$. Besides BM itself, a noteworthy case of the GM process is the Ornstein–Uhlenbeck (OU) process; in fact, any continuous GM process can be represented in terms of an OU process (see, e.g., [17]).

Then, the fractional integral of a GM process $Y(t)$ is defined by:

$$X_\alpha(t) = \frac{1}{\Gamma(\alpha)} \int_0^t (t-s)^{\alpha-1} Y(s) ds. \quad (36)$$

However, the covariance and variance function of $X_\alpha(t)$ are far more complicated than in the case of FIBM (see [1]).

We conclude with a remark concerning the FPT of OU process.

Remark 3. *Let $Z(t)$ be OU process, starting from $Z(0) = z > 0$, driven by the SDE*

$$dZ(t) = -\mu Z(t)dt + \sigma dB(t),$$

for some $\mu, \sigma > 0$. One has

$$Z(t) = e^{-\mu t}(z + B(\rho(t))),$$

where

$$\rho(t) = \frac{\sigma^2}{2\mu}\left(e^{2\mu t} - 1\right).$$

Set now $\tau_Z(z) = \inf\{t > 0 : Z(t) = 0\}$, *then* $\tau_Z(z) = \inf\{t > 0 : B(\rho(t)) = z\}$, *where equality is meant in distribution, so* $\rho(\tau_Z(z)) \equiv \tau^B(z) = \inf\{s > 0 : B(s) = z\}$. *Then, from* (5), *one gets* $P(\tau_Z(z) \leq t) = 2(1 - \Phi(z/\sqrt{\rho(t)})$, *and* $\lim_{t\to+\infty} P(\tau_Z(z) > t)\sqrt{\rho(t)} = z\sqrt{\frac{2}{\pi}}$. *From this, it follows that the first passage time of* $Z(t)$ *through zero has a much lighter tail than that of BM; in particular,*

$$\frac{d}{dt}P(\tau(z) \leq t) = \frac{z}{\sqrt{2\pi}\rho^{3/2}(t)}e^{-z^2/2\rho(t)}\rho'(t) \sim const \cdot e^{-\mu t}, \text{ as } t \to +\infty.$$

Therefore, $E(\tau(z)) < +\infty$, *unlike the case of BM, for which* $E(\tau^B) = +\infty$.

6. Conclusions and Final Remarks

In this paper, we have considered fractionally integrated Brownian motion (FIBM) of order $\alpha \in (0,1]$, that is, $X_\alpha(t) = \frac{1}{\Gamma(\alpha)}\int_0^t (t-s)^{\alpha-1}B(s)ds$. The FIBM is an interesting process, since it appears, e.g., in the framework of diffusion models for neuronal activity (see [1]), where one expects that the inter-spike instants will have a heavy tail distribution, i.e., a power-law decaying rate.

The goal of this paper was to perform a qualitative study of the decaying rate of the tail distribution of τ_S^α, where $\tau_S^\alpha = \inf\{t > 0 : X_\alpha(t) \geq S\}$ is the first-passage time (FPT) of $X_\alpha(t)$ through the barrier $S > 0$. Precisely, we have studied the so-called persistent exponent $\theta = \theta(\alpha)$ of the FPT tail, such that $P(\tau_S^\alpha > t) = t^{-\theta+o(1)}$, as $t \to +\infty$. This study has been carried out by numerical simulation of long enough trajectories of the process $X_\alpha(t)$. In fact, we have estimated $\theta(\alpha)$, as the order α of fractional integration varies in $(0,1]$, and we have showed that it is a non-increasing function of $\alpha \in (0,1]$, with $1/4 \leq \theta(\alpha) \leq 1/2$. This means that the tail of the distribution of τ_S^α becomes heavier and heavier as $\alpha \in (0,1]$ increases. Note that, to our knowledge, none of the known results in the literature regard the theoretical computation of the persistence exponent for FIBM $X_\alpha(t)$, except for $\alpha = 0$ (in the case of BM) and $\alpha = 1$ (in the case of integrated BM). Our numerical estimations confirm that for $\alpha = 0$, one has $\theta(\alpha) = 1/2$, while for $\alpha = 1$, one has $\theta(\alpha) = 1/4$ (see [6,7]).

In particular, we have numerically validated a new conjecture about the analytical expression of the function $\theta = \theta(\alpha), \alpha \in (0,1]$, namely $\theta(\alpha) = \frac{1}{2(\alpha+1)}$, $\alpha \in (0,1]$. Such a numerical validation has been carried out by simulation of long enough trajectories of the process $X_\alpha(t)$ in two ways. In the first one, we have estimated the persistent exponent $\theta(\alpha)$ by using the simulated FPT density obtained for any α. In the second one, we have estimated the persistent exponent by directly calculating $P(\max_{0\leq s\leq t}\int_0^s B(z)dz < 1)$, which is nothing but $P(\tau_1^\alpha > t)$. Both ways confirm our conclusions within the limits of numerical approximation.

In the final part of the paper, we have investigated the self-similarity characteristics of $X_\alpha(t)$ and we have found an upper bound to its covariance function; moreover, we have given some details on the fractionally integrated Gauss–Markov process .

The arguments of this paper allow us, in principle, to also study the decaying rate of the tail distribution (and therefore of the corresponding persistent exponent) of the FPT of the fractional integral of order α of a Gauss–Markov process.

Author Contributions: Conceptualization, M.A.; Methodology, M.A. and E.P.; Software, E.P.; Formal analysis, M.A.; Investigation, E.P.; Data curation, E.P.; Writing—original draft, M.A. and E.P.; Writing—review & editing, M.A. and E.P. All authors have read and agreed to the published version of the manuscript.

Funding: This research is partially supported by MIUR-PRIN 2017, project "Stochastic Models for Complex Systems", no. 2017JFFHSH, and INdAM-GNCS.

Institutional Review Board Statement: Not applicable.

Informed Consent Statement: Not applicable.

Acknowledgments: The authors express particular thanks to the anonymous Reviewers for their helpful comments, leading to an improved presentation.

Conflicts of Interest: The authors declare no conflict of interest.

References

1. Abundo, M.; Pirozzi, E. Fractionally Integrated Gauss–Markov processes and applications. *Commun. Nonlinear Sci. Numer. Simulat.* **2021**, *101*, 105862. [CrossRef]
2. Ascione, G.; Pirozzi, E. On a stochastic neuronal model integrating correlated inputs. *Math. Biosci. Eng.* **2019**, *16*, 5206–5225. [CrossRef] [PubMed]
3. Aurzada, F.; Kilian, M. Asymptotics of the Persistence Exponent of Integrated Fractional Brownian Motion and Fractionally Integrated Brownian Motion. *Theory Probab. Its Appl.* **2022**, *67*, 77–88. [CrossRef]
4. Ascione, G.; Pirozzi, E. Generalized fractional calculus for gompertz-type models. *Mathematics* **2021**, *9*, 2140. [CrossRef]
5. Garrappa, R.; Kaslik, E.; Popolizio, M. Evaluation of fractional integrals and derivatives of elementary functions: Overview and tutorial. *Mathematics* **2019**, *7*, 407. [CrossRef]
6. Goldman, M. On the first passage of the integrated Wiener process. *Ann. Math. Statist.* **1971**, *42*, 2150–2155. [CrossRef]
7. Sinai, Y.G. Statistics of shocks in solution of inviscid Burgers equation. *Comm. Math. Phys.* **1992**, *148*, 601–621. [CrossRef]
8. Profeta, C.; Simon, T. Persistence of integrated stable processes. *Probab. Theory Relat. Fields* **2015**, *62*, 463–485. [CrossRef]
9. Molchan, G.M.; Khokhlov, A. Small values of the maximum for the integral of fractional Brownian motion. *J. Stat. Phys.* **2004**, *114*, 923–946. [CrossRef]
10. Aurzada, F.; Dereich, S. Universality of the asymptotics of the one-sided exit problem for integrated processes. *Ann. L'Insititut Henri-PoincarÉ-Probab. Stat.* **2013**, *49*, 236–251. [CrossRef]
11. Li, W.V.; Shao, Q.-M. Lower tail probabilities for Gaussian processes. *Ann. Probab.* **2004**, *32*, 216–242. [CrossRef]
12. Ascione, G.; Pirozzi, E.; Toaldo, B. On the exit time from open sets of some semi-markov processes. *Ann. Appl. Probab.* **2020**, *30*, 1130–1163. [CrossRef]
13. Klebaner, F.C. *Introduction to Stochastic Calculus with Applications*, 2nd ed.; Imperial College Press: London, UK, 2005.
14. Novikov, A.A. Martingales, Tauberian theorem, and strategies of gambling. *Theor. Probab. Appl.* **1997**, *41*, 716–729. [CrossRef]
15. Groenebom, P.; Jongbloed, G.; Wellner, J.A. Integrated Brownian motion, conditioned to be positive. *Ann. Probab.* **1999**, *27*, 1283–1303. [CrossRef]
16. Haugh M. *Generating Random Variables and Stochastic Processes*; IEOR E4703: Monte Carlo Simulation; Columbia University, New York, NY, USA, 2004.
17. Nobile, A.G.; Pirozzi, E.; Ricciardi, L.M. Asymptotics and evaluations of FPT densities through varying boundaries for Gauss–Markov processes. *Sci. Math. Jpn.* **2008**, *67*, 241–266.

fractal and fractional

MDPI

Article

Approximating the First Passage Time Density of Diffusion Processes with State-Dependent Jumps

Giuseppe D'Onofrio [1,*,†] and Alessandro Lanteri [2,†]

1 Dipartimento di Scienze Matematiche, Politecnico di Torino, 10129 Turin, Italy
2 Department of Economics, Management and Quantitative Methods, Università degli Studi di Milano, 20122 Milan, Italy
* Correspondence: giuseppe_donofrio@polito.it
† These authors contributed equally to this work.

Abstract: We study the problem of the first passage time through a constant boundary for a jump diffusion process whose infinitesimal generator is a nonlocal Jacobi operator. Due to the lack of analytical results, we address the problem using a discretization scheme for simulating the trajectories of jump diffusion processes with state-dependent jumps in both frequency and amplitude. We obtain numerical approximations on their first passage time probability density functions and results for the qualitative behavior of other statistics of this random variable. Finally, we provide two examples of application of the method for different choices of the distribution involved in the mechanism of generation of the jumps.

Keywords: first passage time problem; Jacobi process; simulation algorithm; nonlocal operator; Wright–Fisher model

Citation: D'Onofrio, G., Lanteri, A. Approximating the First Passage Time Density of Diffusion Processes with State-Dependent Jumps. *Fractal Fract.* **2023**, *7*, 30. https://doi.org/10.3390/fractalfract7010030

Academic Editors: Mario Abundo and Bruce Henry

Received: 28 November 2022
Revised: 20 December 2022
Accepted: 23 December 2022
Published: 28 December 2022

1. Introduction

Recently, there has been growing interest in jump diffusion models in many applied areas, ranging from computational neuroscience [1–3] to mathematical biology [4], metrology [5] and queueing theory [6], just to name a few. In particular, they have been popular in financial modeling, starting with the celebrated paper by Merton [7]. Since this, the use of such models has been increasing in real markets and theoretical studies (see, for instance, [8–12]), thanks to their ability to account for some empirically observed effects that otherwise would not be explained by traditional diffusion-based models. A comprehensive discussion on this matter can be found in [13]. Roughly speaking, by choosing the parameters of the jump process appropriately, one can generate a wide variety of dynamics incorporating relevant effects without relying only on a very large amount of noise. In all these application contexts, it is often required to face the problem of the first passage time (FPT) of the process describing the dynamics of the model through a boundary [14,15]. Depending on the context, this crossing is interpreted in a different way, but from a mathematical point of view, its treatment is formally the same. Despite being a classical problem [16], its resolution is non-trivial, and exact analytical results are available only in a few cases even for pure diffusion processes [17,18].

Early attempts to introduce jumps occurring at exponential times can be found in [19–24], where to maintain mathematical tractability, the jumps were assumed to be of a constant amplitude or coming from a fixed distribution. More recently, results on generalized mechanisms of the generation of jumps have appeared, assuming that the jump size depends on the value of the process [25,26] or the state dependence is for both size and frequency [27]. Following this direction, we consider a family of processes with state-dependent jumps whose diffusion part evolves according to a Jacobi (or Wright–Fisher) model. In [28], the authors introduced and studied nonlocal Jacobi operators, which generalized the classical (local) Jacobi operators. Apart from some analytical results obtained in [27], the literature on the FPT problem for these processes is scarce. For this reason, the study of approximations and simulations of the involved quantities constitutes a fundamental tool.

In this paper, we use a discretization scheme for simulating sample paths of these jump diffusion processes with state-dependent jump intensities. At each time step of the algorithm, a downward jump can occur with a probability and amplitude that depend on the distribution characterizing the jump component and the actual state of the process Between each jump epoch, the dynamics of the constructed process are purely diffusive and are simulated using the Milstein's discretization method [29].

From the simulations of the trajectories of the process, we obtain approximations of the density of the FPT through a constant boundary for different choices of the measure describing the distribution of the jumps. In particular, in the case of the Jacobi process with jumps, we consider exponential and Pareto distributions. From the simulations, we observe that despite the presence of only downward jumps, the decay of the tails of the FPT pdfs is fast, as it happens for the diffusion processes without jumps. Moreover, under specific initial conditions, we might observe a bimodal FPT pdf. This behavior suggests the existence of interactions between the two components of the dynamics, resulting in a mixture of two distributions. From the reiteration of the simulation procedure, we also studied the behavior of some quantities related to the FPT, namely the mean and the variance of the FPT as a function of the parameters characterizing the jump component and the average number of jumps, and we observed a non-trivial behavior for the average jump size, which had a non-monotone behavior as a result of the state dependence of the jumps for both frequency and amplitude.

Throughout this paper, we use the description of the process involving an infinitesimal generator. This approach is more convenient due to the fact that the usual formulation using a stochastic differential equation is made less intuitive by the presence of possibly complicated random measures. Moreover, the simulation strategy adopted here only needs the knowledge of a distribution Π, which will be defined later, and the values of the drift and diffusion parameters. Even the Bernstein function associated with the generator, which is essential for the calculation of the analytical results, does not need to be known explicitly.

The algorithm that we present is specialized for a non-local Jacobi operator, but since the continuous part of the trajectory is constructed using a classical discretization scheme, the procedure can be applied to other non-local perturbations of classical operators, as long as it is proven that the resulting operator is still the generator of a Markov process.

This paper is structured as follows. In Section 2, we introduce the Jacobi processes with jumps and the qualitative behavior of their trajectories. In Section 3, we present the problem of the FPT through the analytical results available in the literature. Section 4 is devoted to the description of the numerical procedure that simulates the sample paths of the process. In Sections 5 and 6, we provide two examples of use of the discretization scheme for different choices of the distribution involved in the mechanism of generation of the jumps. Finally, in Section 7, we discuss the obtained results and highlight possible future works.

2. Jacobi Processes with Jumps and Their First Passage Time Problems

Let us denote with $Y = (Y_t)_{t\geq 0}$ the generalized Jacobi process with jumps introduced in [28]. For this kind of Levy-type process, the best description is given in terms of infinitesimal generators, which are functional operators whose terms contain the drift part of the process, the diffusion component and the contribution of the jumps using the integral with respect to a random measure (for a complete description of Feller semigroups and generators, see, for instance, [30]). The process Y is a Feller process on $[0,1]$, whose infinitesimal generator is given for a smooth function f on $[0,1]$ by

$$\mathcal{J}_Y f(y) = \mathbb{J}f(y) + \underbrace{\int_0^\infty (f(e^{-r}y) - f(y))\frac{\Pi(dr)}{y}}_{\text{jump part}} \tag{1}$$

where $\mathbb{J}$ is the classical Jacobi operator

$$\mathbb{J}f(y) = \underbrace{\frac{\sigma^2}{2}y(1-y)f''(y)}_{\text{diffusion part}} - \underbrace{(\lambda y - \mu)f'(y)}_{\text{drift part}}$$

with $\sigma^2 > 0$, where Π is a finite, nonnegative Radon measure on $\mathbb{R}_+$ with $\hbar = \int_0^\infty r\Pi(dr) < \infty$.

Throughout this paper, we impose the following assumption that guarantees that $y = 0$ is an entrance boundary:

$$\mu > \hbar + \frac{\sigma^2}{2}. \tag{2}$$

The latter condition is a standing assumption in [28] and extends the Feller classification of boundaries in the presence of jumps. The results presented here could be obtained with an arbitrary set of parameters satisfying these conditions.

In [28], it is shown that $\mathcal{J}_Y$, which is obtained as a nonlocal perturbation of the generator of the classical Jacobi process, is indeed the generator of a Markov process on $[0,1]$ with càdlàg trajectories. The hypotheses on the Π measure guarantee that $\mathcal{J}_Y$ satisfies the positive maximum principle which, together with the Hille–Yosida–Ray theorem for Markov generators, ensures that $\mathcal{J}_Y$ is the infinitesimal generator of a Markov semigroup on $C^1[0,1]$ (for more technical details, see [28]). As a consequence, the jumps are only downward, and both the amplitude and the intensity of the jumps are state-dependent. In fact, the process jumps from state y to state $e^{-r}y$ at a frequency given by $\Pi(dr)/y$, which is inversely proportional to the achieved state. When the process is close to the lower boundary (i.e., zero), the average number of jumps is high, but the corresponding jump size is small, as the support of the distribution of the amplitude of the jumps is $[0,y]$. Conversely, for higher values of the state of the process, the probability of jumping becomes smaller, whereas the average jump size depends on Π. See Figure 1 for an example of two possible paths of the process under investigation.

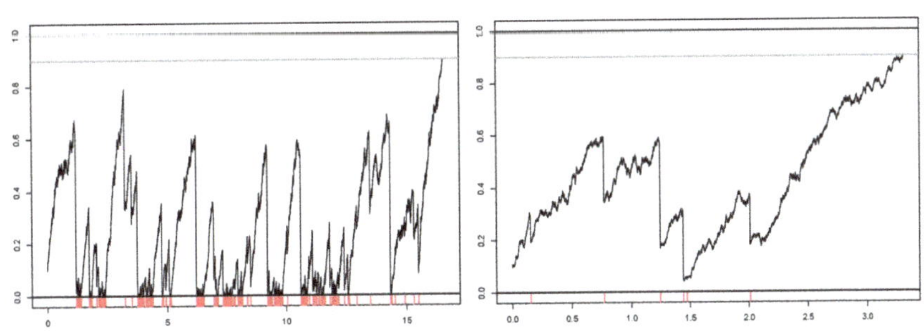

Figure 1. Examples of trajectories of Y for $\alpha = 1$ (**left**) and $\alpha = 2$ (**right**) while fixing the other parameters. At the bottom of each plot, each red vertical segment represents the time point of a jump.

Since this process can perform a finite number of jumps in a finite time, we can derive a path interpretation of this Markov process (see [31]). The stochastic process Y starts from y_0 by undergoing the same dynamics of the classical Jacobi process until a random time $\mathcal{T}$, at which the process performs a downward jump. The survival probability up to time t of $\mathcal{T}$ is given by

$$\mathbb{P}(\mathcal{T} > t) = e^{-\frac{\Pi(\mathbb{R}^+)}{Y_t}}. \tag{3}$$

After a jump, the process restarts from the new position, undergoing the diffusion dynamics until the next jump.

Different choices of Π allow different sizes for the jumps. The more mass Π concentrates around zero, the smaller, in principle, the amplitude of the jumps is. If Π admits large values with high probability, then the state of the process can almost be set to zero after the jump.

Bernstein and Bernstein–Gamma Functions

In this section, we recall a few definitions and results that will be useful in the following.

We recall that a function $\phi : [0,\infty) \to [0,\infty)$ is a Bernstein function if it is infinitely differentiable on $\mathbb{R}_+$ and $(-1)^{n+1}\frac{d^n}{du^n}\phi(u) \geq 0$ for all $n = 1, 2, \ldots$ and $u \geq 0$ [32].

We observe that $\mathcal{J}_Y$ is uniquely determined by σ^2, Π, μ and λ. In particular, by fixing λ, the triplet (σ^2, Π, μ) constitutes a Lévy triplet of the Bernstein function ϕ defined, for $u \geq 0$, by

$$\phi(u) = u + \left(\frac{2}{\sigma^2}\mu - \frac{2}{\sigma^2}\hbar - 1\right) + \frac{2}{\sigma^2}\int_0^\infty (1 - e^{-ur})\overline{\Pi}(r)dr \tag{4}$$

where $\overline{\Pi}(r) = \int_r^\infty \Pi(du)$ such that for a fixed λ, there is a one-to-one correspondence between ϕ and $\mathcal{J}_Y$ (see [27] for more details).

In [33], the authors wrote W_ϕ for the solution in the space of positive definite functions for the recurrence equation $W_\phi(z+1) = \phi(z)W_\phi(z)$, with $W_\phi(1) = 1$, $z \in \mathbb{C}$ and $\mathrm{Re}(z) > 0$. For any $n \in \mathbb{N}$, we set for the Bernstein-Gamma function

$$W_\phi(n+1) = \prod_{k=1}^{n} \phi(k) \tag{5}$$

with the convention $\prod_{k=1}^{0} \phi(k) = 1$. Note that the gamma function appears as a special case of the Bernstein–gamma function W_ϕ for $\phi(n) = n$.

Using this function, it is possible to introduce the mapping

$$_2F_1(a, b, \phi; x) = \sum_{n=0}^{\infty} \frac{(a)_n (b)_n}{n!} \frac{x^n}{W_\phi(n+1)} \tag{6}$$

with $(a)_n = \frac{\Gamma(a+n)}{\Gamma(a)}$, $n \in \mathbb{N} \cup \{0\}$ and $a \in \mathbb{C}$, which generalizes the Gauss hypergeometric function that appears as a special case for $\Pi \equiv 0$ (see [27] for more details).

We are now ready to formulate the problem of the first passage time of Y through a constant boundary and resume the existing analytical results.

3. The First Passage Time Problem

Let us consider the evolution of the stochastic process Y in the presence of a constant threshold $S \in (0,1)$. We are interested in the random time in which the process reaches the threshold S for the first time (i.e., the random variable):

$$T_S = \inf\{t > 0;\ Y_t \geq S | Y_0 = y_0 < S\}. \tag{7}$$

The direct problem of the first passage time consists mainly of finding the distribution of T_S. Although it is a classical and easy-to-state problem, its solution is, for most of the stochastic processes, not available [15,16]. An analytical closed-form expression for the probability density function $g(t)$ of T_S is not known even for the classical Jacobi process [34]. Often, it is convenient to evaluate the Laplace transform of $g(t)$ in order to obtain information on the distribution, the probability of crossing the threshold and the moments of T_S. For the Jacobi process with jumps, the Laplace transform of $g(t)$ is known to be [27], for any $0 < y_0 < S < 1$ and $q > 0$, the following:

$$\mathbb{E}_{y_0}\left[e^{-qT_S}\right] = \frac{{}_2F_1(\kappa(q),\theta(q);\phi;y_0)}{{}_2F_1(\kappa(q),\theta(q);\phi;S)}, \tag{8}$$

involving the mapping defined in Equation (6), where ϕ is a Bernstein function and $\kappa(q)$ and $\theta(q)$ are solutions to the system

$$\begin{cases} \kappa(q)\theta(q) = \frac{2q}{\sigma^2} \\ \kappa(q) + \theta(q) + 1 = \frac{2\lambda}{\sigma^2}. \end{cases} \tag{9}$$

In principle, the moments of T_S of any order can be computed using derivatives of $\mathbb{E}_{y_0}\left[e^{-qT_S}\right]$ when they exist. The first moment is known to have the following explicit analytical expression [27]:

$$\mathbb{E}_{y_0}[T_S] = \frac{2}{\sigma^2}\sum_{n=0}^{\infty}\frac{(2\lambda/\sigma^2)_n}{n+1}\frac{S^{n+1}-y_0^{n+1}}{W_\phi(n+2)}. \tag{10}$$

However, the dependence of $\mathbb{E}_{y_0}[T_S]$ on the parameters of the process is non-trivial since the contribution of the drift is hidden in the function W_ϕ, which merges the contribution of the deterministic and diffusion components.

Using terminology that comes from the context of computational neuroscience, we distinguish between two possible regimes to characterize the tendency of the process to cross the barrier. If the asymptotic mean value of the process (The process has a stationary distribution, which is a generalized beta distribution.) is larger than S, then the process is in the so-called suprathreshold regime. In the classical case, in this regime, the crossings are regular, and the dynamics are driven mainly by the drift part. If the asymptotic mean is smaller than S, then the process is said to be in the subthreshold regime, and the noise plays a prominent role in the crossing of the threshold. The Jacobi process with jumps is in the suprathreshold regime if

$$\mu > a\lambda + \int_0^\infty e^{-r}\overline{\Pi}(r)dr. \tag{11}$$

The derivation of higher-order moments from the Laplace transform is impractical, involving at least second derivatives of the generalized Gaussian hypergeometric function in Equation (6) with respect to the parameters. For this reason, it is fundamental to construct an algorithm to simulate trajectories for the family of one-dimensional jump diffusion processes, with the state-dependent intensity generated by the functional in Equation (1) for different choices of Π.

4. The Discretization Scheme for the First Passage Time

We use a discretization scheme for simulating the sample paths of jump diffusion processes with state-dependent jump intensities. Due to the dependence of the jumps on the current state of the process, in terms of both frequency and amplitude, the times when the jumps occur cannot be drawn in advance in the simulation. For this reason, at each time step of the algorithm, a value for r is sampled from the distribution $\Pi(dr)$, and according to the probability in Equation (3), a jump from Y_t to $e^{-r}Y_t$ may occur. Then, the trajectory moves according to the diffusion from state $e^{-r}Y_t$ if there was a jump; otherwise, it moves according to the diffusion from state Y_t. Between the jump epochs, the dynamics of the constructed process are purely diffusive and are simulated using Milstein's discretization method, which is a generalization of the Euler–Maruyama scheme used for stochastic processes with multiplicative noise [29].

When the intensity of the point process driving the jump component is state-dependent the error generated in the construction of the continuous component can be amplified in the simulation of the jumps and depends on the size of the time step. However, this algorithm is related to the discretization schemes for which it was proven in [35] that the method converges and the weak convergence order equals the order of the adopted time step. This means that the simulation error is of the same order as that of the discretization schemes for pure diffusions.

To avoid discretization errors, one can think of using an exact algorithm that simulates directly the hitting times without constructing the whole paths, as in [10] (see also [36,37]) For the special case of the Jacobi diffusion see [38].

However, the lack of knowledge for many functions and properties concerning these processes and the difficult implementation of the generalized functions involved in the transition density and the stationary distribution of the process may prevent the use of these exact strategies.

The proposed discretization strategy is light and very simple to implement, and it has an advantage: we can simulate the process just from the distribution Π without knowing explicitly the associated Bernstein function ϕ. Moreover, it relies on the study of the process in terms of its generator. Using generators in this context is more convenient with respect to the usual definition of the process as solution to a stochastic differential equation (SDE). Indeed, for a state-dependent jump process, the theory of SDEs is still incomplete and involves integrals with respect to some random measures that make both the numerical implementation and the interpretation of the dynamics hard.

In this paper, we focus mostly on the simulation of the paths of the process in order to answer the problem of the first passage time. For this reason, we construct the trajectory until it reaches the level S, and we record the time of this crossing. We repeat this procedure n times, and we use the collected FPT times to find an approximation of the FPT density and other statistics for which it is not possible to have analytical results.

In Algorithm 1 we illustrate a scheme of the sampling procedure we use to draw an FPT from the process.

Algorithm 1 Sampling FPT

Require: $y_0, S, dt, \sigma^2, \lambda, \mu$
Ensure: FPT sample t_S
  while $y_i < S$ **do**
    $r \sim \Pi(dr)$
    $j \sim Be\left(1 - exp(-\frac{r}{y_{i-1}}dt)\right)$
    if $j = 1$ **then** ▷ a jump occurs
      $y^* \leftarrow e^r y_{i-1}$
    else if $j = 0$ **then** ▷ a jump does not occur
      $y^* \leftarrow y_{i-1}$
    end if
    $g(y^*) \leftarrow \sqrt{\frac{\sigma^2}{2} y^*(1-y^*)}$
    $y_i \leftarrow y^* - (\lambda y^* - \mu)dt + g(y^*)\Delta W_i + \frac{1}{2} g(y^*)(g(y^*))'((\Delta W_i)^2 - dt)$ ▷ diffusion
  end while
  $t_S = i$

5. Example: Exponential Distribution

We consider a parametric family of non-local Jacobi operators of the form in Equation (1), for which

$$\overline{\Pi}(r) = \int_r^\infty \Pi(du) = e^{-\alpha r}, \quad r > 0 \tag{12}$$

where

$$\Pi(dr) = \alpha e^{-\alpha r} dr \tag{13}$$

is an exponential probability density function with a parameter α. In particular, we will choose $\alpha \geq 1$ throughout the paper as in [28].

Moreover, to guarantee from the assumption in Equation (2) that $y = 0$ is an entrance boundary, we have

$$\mu > \frac{\sigma^2}{2} + \frac{1}{\alpha}. \tag{14}$$

The presence in the last inequality of a positive term $1/\alpha$ suggests that the maximum noise amplitude has to be smaller than in the classical case. A large value of σ can lead the process across the lower boundary, a condition that we want to avoid. Unfortunately, classical approximation schemes cannot preserve the properties of the boundaries independent of the choice of the time discretization step, even if the theoretical assumption in Equation (14) is satisfied [39,40]. In particular, for the Jacobi process, even the splitting methods do not preserve the boundary behavior, and other strategies such as the balanced implicit split step (BISS) method, which is able to preserve the boundary structure, are lacking in accuracy (see [41] for an extensive discussion).

Finally, in this case, the regime is a suprathreshold if

$$\mu > a\lambda + \frac{1}{1+\alpha}, \tag{15}$$

We observe that, as expected, the downward jumps make the asymptotic mean of Y, $\frac{\mu}{\lambda} - \frac{1}{1+\alpha}$, smaller than that of the classical Jacobi process (μ/λ):

Remark 1. *The integro-differential operator $\mathcal{J}_Y$ from Equation* (1) *takes the form*

$$\mathcal{J}f(y) \quad = \quad \frac{\sigma^2}{2}y(1-y)f''(y) - (\lambda y - \mu)f'(y) - \int_0^1 (f(r) - f(y))\frac{r^\alpha}{y^{\alpha+1}}dr. \tag{16}$$

Moreover, $\hbar = 1/\alpha$, and simple algebra yields to the explicit expression of the corresponding Bernstein function:

$$\phi(u) = u + \frac{2}{\sigma^2}\left(\mu - \frac{1}{u+\alpha}\right) - 1. \tag{17}$$

However, the application of the proposed discretization scheme does not require the knowledge of the explicit expression of the infinitesimal operator nor of the Bernstein function ϕ. This constitutes a great advantage when working with distributions whose expression prevents easy calculation of the involved quantities.

We want to investigate the behavior of the FPT density and other statistical quantities of this random variable under a change in the parameters characterizing the distribution Π. Precise analysis of these moments is made difficult by the presence of the generalized hypergeometric function in the expression of the Laplace transform (Equation (8)), while a formal analytical study on the FPT pdf is prevented by the lack of explicit results. For these reasons, the information obtained from the simulations are of great importance. In the following, we show some of the qualitative statistical behaviors of T_S using the simulation scheme described in the previous section.

Let us consider the jump diffusion process Y generated by the non-local operator in Equation (1), with Π given in Equation (13).

In Figure 1, we show two possible realizations of Y until it reaches the threshold S for the first time. We can see the impact of different choices of α on the trajectories of the considered stochastic process. A lower value of α implies that possibly larger values r are sampled from Π, resulting in a lower frequency of the jumps in principle and a higher size for them. However, the state dependence of the jumps makes the jump generation mechanism more complex. Moreover, these pictures display how the jumps were more dense when the process was close to the lower boundary, as expected from the probability in Equation (3).

In [27], it was shown, using Equation (10), that the mean FPT decreases as α increases. This is due to the fact that for large values of α, a small value r is most likely sampled from Π, affecting the probability (Equation (3)). Using simulations, we observed the same behavior for the variance of T_S as a function of α (not shown). This is explained equivalently by the fact that the average number of jumps decreased as α increased (see Figure 2). However, the dynamics was not as simple as it may appear due to the presence of state-dependent jumps. Indeed, it is interesting to observe the behavior of the average length of the jumps as a function of α in Figure 2. Up to some value of α, the average size of the jumps increased with α. This behavior could sound counterintuitive, since lower values of α imply, on average, a higher gap between y and $e^{-r}y$. However, we have to take into account that jumps were more frequent when y was small, and since the support of the distribution of the amplitude of the jumps was $[0, y]$, most of the jumps were short. Therefore, roughly speaking, even if there is a large jump that pushes the process to the lower boundary, it will be compensated for by many small jumps close to the zero level. Even for higher values of α, we observed that the average length of the jumps started to decrease. In this case, very large jumps occurred with a small probability, so on average, the size of the jumps decreased.

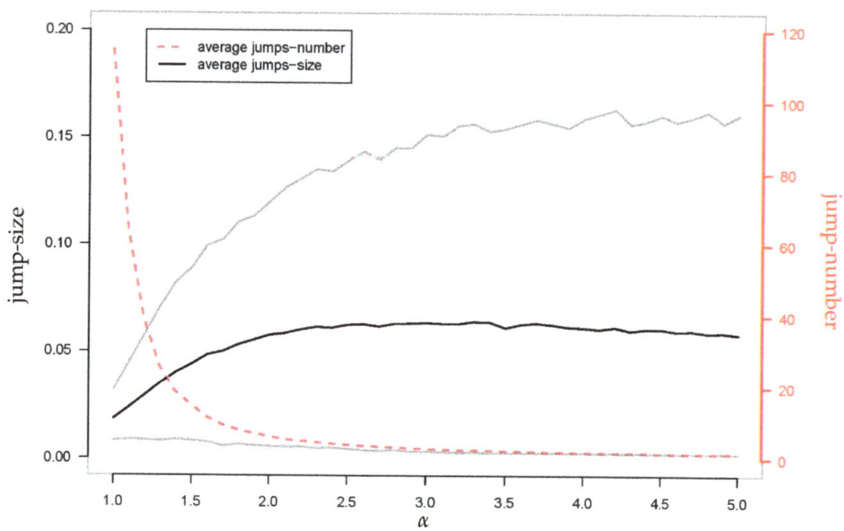

Figure 2. Average jump lengths (black solid line) and average number of jumps (red dashed line) as a function of α. Confidence bands for the jump lengths (gray solid line) were obtained for the 5th and the 95th percentiles. The averages were taken over 5000 simulated sample paths of Y from Equation (1) for $\mu = 1.1, \sigma = 0.1, \lambda = 1.1, x_0 = 5 \times 10^{-4}, S = 0.9$ and time step $dt = 0.01$.

If a diffusion process admits a stationary distribution, then the corresponding FPT pdf is known to have asymptotically, for large times and large boundaries, an approximately exponential distribution whose mean is related to the average first passage time from the origin to the boundary (see [42] for the Ornstein–Uhlenbeck process, ref. [43] for one-dimensional diffusion processes and [44] for Gauss–Markov processes). Since the presence of downward jumps decreases the asymptotic mean of the process with respect to the case without jumps and increases the mean FTP, it is natural to ask whether the FPT pdf tail decays slower.

In Figure 3, we show approximations of the FPT pdfs obtained from histograms of 5×10^4 simulated first passage times of Y through S for 6 different choices of α. As expected, we can appreciate that the threshold S was more likely to be crossed earlier for larger values of α, but the tail decay remained qualitatively the same.

To measure the tail decay, we selected the densities $\tilde{g}$ of the histograms after the third quartile, which were the heights of the bars on the right-hand side of the histogram that summed to a probability approximately equal to 0.25, and fit them with two different models. More specifically, we used an exponential function and a power law function, defined by

$$M_e: \quad h = \beta_0 + \beta_1 e^{-\beta_2 t} \tag{18}$$

$$M_p: \quad h = \beta_0 + \beta_1 t^{-\beta_2}. \tag{19}$$

The estimation of the parameters $(\beta_0, \beta_1, \beta_2)$ for the two models was performed using non-linear optimization software.

In the legends of Figure 3, we display the logarithm of the mean squared error (MSE) as a measure of goodness of fit for both curves. We can see that although both curves approximated the tails well, the approximation error of the exponential function was always the smallest. This might indicate that the tails of the FPT pdf had an exponential decay.

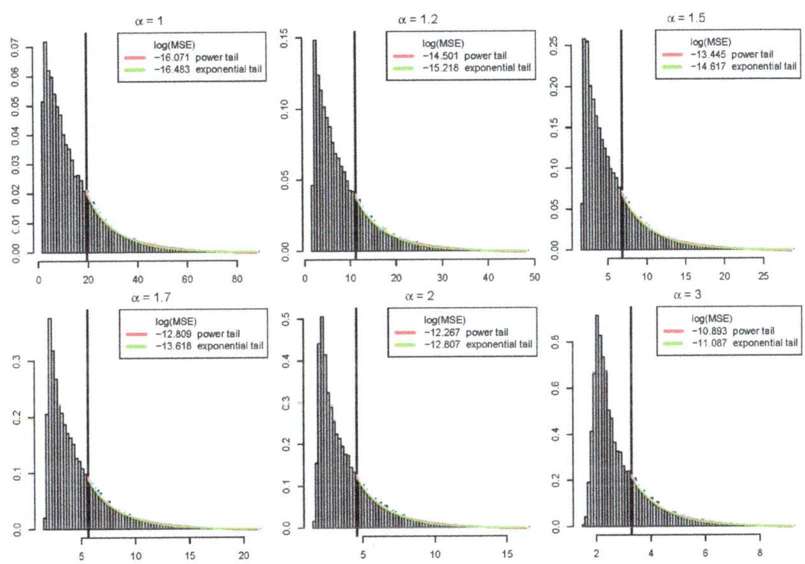

Figure 3. Histograms of FPT data obtained from 5×10^4 simulations of trajectories of Y for 6 different choices of α. Other parameters were $\mu = 1.1$, $\sigma = 0.1$, $\lambda = 1.1$, $y_0 = 5 \times 10^{-5}$, $S = 0.9$ and $dt = 0.01$. The black vertical line indicates the third quartile, marking the beginning of the tail of the approximated FPT pdf. The legend shows the MSE in logarithmic scale of the estimation of the tails made with exponential and power curves.

Interestingly, the FPT distribution could show a bimodal behavior when the starting point y_0 was close to the threshold S. This behavior followed from the fact that if no jump occurred in the very first moments, then the positive drift quickly pushed the process toward S (first peak). On the other hand, if a large jump took place before the process was absorbed right away in S, then the process would take a longer time to reach the threshold with a distribution showing a longer tail (second peak). In other words, one could see this distribution as a mixture of two distributions of the first passage time: one related to a Jacobi process without jumps and the other related to a Jacobi process with jumps and a starting point y_0 far from S, where the weights of such a mixture depend on the probability of the process to perform a jump before the drift pushes it across the threshold

S. An example of this behavior can be appreciated in Figure 4, where to better stress the bimodality of the distribution, we display an histogram of the logarithm of the FPT in a setting where y_0 is close to S.

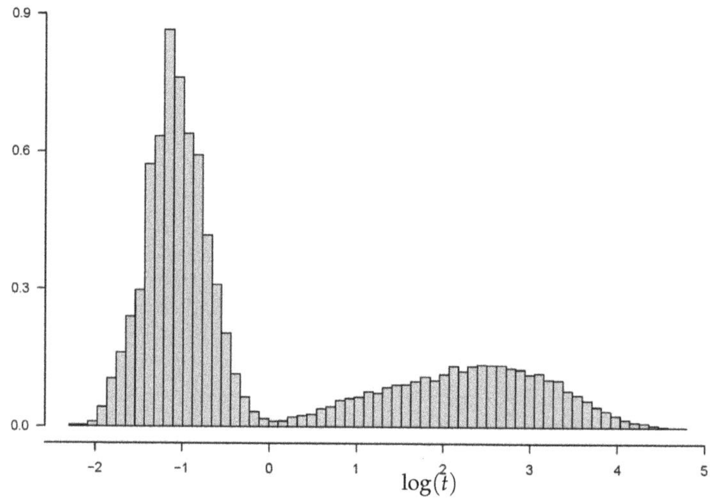

Figure 4. Histogram of the logarithm of FPT data obtained from 5000 simulations of trajectories of Y, with $\alpha = 1$ and $y_0 = 0.85$. All other parameters were chosen as in Figure 3.

6. Example: Pareto Distribution

We consider a parametric family of non-local Jacobi operators as in Equation (1) for which

$$\overline{\Pi}(r) = \int_r^{\infty} \Pi(du) = \begin{cases} \left(\frac{\eta}{r}\right)^{\theta} & \text{if} \quad r \geq \eta, \\ 1 & \text{if} \quad r < \eta \end{cases}$$

where

$$\Pi(dr) = \begin{cases} \frac{\theta \eta^{\theta}}{r^{\theta+1}} dr & \text{if } r \geq \eta \\ 0 & \text{if } r < \eta \end{cases} \tag{20}$$

is a Pareto Type I probability density function with a shape parameter $\theta > 0$ and location parameter $\eta > 0$. In order to match the assumption that $\hbar = \int_0^{\infty} r\Pi(dr) < \infty$, we will choose $\theta > 1$ throughout this paper. Moreover, to guarantee that $y = 0$ is an entrance boundary, from Equation (2), we have

$$\mu > \frac{\theta \eta}{\theta - 1} + \frac{\sigma^2}{2}. \tag{21}$$

In this case, the regime is a suprathreshold if, for $r \geq \eta$, the following is true:

$$\mu > S\lambda + \int_0^{\infty} e^{-r} \overline{\Pi}(dr) = S\lambda + \int_0^{\infty} e^{-r} \left(\frac{\eta}{r}\right)^{\theta} dr = S\lambda + \eta^{\theta} \Gamma(1 - \theta), \tag{22}$$

where the last equality involving the integral representation of the gamma function holds only for $\mathrm{Re}(\theta) > 1$, which is a case that cannot be considered here.

An analytical expression of the moments of T_S in the case of the Pareto distribution is not known and neither is the expression of the Bernstein function associated with the non-local operator. Using the discretization scheme, we can find an estimation of the mean and variance of T_S as a function of the two parameters characterizing the Pareto distribution.

In Figure 5, we show the behavior of the variance of the FPT when tuning simultaneously the scale and location parameters θ and η, respectively. The variance increases with the location parameter η. In fact, the support of the Pareto distribution is the interval $[\eta, \infty]$, meaning that a large value $r \in [\eta, \infty]$ will be sampled by the algorithm if η is large. At the same time, the variance of T_S decreases as θ increases due to the shape of the distribution for large values of θ that favour small values of r. In order to match Equation (21) for all the couples (θ, η), the chosen drift is relatively strong ($\mu = 10.5$), which explains why the resulting variance was small in all the considered cases. Qualitatively, the same behavior can be observed for the mean FPT (not shown here), where the mean FPT increases with η and decreases as θ increases.

Figure 5. Variance of the FPT as a function of the scale and location parameters of the Pareto distribution. The heatmap was obtained by simulating 2000 FPTs for each couple of parameters. The trajectories were obtained from Equations (1) and (20) with $\mu = 10.5$, $\sigma = 0.1$, $\lambda = 1.1$, $x_0 = 5 \times 10^{-3}$, $S = 0.9$ and $dt = 0.01$.

Under the assumption of $\theta > 2$, which guarantees finite variance for the Pareto distribution, we consider the two following parameter choices:

Case 1: $\theta = 1 + \sqrt{2}, \quad \eta = \frac{\sqrt{2}}{1+\sqrt{2}}$;

Case 2: $\theta = 1 + \sqrt{2}, \quad \eta = \frac{\sqrt{2}}{2(1+\sqrt{2})}$.

These choices guarantee expected values equal to 1 and $1/2$, respectively, and variances equal to the square of the means, as in the exponential case, allowing a comparison between the two examples.

In Figure 6, we consider the histograms of the FPT for the two cases. As performed in the previous section, we applied the fitting models of Equations (18) and (19) to the tails of the distributions. Additionally, in this case, both models fit well, but the use of an exponential curve resulted in a lower MSE. The Pareto distribution is a classical example of a heavy-tailed probability distribution, meaning that large values of r in the mechanism of generation of jumps can be chosen by the algorithm. It could be natural to expect that the shape of the FPT pdf could be stretched by the heavy tails of the Pareto distribution. However, for our parameter choice, the tail of the Pareto distribution became fatter than the one of the exponential distribution only for values with negligible probabilities.

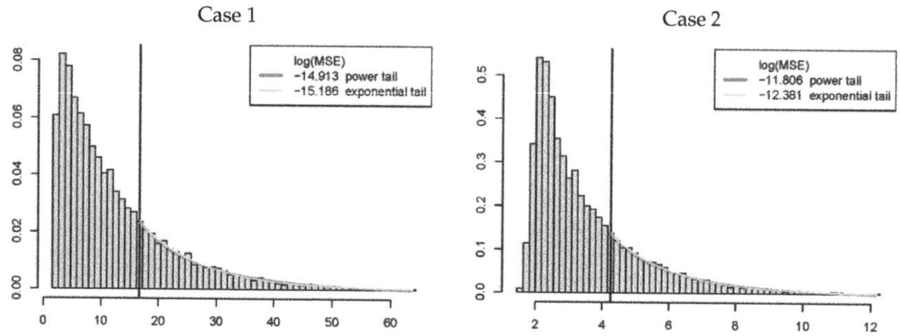

Figure 6. Histograms of 2×10^4 simulated FPT data. The parameters of the Pareto distribution were those of Case 1 (**left**) and Case 2 (**right**). All other parameters were chosen as in Figure 3. The black vertical line indicates the third quartile, marking the beginning of the tail of the approximated FPT pdf. The legend shows the MSE in a logarithmic scale of the estimation of the tails made with exponential and power curves.

7. Discussion

Due to the lack of analytical results regarding the FPT of jump diffusion processes, for which the jumps are state-dependent in terms of both frequency and amplitude, the use of simulations is crucial. Using a discretization scheme, we simulated the trajectories of these processes, and we studied the problem of their passage times through a constant boundary. The method is specialized for a Jacobi process with jumps but can be used for any Markov process whose infinitesimal generator is obtained as a non-local perturbation of a classical operator. We obtained approximations on the FPT pdf for different choices of the measure describing the distribution of the jumps. In particular, in the case of the Jacobi process with jumps, we considered exponential and Pareto distributions. From the simulations, we observed that despite the presence of only downward jumps, the decay of the tails of the FPT pdfs was fast, as happened for the diffusion processes without jumps. This was a consequence of the main assumption in Equation (2) for the drift of the process, which guaranteed the Markov property. Analytical results for the pdf's tails' decay could be obtained by Tauberian theorems for Laplace transforms, but the presence of the generalized functions in Equation (8) prevented straightforward calculations.

Another interesting feature that might be observed is the multimodality of the FPT pdf, which can appear even if the drift part of the process is non-periodic. The effect is more visible if the starting point of the process y_0 is close enough to the boundary S. In this case, the first peak of the FPT pdf is determined by the drift part of the process, and a second bump is visible, suggesting the existence of interactions between the components of the dynamics, resulting in a mixture of two distributions. In [2], a similar behavior was observed but only in the presence of both positive and negative jumps.

From iterations of the simulation procedure, we also studied the behavior of some moments of the FPT. In particular, we studied the variance of the FPT as a function of the parameters characterizing the jump component in the case of both one- and two-parameter distributions. Finally, we studied the average number of jumps as a function of the parameter of Π, and we observed a non-trivial behavior of the average jump size, which had a non-monotone behavior as a result of the state dependence of the jumps in terms of both frequency and amplitude.

We stress that we used the infinitesimal generator in Equation (1), but in the application context, more specific operators defined on different intervals can be used. The same result follows if one can identify a homeomorphism between the new semigroups and the one of Jacobi processes with jumps on (0, 1) defined in Equation (1). For example, this was performed in the framework of mathematical neuroscience in [27], taking advantage of the intertwining approach.

Author Contributions: Conceptualization, G.D. and A.L.; Methodology, G.D. and A.L.; Software, G.D. and A.L.; Investigation, G.D. and A.L.; Writing—original draft, G.D. and A.L.; Writing—review & editing, G.D. and A.L.; Funding acquisition, G.D. All authors have read and agreed to the published version of the manuscript.

Funding: This work was supported by the Gruppo Nazionale per l'Analisi Matematica, la Probabilità e le loro Applicazioni (GNAMPA—INdAM).

Data Availability Statement: The codes used during this study are available from the authors on reasonable request.

Conflicts of Interest: The authors declare no conflict of interest.

Abbreviations

The following abbreviations are used in this manuscript:

FPT	First passage time
pdf	Probability density function
MSE	Mean squared error
SDE	Stochastic differential equation

References

1. Jahn, P.; Berg, R.W.; Hounsgaard, J.; Ditlevsen, S. Motoneuron membrane potentials follow a time inhomogeneous jump diffusion process. *J. Comput. Neurosci.* **2011**, *31*, 563–579. [CrossRef] [PubMed]
2. Sirovich, R.; Sacerdote, L. Noise induced phenomena in jump diffusion models for single neuron spike activity. *IEEE Int. Jt. Conf. Neural Netw.* **2004**, *4*, 3025–3028.
3. Sirovich, R.; Sacerdote, L.; Villa, A.E.P. Cooperative behavior in a jump diffusion model for a simple network of spiking neurons. *Math. Biosci. Eng.* **2013**, *11*, 385–401. [CrossRef] [PubMed]
4. Giorno, V.; Román-Román, P.; Spina, S.; Torres-Ruiz, F. Estimating a non-homogeneous Gompertz process with jumps as model of tumor dynamics. *Comput. Stat. Data Anal.* **2017**, *107*, 18–31. [CrossRef]
5. Zucca, C.; Tavella, P. A mathematical model for the atomic clock error in case of jumps. *Metrologia* **2015**, *52*, 514. [CrossRef]
6. Dharmaraja, S.; Di Crescenzo, A.; Giorno, V.; Nobile, A.G. A continuous-time Ehrenfest model with catastrophes and its jump-diffusion approximation. *J. Stat. Phys.* **2015**, *161*, 326–345. [CrossRef]
7. Merton, R.C. Option pricing when underlying stock returns are discontinuous. *J. Financ. Econ.* **1976**, *3*, 125–144. [CrossRef]
8. Briani, M., Caramellino, L.; Terenzi, G. Convergence Rate of Markov Chains and Hybrid Numerical Schemes to Jump-Diffusion with Application to the Bates Model. *SIAM J. Numer. Anal.* **2021**, *59*, 477–502. [CrossRef]
9. Brignone, R.; Sgarra, C. Asian options pricing in Hawkes-type jump-diffusion models. *Ann. Financ.* **2020**, *16*, 101–119. [CrossRef]
10. Casella, B.; Roberts, G.O. Exact simulation of jump-diffusion processes with Monte Carlo applications. *Methodol. Comput. Appl. Probab.* **2011**, *13*, 449–473. [CrossRef]
11. Kou, S.G. A jump-diffusion model for option pricing. *Manag. Sci.* **2002**, *48*, 1086–1101. [CrossRef]
12. Øksendal, B.; Sulem, A. *Applied Stochastic Control of Jump Diffusions*; Springer: Berlin/Heidelberg, Germany, 2007.
13. Tankov, P. *Financial Modelling with Jump Processes*; Chapman and Hall/CRC: Boca Raton, FL, USA, 2003.
14. Metzler, R.; Redner, S.; Oshanin, G. *First-Passage Phenomena and Their Applications*; World Scientific: Singapore, 2014.
15. Redner, S. *A Guide to First-Passage Processes*; Cambridge University Press: Cambridge, UK, 2001.
16. Karlin, S.; Taylor, H.E. *A Second Course in Stochastic Processes*; Elsevier: Amsterdam, The Netherlands, 1981.
17. Buonocore, A.; Caputo, L.; D'Onofrio, G.; Pirozzi, E. Closed-form solutions for the first-passage-time problem and neuronal modeling. *Ric. Mat.* **2015**, *64*, 421–439. [CrossRef]
18. Ricciardi, L.M.; Di Crescenzo, A.; Giorno, V.; Nobile, A.G. An outline of theoretical and algorithmic approaches to first passage time problems with applications to biological modeling. *Math. Jpn.* **1999**, *50*, 247–322.
19. Abundo, M. On first-passage times for one-dimensional jump-diffusion processes. *Probab. Math. Stat.* **2000**, *20*, 399–423.
20. Abundo, M. On the first hitting time of a one-dimensional diffusion and a compound Poisson process. *Methodol. Comput. Appl. Probab.* **2010**, *12*, 473–490. [CrossRef]
21. Atiya, A.F.; Metwally, S.A. Efficient estimation of first passage time density function for jump-diffusion processes. *SIAM J. Sci. Comput.* **2005**, *26*, 1760–1775. [CrossRef]
22. Herrmann, S.; Massin, N. Exact simulation of the first passage time through a given level of jump diffusions. *Math. Comput. Simul.* **2023**, *203*, 553–576. [CrossRef]
23. Kou, S.G.; Wang, H. First passage times of a jump diffusion process. *Adv. Appl. Probab.* **2003**, *35*, 504–531. [CrossRef]
24. Xie, J.; Cui, Z.; Zhang, Z. Some new infinite series expansions for the first passage time densities in a jump diffusion model with phase-type jumps. *Appl. Math. Comput.* **2022**, *429*, 127251. [CrossRef]

25. Lefebvre, M. The ruin problem for a Wiener process with state-dependent jumps. *J. Appl. Math. Stat. Inform.* **2020**, *16*, 13–23. [CrossRef]
26. Lefebvre, M. First-passage problems for diffusion processes with state-dependent jumps. *Commun. Stat.-Theory Methods* **2022**, *51*, 2908–2918. [CrossRef]
27. D'Onofrio, G.; Patie, P.; Sacerdote, L. Jacobi processes with jumps as neuronal models: A first passage time analysis. *arXiv* **2022**, arXiv:2205.08237.
28. Cheridito, P.; Patie, P.; Srapionyan, A.; Vaidyanathan, A. On non-local ergodic Jacobi semigroups: Spectral theory, convergence-to-equilibrium and contractivity. *J. l'Ecole Polytech.-Math.* **2021**, *8*, 331–378. [CrossRef]
29. Higham, D.J. An algorithmic introduction to numerical simulation of stochastic differential equations. *SIAM Rev.* **2001**, *43*, 525–546. [CrossRef]
30. Ethier, S.N.; Kurtz, T.G. *Markov Processes: Characterization and Convergence*; John Wiley & Sons: Hoboken, NJ, USA, 2009.
31. Bass R.F. Adding and subtracting jumps from Markov processes. *Trans. Am. Math. Soc.* **1979**, *255*, 363–376. [CrossRef]
32. Schilling, R.L.; Song, R.; Vondraček, Z. *Bernstein Functions*; Volume 37 of de Gruyter Studies in Mathematics; Theory and Applications; Walter de Gruyter & Co.: Berlin, Germany, 2010.
33. Patie, P.; Savov, M. Bernstein-gamma functions and exponential functionals of Lévy processes. *Electron. J. Probab.* **2018**, *23*, 1–101. [CrossRef]
34. D'Onofrio, G.; Tamborrino, M.; Lansky, P. The Jacobi diffusion process as a neuronal model. *Chaos* **2018**, *28*, 103119. [CrossRef]
35. Glasserman, P.; Merener, N. Convergence of a discretization scheme for jump-diffusion processes with state–dependent intensities. *Proc. R. Soc. Lond. Ser. A Math. Phys. Eng. Sci.* **2004**, *460*, 111–127. [CrossRef]
36. Beskos, A.; Roberts, G.O. Exact simulation of diffusions. *Ann. Appl. Probab.* **2005**, *15*, 2422–2444. [CrossRef]
37. Herrmann, S.; Zucca, C. Exact simulation of the first-passage time of diffusions. *J. Sci. Comput.* **2019**, *79*, 1477–1504. [CrossRef]
38. Jenkins, P.A.; Spano, D. Exact simulation of the Wright–Fisher diffusion. *Ann. Appl. Probab.* **2017**, *27*, 1478–1509. [CrossRef]
39. Tubikanec, I.; Tamborrino, M.; Lansky, P.; Buckwar, E. Qualitative properties of different numerical methods for the inhomogeneous geometric Brownian motion. *J. Comput. Appl. Math.* **2022**, *406*, 113951. [CrossRef]
40. Dangerfield, C.E.; Kay, D.; Macnamara, S.; Burrage, K. A boundary preserving numerical algorithm for the Wright–Fisher model with mutation. *BIT Numer. Math.* **2012**, *52*, 283–304. [CrossRef]
41. Eder, J. Boundary Behaviour of Pearson Diffusion Processes and Numerical Splitting Methods Preserving Them. Ph.D. Thesis, Universität Linz, Linz, Austria, 2020.
42. Nobile, A.G.; Ricciardi, L.M.; Sacerdote, L. Exponential trends of Ornstein–Uhlenbeck first-passage-time densities. *J. Appl. Probab.* **1985**, *22*, 360–369. [CrossRef]
43. Giorno, V.; Nobile, A.G.; Ricciardi, L.M. On the asymptotic behaviour of first-passage-time densities for one-dimensional diffusion processes and varying boundaries. *Adv. Appl. Probab.* **1990**, *22*, 883–914. [CrossRef]
44. Nobile, A.G.; Pirozzi, E.; Ricciardi, L.M. Asymptotics and evaluations of FPT densities through varying boundaries for Gauss-Markov processes. *Sci. Math. Jpn.* **2008**, *67*, 241–266.

fractal and fractional

MDPI

Article

On the Absorbing Problems for Wiener, Ornstein–Uhlenbeck, and Feller Diffusion Processes: Similarities and Differences

Virginia Giorno † and Amelia G. Nobile *,†

Dipartimento di Informatica, Università degli Studi di Salerno, Via Giovanni Paolo II n. 132, Fisciano, 84084 Salerno, Italy
* Correspondence: nobile@unisa.it
† These authors contributed equally to this work.

Abstract: For the Wiener, Ornstein–Uhlenbeck, and Feller processes, we study the transition probability density functions with an absorbing boundary in the zero state. Particular attention is paid to the proportional cases and to the time-homogeneous cases, by obtaining the first-passage time densities through the zero state. A detailed study of the asymptotic average of local time in the presence of an absorbing boundary is carried out for the time-homogeneous cases. Some relationships between the transition probability density functions in the presence of an absorbing boundary in the zero state and between the first-passage time densities through zero for Wiener, Ornstein–Uhlenbeck, and Feller processes are proven. Moreover, some asymptotic results between the first-passage time densities through zero state are derived. Various numerical computations are performed to illustrate the role played by parameters.

Keywords: Wiener process; Ornstein–Uhlenbeck process; Feller process; asymptotic average of the local time; first-passage time and its moments

Citation: Giorno, V.; Nobile, A.G. On the Absorbing Problems for Wiener, Ornstein–Uhlenbeck, and Feller Diffusion Processes: Similarities and Differences. *Fractal Fract.* **2023**, *7*, 11. https://doi.org/10.3390/fractalfract7010011

Academic Editor: Mario Abundo

Received: 2 November 2022
Revised: 19 December 2022
Accepted: 20 December 2022
Published: 24 December 2022

1. Introduction and Background

Diffusion models are widely used to describe dynamical systems in economics, finance, biology, genetics, physics, engineering, neuroscience, queueing, and other fields (cf. Bailey [1], Ricciardi [2], Gardiner [3], Stirzaker [4], Janssen et al. [5], Pavliotis [6]). In various applications, it is useful to consider diffusion processes with linear infinitesimal drift and linear infinitesimal variance. This class incorporates Wiener, Ornstein–Uhlenbeck, and Feller diffusion processes. In population dynamics, these processes can be used to describe the growth of a population and the zero state represents the absorbing extinction threshold. With this aim, we study the absorbing problem for linear diffusion processes.

In the remaining part of this section, we shall briefly review some background results on the absorbing problems that will be used in the next sections for Wiener, Ornstein–Uhlenbeck and Feller diffusion processes.

Let $\{\mathcal{Z}(t), t \geq t_0\}$ be a time-inhomogeneous diffusion (TNH-D) process with state-space $\mathcal{D} = (r_1, r_2)$, which satisfies the stochastic differential equation

$$d\mathcal{Z}(t) = \zeta_1[\mathcal{Z}(t), t]\, dt + \sqrt{\zeta_2[\mathcal{Z}(t), t]}\, dW(t), \qquad \mathcal{Z}(t_0) = x_0,$$

with $\zeta_1(x,t)$ and $\zeta_2(x,t)$ denoting, respectively, the infinitesimal drift and the infinitesimal variance of $\mathcal{Z}(t)$ and where $W(t)$ is a standard Brownian motion. Often, $\mathcal{D} = (-\infty, +\infty)$, with $\pm\infty$ unattainable endpoints, but in some cases $\mathcal{Z}(t)$ is confined to the state space $\mathcal{D} = (0, +\infty)$ and in the zero state is imposed an absorbing condition.

When the endpoints $\pm\infty$ of $\mathcal{D}$ are unattainable boundaries, the transition probability density function (PDF) $f_{\mathcal{Z}}(x,t|x_0,t_0)$ is the solution of the backward Kolmogorov equation (cf. Dynkin [7])

$$\frac{\partial f_{\mathcal{Z}}(x,t|x_0,t_0)}{\partial t_0}+\zeta_1(x_0,t_0)\,\frac{\partial f_{\mathcal{Z}}(x,t|x_0,t_0)}{\partial x_0}+\frac{1}{2}\,\zeta_2(x_0,t_0)\,\frac{\partial^2 f_{\mathcal{Z}}(x,t|x_0,t_0)}{\partial x_0^2}=0, \tag{1}$$

with the initial delta condition $\lim_{t_0\uparrow t} f_{\mathcal{Z}}(x,t|x_0,t_0)=\delta(x-x_0)$. In the backward Kolmogovov Equation (1), the forward variables x and t are constant and enter only through the initial and boundary conditions.

We remark that the PDF $f_{\mathcal{Z}}(x,t|x_0,t_0)$ is also solution of a forward Kolmogorov equation, also known as the Fokker–Planck equation (cf. Dynkin [7]), in which the backward variables x_0 and t_0 are essentially constant. In this paper, we choose to use the Kolmogorov backward equation because we will address absorption problems. Indeed, if one is interested to the first-passage time distribution through a fixed state S as a function of the initial position x_0, then the backward Kolmogorov equation provides the most appropriate method (cf. Cox and Miller [8]).

For a diffusion process $\mathcal{Z}(t)$, the first-passage time (FPT) problem can be reduced to estimate the density of the random variable

$$T_{\mathcal{Z}}(S|x_0,t_0)=\begin{cases}\inf_{t\ge t_0}\{t:\mathcal{Z}(t)\ge S\}, & \mathcal{Z}(t_0)=x_0<S,\\ \inf_{t\ge t_0}\{t:\mathcal{Z}(t)\le S\}, & \mathcal{Z}(t_0)=x_0>S,\end{cases}$$

which describes the FPT of $\mathcal{Z}(t)$ through the state S starting from $\mathcal{Z}(t_0)=x_0\neq S$.

The FPT problem plays an important role in various biological applications. For instance, in the context of population dynamics the FPT problem is suitable to model population's extinction or persistence (see Bailey [1], Ricciardi [2], Allen [9,10]).

Let $g_{\mathcal{Z}}(S,t|x_0,t_0)=dP\{T_{\mathcal{Z}}(S|x_0,t_0)\le t\}/dt$ be the FPT density, being $P\{T_{\mathcal{Z}}(S|x_0,t_0)\le t\}$ the distribution function of the random variable $T_{\mathcal{Z}}(S|x_0,t_0)$. If the endpoints of $\mathcal{D}$ are unattainable boundaries, the densities $f_{\mathcal{Z}}(x,t|x_0,t_0)$ and $g_{\mathcal{Z}}(S,t|x_0,t_0)$ are related by the following renewal equation (cf. Blake and Lindsey [11]):

$$f_{\mathcal{Z}}(x,t|x_0,t_0)=\int_{t_0}^{t} g_{\mathcal{Z}}(S,\tau|x_0,t_0)\,f_{\mathcal{Z}}(x,t|S,\tau)\,d\tau, \qquad (x_0<S\le x)\text{ or }(x\le S<x_0). \tag{2}$$

Equation (2) indicates that any sample path that reaches $x\ge S$ $[x\le S]$, after starting from $x_0<S$ $[x_0>S]$ at time t_0, must necessarily cross S for the first time at some intermediate instant $\tau\in(t_0,t)$.

For diffusion processes, closed form expressions for FPT densities through constant boundaries are not available, except in some special cases (see Ricciardi et al. [12], Ding and Rangarajan [13], Molini et al. [14], Giorno and Nobile [15], Masoliver [16]). In particular, closed form expressions are available in the following cases: (i) the Wiener process through an arbitrary constant boundary; (ii) the Ornstein–Uhlenbeck process through the boundary in which the drift vanishes; and (iii) the Feller process through the zero state. In the literature many efforts have been devoted to determining the asymptotic behavior of FPT density and its moments for large boundaries or large times and to search efficient numerical and simulation methods to estimate the FPT densities (cf. Ricciardi et al. [12], Linetsky [17]). Furthermore, the FPT problems play a relevant role also in the context of fractional processes (see, for instance, Guo et al. [18], Wiese [19], Abundo [20], Leonenko and Pirozzi [21]).

For a TNH-D process $\mathcal{Z}(t)$ confined to interval $(0,+\infty)$, with 0 absorbing boundary and $+\infty$ unattainable boundary, we denote with

$$a_{\mathcal{Z}}(x,t|x_0,t_0)=\frac{\partial}{\partial x}P\{\mathcal{Z}(t)\le x;\ \mathcal{Z}(\theta)>0,\forall\theta<t|\mathcal{Z}(t_0)=x_0\},\quad x>0,x_0>0$$

the PDF of $\mathcal{Z}(t)$ with an absorbing condition in the zero state. The PDF $a_{\mathcal{Z}}(x,t|x_0,t_0)$ satisfies the Kolmogorov Equation (1) with the initial condition $\lim_{t_0\uparrow t} a_{\mathcal{Z}}(x,t|x_0,t_0) = \delta(x-x_0)$ and the absorbing condition $\lim_{x_0\downarrow 0} a_{\mathcal{Z}}(x,t|x_0,t_0) = 0$.

The densities $f_{\mathcal{Z}}(x,t|x_0,t_0)$, $g_{\mathcal{Z}}(0,t|x_0,t_0)$, and $a_{\mathcal{Z}}(x,t|x_0,t_0)$ are related by the following integral equations (cf. Siegert [22]):

$$a_{\mathcal{Z}}(x,t|x_0,t_0) = f_{\mathcal{Z}}(x,t|x_0,t_0) - \int_{t_0}^{t} g_{\mathcal{Z}}(0,\theta|x_0,t_0)\, f_{\mathcal{Z}}(x,t|0,\theta)\, d\theta, \quad x_0>0, x>0, \tag{3}$$

$$\int_0^{+\infty} a_{\mathcal{Z}}(x,t|x_0,t_0)\, dx + \int_{t_0}^{t} g_{\mathcal{Z}}(0,\theta|x_0,t_0)\, d\theta = 1, \qquad x_0>0. \tag{4}$$

In the context of population dynamics, the first integral in (4) gives the survival probability, i.e., the probability that the trajectories of the process $\mathcal{Z}(t)$ are not absorbed in the zero state in (t_0,t). Moreover, from (4) one obtains the FPT density

$$g_{\mathcal{Z}}(0,t|x_0,t_0) = -\frac{\partial}{\partial t}\int_0^{+\infty} a_{\mathcal{Z}}(x,t|x_0,t_0)\, dx, \qquad x_0>0, \tag{5}$$

and the ultimate FPT probability of $\mathcal{Z}(t)$ through the zero-state

$$P_{\mathcal{Z}}(0|x_0,t_0) = \int_{t_0}^{+\infty} g_{\mathcal{Z}}(0,\tau|x_0,t_0)\, d\tau = 1 - \lim_{t\to+\infty}\int_0^{+\infty} a_{\mathcal{Z}}(x,t|x_0,t_0)\, dx, \quad x_0>0. \tag{6}$$

In population dynamics, $g_{\mathcal{Z}}(0,t|x_0,t_0)$ in Equation (5) represents the density of the time required to reach the zero state for the first time (extinction density); instead, $P_{\mathcal{Z}}(0|x_0,t_0)$ in Equation (6) provides the probability that the population will become extinct sooner or later.

For a TNH-D process $\mathcal{Z}(t)$, the *local time* $\mathcal{L}(t,x|t_0)$ at an interior state $x \in \mathcal{D}$ is a random variable defined as (cf. Karlin and Taylor [23], Aït-Sahalia and Park [24]):

$$\mathcal{L}(t,x|t_0) = \lim_{\varepsilon\downarrow 0}\frac{1}{2\varepsilon}\int_{t_0}^{t} 1\{|\mathcal{Z}(\theta)-x|\le\varepsilon\}\, d\theta, \qquad t>t_0, \tag{7}$$

where, for $\varepsilon>0$, we have set

$$1\{|\mathcal{Z}(\theta)-x|\le\varepsilon\} = \begin{cases} 1, & |\mathcal{Z}(\theta)-x|\le\varepsilon, \\ 0, & \text{otherwise.}\end{cases}$$

The asymptotic average of the local time in the presence of an absorbing boundary in the zero state, for $x>0$ and $x_0>0$ is:

$$L_{\mathcal{Z}}(x|x_0,t_0) = \lim_{t\to+\infty} E[\mathcal{L}(t,x|t_0)|\mathcal{Z}(t_0)=x_0] = \int_{t_0}^{+\infty} a_{\mathcal{Z}}(x,\theta|x_0,t_0)\, d\theta. \tag{8}$$

For a time-homogeneous diffusion (TH-D) process $\mathcal{Z}(t)$ one has $\zeta_1(x,t)=\zeta_1(x)$ and $\zeta_2(x,t)=\zeta_2(x)$ for all t. In this case, the classification of the endpoints of the state space $\mathcal{D}$, due to Feller [25,26], is based on integrability properties of the functions

$$h_{\mathcal{Z}}(x) = \exp\Big\{-2\int^x \frac{\zeta_1(u)}{\zeta_2(u)}\, du\Big\}, \qquad s_{\mathcal{Z}}(x) = \frac{2}{\zeta_2(u)\, h_{\mathcal{Z}}(u)}, \qquad x\in\mathcal{D}, \tag{9}$$

called scale function and speed density, respectively. Such functions allow us to determine the FPT moments for TH-D processes thanks to the Siegert formula (cf. Siegert [22]). Specifically, if $\mathcal{Z}(t)$ is a TH-D process with state space $\mathcal{D}=(r_1,r_2)$, for $n=1,2,\ldots$ it results in

- for $x_0<S$, if $P_{\mathcal{Z}}(S|x_0) = \int_0^{+\infty} g_{\mathcal{Z}}(S,t|x_0)\, dt = 1$ and if $\int_{r_1}^{z} s_{\mathcal{Z}}(u)\, du$ converges one has:

$$t_n^{(\mathcal{Z})}(S|x_0) = \int_0^{+\infty} t^n\, g_{\mathcal{Z}}(S,t|x_0)\, dt = n \int_{x_0}^{S} dz\, h_{\mathcal{Z}}(z) \int_{r_1}^{z} s_{\mathcal{Z}}(u)\, t_{n-1}(S|u)\, du, \quad x_0 < S, \tag{10}$$

- for $x_0 > S$, if $P_{\mathcal{Z}}(S|x_0) = 1$, and if $\int_z^{r_2} s_{\mathcal{Z}}(u)\, du$ converges one has

$$t_n^{(\mathcal{Z})}(S|x_0) = \int_0^{+\infty} t^n\, g_{\mathcal{Z}}(S,t|x_0)\, dt = n \int_{S}^{x_0} dz\, h_{\mathcal{Z}}(z) \int_{z}^{r_2} s_{\mathcal{Z}}(u)\, t_{n-1}(S|u)\, du, \quad x_0 > S, \tag{11}$$

with $t_0^{(\mathcal{Z})}(S|x_0) = P_{\mathcal{Z}}(S|x_0)$.

In the sequel, for the FPT of TH-D process $\mathcal{Z}(t)$ we denote by

$$\mathrm{Var}^{(\mathcal{Z})}(S|x_0) = t_2^{(\mathcal{Z})}(S|x_0) - [t_1^{(\mathcal{Z})}(S|x_0)]^2, \qquad \mathrm{Cv}^{(\mathcal{Z})}(S|x_0) = \frac{\sqrt{\mathrm{Var}^{(\mathcal{Z})}(S|x_0)}}{t_1^{(\mathcal{Z})}(S|x_0)},$$

$$\Sigma^{(\mathcal{Z})}(S|x_0) = \frac{t_3^{(\mathcal{Z})}(S|x_0) - 3\,t_1^{(\mathcal{Z})}(S|x_0)\, t_2^{(\mathcal{Z})}(S|x_0) + 2\,[t_1^{(\mathcal{Z})}(S|x_0)]^3}{[\mathrm{Var}^{(\mathcal{Z})}(S|x_0)]^{3/2}}.$$

the variance, the coefficient of variation, and the skewness, respectively.

For a TH-D process in $(0,+\infty)$, with 0 absorbing boundary, if $x_0 > 0$ and $x > 0$ the asymptotic average of the local time is (cf. Giorno and Nobile [27]):

$$L_{\mathcal{Z}}(x|x_0) = \begin{cases} s_{\mathcal{Z}}(x) \int_0^{x_0 \wedge x} h_{\mathcal{Z}}(z)dz, & +\infty \text{ unattainable, nonattracting,} \\ s_{\mathcal{Z}}(x) P_{\mathcal{Z}}(0|x \vee x_0) \int_0^{x_0 \wedge x} h_{\mathcal{Z}}(z)dz, & +\infty \text{ unattainable, attracting,} \end{cases} \tag{12}$$

where $x_0 \wedge x = \min(x_0, x)$ and $x_0 \vee x = \max(x_0, x)$.

For a TH-D process $\mathcal{Z}(t)$, in the sequel we denote by

$$\varphi_\lambda^{(\mathcal{Z})}(x|x_0) = \int_0^{+\infty} e^{-\lambda t} \varphi_{\mathcal{Z}}(x,t|x_0)\, dt$$

the Laplace transform (LT) of the function $\varphi_{\mathcal{Z}}(x,t|x_0)$.

Plan of the Paper

In Section 2, we consider the time-inhomogeneous Wiener (TNH-W) process $X(t)$, with infinitesimal drift and infinitesimal variance $A_1(t) = \beta(t)$ and $A_2(t) = \sigma^2(t)$, respectively. For $\beta(t) = \gamma\sigma^2(t)$, with $\gamma \in \mathbb{R}$, we determine the PDF $a_X(x,t|x_0,t_0)$ and the FPT density $g_X(0,t|x_0,t_0)$. Furthermore, for the time-homogeneous Wiener (TH-W) process, the FPT moments through a boundary $S \in \mathbb{R}$ and the asymptotic average of the local time are studied.

In Section 3, we take into account the time-inhomogeneous Ornstein–Uhlenbeck (TNH-OU) process $Y(t)$, with infinitesimal drift and infinitesimal variance $B_1(x,t) = \alpha(t)\,x + \beta(t)$ and $B_2(t) = \sigma^2(t)$, respectively. For $\beta(t) = \gamma\sigma^2(t)\, e^{-A(t|0)}$, with $\gamma \in \mathbb{R}$ and $A(t|0) = \int_0^t \alpha(u)\, du$, we determine $a_Y(x,t|x_0,t_0)$ and $g_Y(0,t|x_0,t_0)$. Moreover, for the TH-OU process, the FPT mean through a constant boundary and the asymptotic average of the local time are evaluated.

In Section 4, we consider the time-inhomogeneous Feller (TNH-F) process $Z(t)$ with infinitesimal drift and infinitesimal variance $C_1(x,t) = \alpha(t)\,x + \beta(t)$ and $C_2(x,t) = 2\,r(t)\,x$, respectively, with an absorbing boundary in the zero-state. For $\beta(t) = \xi\, r(t)$, with $0 \leq \xi < 1$, we obtain $a_Z(x,t|x_0,t_0)$ and $g_Z(0,t|x_0,t_0)$. Furthermore, for the TH-F process, the FPT mean through a constant boundary and the asymptotic average of the local time are examined.

We remark that time-inhomogeneous Wiener, Ornstein–Uhlenbeck and Feller diffusion processes are used in biological systems to model the growth of a population. In such a context, $\alpha(t)$ represents the growth intensity function and $\beta(t)$ denotes the immigration/emigration intensity function. The functions $\sigma^2(t)$ (in Wiener and Ornstein–Uhlenbeck processes) and $r(t)$ (in the Feller process) are the noise intensity functions and take into account the environmental fluctuations.

In Sections 2–4, by using Siegert Formulas (10) and (11), extensive computation are performed with MATHEMATICA to obtain the mean, the variance, the coefficient of variation, and the skewness of FPT for the TH-W, TH-OU, and TH-F processes for various choices of parameters. For these processes, some considerations on the asymptotic average of the local time in the presence of an absorbing boundary in the zero state are also made.

In Section 5, for $\beta(t) = r(t)/2$, some relationships between the PDF in the presence of an absorbing boundary in the zero state and between the FPT densities through zero for Wiener, Ornstein–Uhlenbeck and Feller processes are proved. Moreover, for $\beta(t) = \xi\, r(t)$ $(0 < \xi < 1)$ some asymptotic results for large times between the FPT densities are provided.

2. Wiener-Type Diffusion Process

Let $\{X(t), t \geq t_0\}$, $t_0 \geq 0$, be a TNH-W process, having infinitesimal drift and infinitesimal variance

$$A_1(t) = \beta(t), \qquad A_2(t) = \sigma^2(t), \tag{13}$$

with the state space $\mathbb{R}$, where $\beta(t) \in \mathbb{R}$ and $\sigma(t) > 0$ are continuous functions.

The Wiener process arises as the mathematical limit of other stochastic processes, such as random walks (see Knight [28]). This process has been originally used in physics to model the motion of particles suspended in a fluid and it is still used as a mathematical model for various random phenomena in applied mathematics, economics, quantitative finance, evolutionary biology, and physics.

For $t \geq t_0$, the PDF of $X(t)$ is normal,

$$f_X(x,t|x_0,t_0) = \frac{1}{\sqrt{2\pi V_X(t|t_0)}}\,\exp\Big\{-\frac{\big[x - M_X(t|x_0,t_0)\big]^2}{2\,V_X(t|t_0)}\Big\}, \qquad x, x_0 \in \mathbb{R}, \tag{14}$$

with

$$M_X(t|x_0,t_0) = x_0 + \int_{t_0}^t \beta(u)\,du, \qquad V_X(t|t_0) = \int_{t_0}^t \sigma^2(u)\,du.$$

We now consider the TNH-W process $X(t)$, having infinitesimal moments given in (13), restricted to the state space $(0,+\infty)$ with 0 absorbing boundary; we denote by $a_X(x,t|x_0,t_0)$ its PDF. For the Wiener process $X(t)$ in the presence of an absorbing boundary in the zero state, we analyze two cases: the proportional case with $\beta(t) = \gamma\,\sigma^2(t)$, being $\gamma \in \mathbb{R}$ and $\sigma(t) > 0$, and the time-homogeneous case.

2.1. Proportional Case for the Wiener Process

Proposition 1. *Let $\beta(t) = \gamma\,\sigma^2(t)$, with $\gamma \in \mathbb{R}$ and $\sigma(t) > 0$ in (13). For the TNH-W process $X(t)$ one has*

$$a_X(x,t|x_0,t_0) = f_X(x,t|x_0,t_0) - e^{2\gamma x} f_X(-x,t|x_0,t_0), \qquad x > 0, x_0 > 0, \tag{15}$$

with $f_X(x,t|x_0,t_0)$ given in (14).

Proof. If $\beta(t) = \gamma\,\sigma^2(t)$, from (14) the following symmetry relation holds,

$$f_X(x,t|0,t_0) = e^{2\gamma x}\, f_X(-x,t|0,t_0), \qquad x \in \mathbb{R},$$

so that from (3) one has

$$a_X(x,t|x_0,t_0) = f_X(x,t|x_0,t_0) - e^{2\gamma x}\int_{t_0}^t g_X(0,\tau|x_0,t_0)\, f_X(-x,t|0,\tau)\,d\tau, \qquad x_0 > 0, x > 0. \tag{16}$$

Hence, by virtue of the renewal Equation (2), Equation (15) follows from (16). □

From (15), for $\beta(t) = \gamma\,\sigma^2(t)$, with $\gamma \in \mathbb{R}$ and $\sigma(t) > 0$, one explicitly obtains

$$a_X(x,t|x_0,t_0) = \frac{1}{\sqrt{2\pi V_X(t|t_0)}} \left[\exp\left\{-\frac{\left[x - x_0 - \gamma\,V_X(t|t_0)\right]^2}{2\,V_X(t|t_0)}\right\} - e^{2\gamma\,x}\exp\left\{-\frac{\left[x + x_0 + \gamma\,V_X(t|t_0)\right]^2}{2\,V_X(t|t_0)}\right\}\right], \qquad x_0 > 0, x > 0. \tag{17}$$

We note that Equation (17) for $t_0 = 0$ is in agreement with Equation (25) in Molini et al. [14].

Proposition 2. *Under the assumptions of Proposition 1, for the TNH-W process $X(t)$ one has*

$$g_X(0,t|x_0,t_0) = \frac{x_0\,\sigma^2(t)}{\sqrt{2\pi\,\left[V_X(t|t_0)\right]^3}}\,\exp\left\{-\frac{\left[x_0 + \gamma\,V_X(t|t_0)\right]^2}{2V_X(t|t_0)}\right\}, \qquad x_0 > 0. \tag{18}$$

Moreover, if $\lim_{t\to+\infty} V_X(t|t_0) = +\infty$, the ultimate FPT probability of $X(t)$ through zero is

$$P_X(0|x_0,t_0) = \int_{t_0}^{+\infty} g_X(0,t|x_0,t_0)\,dt = \begin{cases} 1, & \gamma \le 0, \\ e^{-2\gamma\,x_0}, & \gamma > 0, \end{cases} \qquad x_0 > 0. \tag{19}$$

Proof. For $x_0 > 0$, from (17) one obtains

$$\int_0^{+\infty} a_X(x,t|x_0,t_0)\,dx = \frac{1}{2}\left[1 + \mathrm{Erf}\left(\frac{x_0 + \gamma\,V_X(t|t_0)}{\sqrt{2\,V_X(t|t_0)}}\right) - e^{-2\gamma x_0}\,\mathrm{Erfc}\left(\frac{x_0 - \gamma\,V_X(t|t_0)}{\sqrt{2\,V_X(t|t_0)}}\right)\right], \tag{20}$$

where $\mathrm{Erf}(x) = (2/\sqrt{\pi})\int_0^x e^{-z^2}\,dz$ denotes the error function and $\mathrm{Erfc}(x) = 1 - \mathrm{Erf}(x)$ is the complementary error function. Hence, due to (5) and recalling (20), Equation (18) follows. Finally, if $\lim_{t\to+\infty} V_X(t|t_0) = +\infty$, Equation (19) follows, making use of (20) in (6) and by noting that

$$\lim_{t\to+\infty} \mathrm{Erf}\left(\frac{x_0 + \gamma\,V_X(t|t_0)}{\sqrt{2\,V_X(t|t_0)}}\right) = \begin{cases} -1, & \gamma < 0, \\ 0, & \gamma = 0, \\ 1, & \gamma > 0, \end{cases} \qquad \lim_{t\to+\infty} \mathrm{Erfc}\left(\frac{x_0 - \gamma\,V_X(t|t_0)}{\sqrt{2\,V_X(t|t_0)}}\right) = \begin{cases} 0, & \gamma < 0, \\ 1, & \gamma = 0, \\ 2, & \gamma > 0, \end{cases}$$

for any x_0. □

Equation (19) shows that if $\beta(t) = \gamma\sigma^2(t)$, with $\gamma \in \mathbb{R}$ and $\sigma(t) > 0$ in (13), the first-passage for the Wiener process through zero is a sure event for $\gamma > 0$ and $x_0 > 0$.

2.2. Time-Homogeneous Case for the Wiener Process

We consider the TH-W process, obtained from (13) by setting $\beta(t) = \beta$ and $\sigma^2(t) = \sigma^2$, with $\beta \in \mathbb{R}$ and $\sigma > 0$. When $\beta > 0$ ($\beta < 0$) the end point $-\infty$ is a nonattracting (attracting) natural boundary and the end point $+\infty$ is an attracting (nonattracting) natural boundary. Instead, for $\beta = 0$ the end points $-\infty$ and $+\infty$ are nonattracting natural boundaries. The scale function and the speed density, defined in (9) for the TH-W process $X(t)$ are

$$h_X(x) = \exp\left\{-\frac{2\beta}{\sigma^2}\,x\right\}, \qquad s_X(x) = \frac{2}{\sigma^2}\,\exp\left\{\frac{2\beta}{\sigma^2}\,x\right\}, \tag{21}$$

respectively.

The FPT density of the TH-W process $X(t)$ through the constant boundary S starting from x_0 is

$$g_X(S,t|x_0) = \frac{|S - x_0|}{\sigma\sqrt{2\pi\,t^3}}\,\exp\left\{-\frac{(S - x_0 - \beta\,t)^2}{2\sigma^2\,t}\right\}, \qquad S \neq x_0 \tag{22}$$

and the ultimate FPT probability is

$$P_X(S|x_0) = \int_0^{+\infty} g_X(S,t|x_0)\, dt = \begin{cases} 1, & \beta = 0 \text{ or } \beta(S-x_0) > 0, \\ \exp\left\{\frac{2\beta(S-x_0)}{\sigma^2}\right\}, & \beta(S-x_0) < 0. \end{cases} \quad (23)$$

For $\beta(S-x_0) > 0$, the FPT moments of the TH-W process $X(t)$ are finite and from (22) one has

$$t_n^{(X)}(S|x_0) = \frac{2\,|S-x_0|}{\sigma\sqrt{2\pi}} \left(\frac{S-x_0}{\beta}\right)^{n-1/2} \exp\left\{\frac{\beta(S-x_0)}{\sigma^2}\right\} K_{n-1/2}\left[\frac{\beta(S-x_0)}{\sigma^2}\right], \quad n = 1,2,\ldots$$

where $K_\nu(z)$ denotes the modified Bessel function of the third kind, which can be expressed in terms of the modified Bessel function of first kind $I_\nu(z)$ (see Abramowitz and Stegun [29], p. 375, n. 9.6.2),

$$K_\nu(z) = \frac{\pi}{2}\,\frac{I_{-\nu}(z) - I_\nu(z)}{\sin(\nu\,\pi)}, \qquad I_\nu(z) = \sum_{k=0}^{+\infty} \frac{1}{k!\,\Gamma(\nu+k+1)} \left(\frac{z}{2}\right)^{2k+\nu}, \quad (24)$$

where $\Gamma(\nu) = \int_0^{+\infty} y^{\nu-1} e^{-y}\, dy$, with $\mathrm{Re}\,\nu > 0$, is the Euler gamma function.

In particular, for $\beta(S-x_0) > 0$ the first three FPT moments of the TH-W process $X(t)$ are

$$t_1^{(X)}(S|x_0) = \frac{S-x_0}{\beta}, \qquad t_2^{(X)}(S|x_0) = \left(\frac{S-x_0}{\beta}\right)^2 \left\{1 + \frac{\sigma^2}{\beta\,(S-x_0)}\right\},$$
$$t_3^{(X)}(S|x_0) = \left(\frac{S-x_0}{\beta}\right)^3 \left\{1 + \frac{3\,\sigma^2}{\beta\,(S-x_0)} + \frac{3\,\sigma^4}{\beta^2\,(S-x_0)^2}\right\}.$$

In Tables 1 and 2, the mean $t_1^{(X)}(S|x_0)$, the variance $\mathrm{Var}^{(X)}(S|x_0)$, the coefficient of variation $\mathrm{Cv}^{(X)}(S|x_0)$, and the skewness $\Sigma^{(X)}(S|x_0)$ of the FPT are listed for $x_0 = 4, \sigma = 1$ and some choices of β and S.

Table 1. For the Wiener process, with $A_1(x) = \beta$ and $A_2(x) = 1$, the mean, the variance, the coefficient of variation, and the skewness of FPT are listed for $x_0 = 4, \beta = 0.1, 0.2$ and for increasing values the boundary $S > x_0$.

	S	$t_1^{(X)}(S\|x_0)$	$\mathrm{Var}^{(X)}(S\|x_0)$	$\mathrm{Cv}^{(X)}(S\|x_0)$	$\Sigma^{(X)}(S\|x_0)$
$\beta = 0.1$	100	960	96,000	0.322749	0.968246
	500	4960	496,000	0.141990	0.425971
	1000	9960	996,000	0.100201	0.300602
	1500	14,960	1,496,000	0.0817587	0.245276
	2000	19,960	1,996,000	0.0707815	0.212344
	2500	24,960	2,496,000	0.0632962	0.189889
	3000	29,960	2,996,000	0.0577736	0.173321
$\beta = 0.2$	100	480	12,000	0.228218	0.684653
	500	2480	62,000	0.100402	0.301207
	1000	4980	124,500	0.0708525	0.212558
	1500	7480	187,000	0.0578122	0.173436
	2000	9980	249,500	0.0500501	0.150150
	2500	12,480	312,000	0.0447572	0.134272
	3000	14,980	374,500	0.0408521	0.122556

Table 2. As in Table 1, with $x_0 = 4, \sigma = 1, \beta = -0.1, -0.2$ and for decreasing values the boundary $S \in [0, x_0)$.

	S	$t_1^{(X)}(S\|x_0)$	$\text{Var}^{(X)}(S\|x_0)$	$\text{Cv}^{(X)}(S\|x_0)$	$\Sigma^{(X)}(S\|x_0)$
$\beta = -0.1$	3.5	5	500	4.47214	13.4164
	3.0	10	1000	3.16228	9.48683
	2.5	15	1500	2.58199	7.74597
	2.0	20	2000	2.23607	6.7082
	1.5	25	2500	2.0	6.0
	1.0	30	3000	1.82574	5.47723
	0.5	35	3500	1.69031	5.07093
	0.0	40	4000	1.58114	4.74342
$\beta = -0.2$	3.5	2.5	62.5	3.16228	9.48683
	3.0	5	125	2.23607	6.7082
	2.5	7.5	187.5	1.82574	5.47723
	2.0	10	250	1.58114	4.74342
	1.5	12.5	312.5	1.41421	4.24264
	1.0	15	375	1.29099	3.87298
	0.5	17.5	437.5	1.19523	3.58569
	0.0	20	500	1.11803	3.3541

As shown in Tables 1 and 2, for the TH-W process $X(t)$ the coefficient of variation and the skewness of the FPT decrease when S moves away from x_0.

Moreover, by setting $\beta(t) = \beta, \sigma^2(t) = \sigma^2$ and $\gamma = \beta/\sigma^2$ in (17), for the TH-W process $X(t)$ one has

$$a_X(x,t|x_0) = \frac{1}{\sqrt{2\pi\sigma^2 t}}\left[\exp\left\{-\frac{(x - x_0 - \beta t)^2}{2\sigma^2 t}\right\} - \exp\left\{\frac{2\beta x}{\sigma^2}\right\}\exp\left\{-\frac{(x + x_0 + \beta t)^2}{2\sigma^2 t}\right\}\right] \tag{25}$$

with $x_0 > 0$ and $x > 0$.

Proposition 3. *For the TH-W process $X(t)$, the asymptotic average of the local time is*

$$\begin{aligned} L_X(x|x_0) &= \int_0^{+\infty} a_X(x,t|x_0)\,dt \\ &= \begin{cases} \frac{1}{|\beta|}\exp\left\{\frac{\beta(x-x_0)}{\sigma^2}\right\}\left[\exp\left\{-\frac{|\beta(x-x_0)|}{\sigma^2}\right\} - \exp\left\{-\frac{|\beta(x+x_0)|}{\sigma^2}\right\}\right], & \beta \neq 0, \\ \frac{|x+x_0|}{\sigma^2} - \frac{|x-x_0|}{\sigma^2} = \frac{2(x_0 \wedge x)}{\sigma^2}, & \beta = 0, \end{cases} \end{aligned} \tag{26}$$

with $x_0 \wedge x = \min(x_0, x)$ and $x_0 \vee x = \max(x_0, x)$.

Proof. Because $+\infty$ is a nonattracting boundary for $\beta \leq 0$ and attracting for $\beta > 0$, Equation (26) follows from (12) making use of (21) and (23). □

From (26), for $\beta \in \mathbb{R}$ and $\sigma > 0$ one has $\lim_{x\downarrow 0} L_X(x|x_0) = 0$ and

$$\lim_{x\uparrow+\infty} L_X(x|x_0) = \begin{cases} 0, & \beta < 0, \\ \frac{2x_0}{\sigma^2}, & \beta = 0, \\ \frac{1-e^{-2\beta x_0/\sigma^2}}{\beta}, & \beta > 0. \end{cases}$$

We note that $L_X(x|x_0)$ tends to zero as x increases if $\beta < 0$, and it approaches a positive value when $\beta \geq 0$.

In Figure 1, the asymptotic average of the local time for the TH-W process $X(t)$ is plotted for $x_0 = 4, \sigma = 1$ and some choices of β. We note that $L_X(x|x_0)$ tends to zero only if $\beta < 0$, otherwise approaches to a positive value.

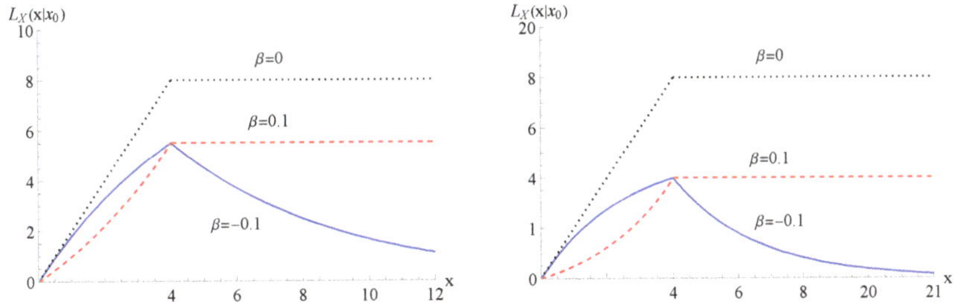

(a) (b)

Figure 1. $L_X(x|x_0)$, given in (26), with $x_0 = 4, \sigma = 1$ and some choices of β. In (**a**) $\beta = -0.1, 0, 0.1$ and in (**b**) $\beta = -0.2, 0, 0.2$.

3. Ornstein–Uhlenbeck-Type Diffusion Process

Let $\{Y(t), t \geq t_0\}$, $t_0 \geq 0$, be a TNH-OU process, having infinitesimal drift and infinitesimal variance

$$B_1(x,t) = \alpha(t)\,x + \beta(t), \qquad B_2(t) = \sigma^2(t), \qquad x \in \mathbb{R}, \tag{27}$$

with state space $\mathbb{R}$, where $\alpha(t) \in \mathbb{R}$, $\beta(t) \in \mathbb{R}$, $\sigma(t) > 0$ are continuous functions. Note that when $\alpha(t) = 0$ for all t, the process $Y(t)$ identifies with the TNH-W process $X(t)$ having infinitesimal moments (13).

Although the Ornstein–Uhlenbeck process has been originally used in physics to model the velocity of a Brownian particle (see Uhlenbeck and Ornstein [30]), it finds many applications in several scientific fields. In particular, the Ornstein–Uhlenbeck process is frequently proposed as a stochastic model for the single neuronal activity (see Ricciardi and Sacerdote [31], Lánský and Ditlevsen [32]). A wide field of applications of the Ornstein–Uhlenbeck process lies also in mathematical finance to model the evolution of the interest rate of financial markets (cf. Vasicek [33], Hull and White [34]).

The PDF of $Y(t)$ is normal,

$$f_Y(x,t|x_0,t_0) = \frac{1}{\sqrt{2\pi V_Y(t|t_0)}} \exp\Big\{-\frac{[x - M_Y(t|x_0,t_0)]^2}{2\,V_Y(t|t_0)}\Big\}, \qquad x, x_0 \in \mathbb{R}, \tag{28}$$

with

$$M_Y(t|x_0,t_0) = x_0\, e^{A(t|t_0)} + \int_{t_0}^{t} \beta(\theta)\, e^{A(t|\theta)}\, d\theta, \qquad V_Y(t|t_0) = \int_{t_0}^{t} \sigma^2(\theta)\, e^{2A(t|\theta)}\, d\theta, \tag{29}$$

being

$$A(t|t_0) = \int_{t_0}^{t} \alpha(\theta)\, d\theta. \tag{30}$$

We now consider the TNH-OU process $Y(t)$, having infinitesimal moments given in (27), restricted to the state space $(0, +\infty)$, with 0 absorbing boundary, and denote by $a_Y(x,t|x_0,t_0)$ its PDF. For the TNH-OU process $Y(t)$ with 0 absorbing boundary, we take into account two cases: the proportional case in which $\beta(t) = \gamma\,\sigma^2(t)\, e^{-A(t|0)}$, with $\gamma \in \mathbb{R}$, $\alpha(t) \in \mathbb{R}$ and $\sigma(t) > 0$, and the time-homogeneous case.

3.1. Proportional Case for the Ornstein-Uhlenbeck Process

Proposition 4. *Let $\beta(t) = \gamma\,\sigma^2(t)\,e^{-A(t|0)}$, with $\gamma \in \mathbb{R}$, $\alpha(t) \in \mathbb{R}$, $\sigma(t) > 0$ in (27) and $A(t|0)$ defined in (30). For the TNH-OU process $Y(t)$ one has*

$$a_Y(x,t|x_0,t_0) = f_Y(x,t|x_0,t_0) - \exp\Big\{2\gamma\,x\,e^{-A(t|0)}\Big\}\,f_Y(-x,t|x_0,t_0), \qquad x > 0, x_0 > 0, \tag{31}$$

with $f_Y(x,t|x_0,t_0)$ given in (28).

Proof. By choosing $\beta(t) = \gamma\,\sigma^2(t)\,e^{-A(t|0)}$, from (28) the following symmetry relation holds,

$$f_Y(x,t|0,t_0) = \exp\Big\{2\gamma\,x\,e^{-A(t|0)}\Big\}\,f_Y(-x,t|0,t_0), \qquad x \in \mathbb{R},$$

so that from (3) one obtains

$$a_Y(x,t|x_0,t_0) = f_Y(x,t|x_0,t_0) - \exp\Big\{2\gamma\,x\,e^{-A(t|0)}\Big\}\int_{t_0}^{t} g_Y(0,\tau|x_0,t_0)\,f_Y(-x,t|0,\tau)\,d\tau \tag{32}$$

for $x_0 > 0$ and $x > 0$. Hence, by virtue of the renewal Equation (2), Equation (31) follows from (32). □

From (31), if $\beta(t) = \gamma\,\sigma^2(t)\,e^{-A(t|0)}$, for $x_0 > 0$ and $x > 0$ one obtains

$$\begin{aligned} a_Y(x,t|x_0,t_0) = \frac{1}{\sqrt{2\pi V_Y(t|t_0)}}&\left[\exp\Big\{-\frac{\big[x - x_0\,e^{A(t|t_0)} - \gamma\,e^{-A(t|0)}\,V_Y(t|t_0)\big]^2}{2\,V_Y(t|t_0)}\Big\}\right. \\ &\left. - \exp\Big\{2\gamma\,x\,e^{-A(t|0)}\Big\}\exp\Big\{-\frac{\big[x + x_0\,e^{A(t|t_0)} + \gamma\,e^{-A(t|0)}\,V_Y(t|t_0)\big]^2}{2\,V_Y(t|t_0)}\Big\}\right]. \end{aligned} \tag{33}$$

Proposition 5. *Under the assumptions of Proposition 4, for the TNH-OU process $Y(t)$ one has*

$$g_Y(0,t|x_0,t_0) = \frac{x_0\,\sigma^2(t)\,e^{A(t|t_0)}}{\sqrt{2\pi\,[V_Y(t|t_0)]^3}}\exp\Big\{-\frac{\big[x_0\,e^{A(t|t_0)} + \gamma\,e^{-A(t|0)}\,V_Y(t|t_0)\big]^2}{2\,V_Y(t|t_0)}\Big\}, \qquad x_0 > 0. \tag{34}$$

Furthermore, if $\lim_{t\to+\infty}[e^{-2A(t|t_0)}V_Y(t|t_0)] = +\infty$, the ultimate FPT probability for $x_0 > 0$ is

$$P_Y(0|x_0,t_0) = \int_{t_0}^{+\infty} g_Y(0,t|x_0,t_0)\,dt = \begin{cases} 1, & \gamma \le 0, \\ \exp\Big\{-2\gamma\,x_0\,e^{-A(t_0|0)}\Big\}, & \gamma > 0. \end{cases} \tag{35}$$

Proof. Recalling (33), one obtains

$$\begin{aligned} \int_0^{+\infty} a_Y(x,t|x_0,t_0)\,dx = \frac{1}{2}&\left[1 + \mathrm{Erf}\Big(\frac{x_0\,e^{A(t|t_0)} + \gamma\,e^{-A(t|0)}\,V_Y(t|t_0)}{\sqrt{2\,V_Y(t|t_0)}}\Big)\right. \\ &\left. - \exp\Big\{-2\gamma\,x_0\,e^{-A(t_0|0)}\Big\}\,\mathrm{Erfc}\Big(\frac{x_0\,e^{A(t|t_0)} - \gamma\,e^{-A(t|0)}\,V_Y(t|t_0)}{\sqrt{2\,V_Y(t|t_0)}}\Big)\right], \qquad x_0 > 0. \end{aligned} \tag{36}$$

By virtue of (5) and recalling (36), Equation (34) follows. Moreover, under the assumption $\lim_{t\to+\infty}[e^{-2A(t|t_0)}V_Y(t|t_0)] = +\infty$, Equation (35) follows, making use of (36) in (6) by noting that

$$\lim_{t\to+\infty} \mathrm{Erf}\left(\frac{x_0\, e^{A(t|t_0)} + \gamma\, e^{-A(t|0)}\, V_Y(t|t_0)}{\sqrt{2\, V_Y(t|t_0)}}\right) = \begin{cases} -1, & \gamma < 0, \\ 0, & \gamma = 0, \\ 1, & \gamma > 0, \end{cases}$$

$$\lim_{t\to+\infty} \mathrm{Erfc}\left(\frac{x_0\, e^{A(t|t_0)} - \gamma\, e^{-A(t|0)}\, V_Y(t|t_0)}{\sqrt{2\, V_Y(t|t_0)}}\right)\Bigg] = \begin{cases} 0, & \gamma < 0, \\ 1, & \gamma = 0, \\ 2, & \gamma > 0, \end{cases}$$

for any x_0. $\square$

3.2. Time-Homogeneous Case for the Ornstein–Uhlenbeck Process

We consider the TH-OU process $Y(t)$, by setting in (27) $\alpha(t) = \alpha$, $\beta(t) = \beta$, $\sigma^2(t) = \sigma^2$, with $\alpha \neq 0$, $\beta \in \mathbb{R}$ and $\sigma > 0$. The end points $-\infty$ and $+\infty$ are nonattracting natural boundaries for $\alpha < 0$ and attracting natural boundaries for $\alpha > 0$. The scale function and the speed density, defined in (9), for the TH-OU process $Y(t)$ are

$$h_Y(x) = \exp\Big\{-\frac{\alpha}{\sigma^2}\left(x^2 + \frac{2\beta}{\alpha}\, x\right)\Big\}, \qquad s_Y(x) = \frac{2}{\sigma^2}\exp\Big\{\frac{\alpha}{\sigma^2}\left(x^2 + \frac{2\beta}{\alpha}\, x\right)\Big\}, \tag{37}$$

respectively. The LT of $f_Y(x,t|x_0)$ is

$$f_\lambda^{(Y)}(x|x_0) = \begin{cases} \dfrac{2^{\frac{\lambda}{|\alpha|}-1}}{\sigma\pi\sqrt{|\alpha|}}\Gamma\left(\frac{\lambda}{2|\alpha|}\right)\Gamma\left(\frac{1}{2}+\frac{\lambda}{2|\alpha|}\right)\exp\Big\{-\frac{|\alpha|}{2\sigma^2}\Big[\left(x+\frac{\beta}{\alpha}\right)^2 - \left(x_0+\frac{\beta}{\alpha}\right)^2\Big]\Big\} \\ \quad\times D_{-\frac{\lambda}{|\alpha|}}\left(-\frac{\sqrt{2|\alpha|}}{\sigma}\Big[x_0 \wedge x + \frac{\beta}{\alpha}\Big]\right) D_{-\frac{\lambda}{|\alpha|}}\left(\frac{\sqrt{2|\alpha|}}{\sigma}\Big[x_0 \vee x + \frac{\beta}{\alpha}\Big]\right), & \alpha < 0, \\ \dfrac{2^{\frac{\lambda}{\alpha}}}{\sigma\pi\sqrt{\alpha}}\Gamma\left(1+\frac{\lambda}{2\alpha}\right)\Gamma\left(\frac{1}{2}+\frac{\lambda}{2\alpha}\right)\exp\Big\{-\frac{\alpha}{2\sigma^2}\Big[\left(x_0+\frac{\beta}{\alpha}\right)^2 - \left(x+\frac{\beta}{\alpha}\right)^2\Big]\Big\} \\ \quad\times D_{-\frac{\lambda}{\alpha}-1}\left(-\frac{\sqrt{2\alpha}}{\sigma}\Big[x_0 \wedge x + \frac{\beta}{\alpha}\Big]\right) D_{-\frac{\lambda}{\alpha}-1}\left(\frac{\sqrt{2\alpha}}{\sigma}\Big[x_0 \vee x + \frac{\beta}{\alpha}\Big]\right), & \alpha > 0, \end{cases} \tag{38}$$

where $D_\nu(z)$ is the parabolic cylinder function defined as (cf. Gradshteyn and Ryzhik [35], p. 1028, no. 9.240). We have

$$D_\nu(z) = 2^{\nu/2} e^{-z^2/4}\left\{\frac{\sqrt{\pi}}{\Gamma\left(\frac{1-\nu}{2}\right)}\Phi\left(-\frac{\nu}{2},\frac{1}{2};\frac{z^2}{2}\right) - \frac{z\sqrt{2\pi}}{\Gamma\left(-\frac{\nu}{2}\right)}\Phi\left(\frac{1-\nu}{2},\frac{3}{2};\frac{z^2}{2}\right)\right\} \tag{39}$$

in terms of Kummer's confluent hypergeometric function

$$\Phi(a,c;x) = 1 + \sum_{n=1}^{+\infty}\frac{(a)_n}{(c)_n}\frac{x^n}{n!},$$

with $(a)_0 = 1$ and $(a)_n = a(a+1)\cdots(a+n-1)$ for $n = 1, 2, \ldots$ In the following, we will make use of the relations (cf. Gradshteyn and Ryzhik [35], p. 1030, no. 9.251 and no. 9.254).

$$D_0(x) = e^{-x^2/4}, \quad D_1(x) = x\, e^{-x^2/4}, \quad D_{-1}(x) = \sqrt{\frac{\pi}{2}}\, e^{x^2/4}\,\mathrm{Erfc}\left(\frac{x}{\sqrt{2}}\right). \tag{40}$$

For the TH-OU process, taking the Laplace transform in (2) and recalling (38), for $x_0 \neq S$ one has

$$
g_\lambda^{(Y)}(S|x_0) = \begin{cases} \exp\Big\{\frac{|\alpha|}{2\sigma^2}\Big[\Big(x_0+\frac{\beta}{\alpha}\Big)^2-\Big(S+\frac{\beta}{\alpha}\Big)^2\Big]\Big\}\,\dfrac{D_{-\frac{\lambda}{|\alpha|}}\Big(\mathrm{sign}(x_0-S)\frac{\sqrt{2|\alpha|}}{\sigma}\Big(x_0+\frac{\beta}{\alpha}\Big)\Big)}{D_{-\frac{\lambda}{|\alpha|}}\Big(\mathrm{sign}(x_0-S)\frac{\sqrt{2|\alpha|}}{\sigma}\Big(S+\frac{\beta}{\alpha}\Big)\Big)}, & \alpha<0,\\[2ex] \exp\Big\{-\frac{\alpha}{2\sigma^2}\Big[\Big(x_0+\frac{\beta}{\alpha}\Big)^2-\Big(S+\frac{\beta}{\alpha}\Big)^2\Big]\Big\}\,\dfrac{D_{-\frac{\lambda}{\alpha}-1}\Big(\mathrm{sign}(x_0-S)\frac{\sqrt{2\alpha}}{\sigma}\Big(x_0+\frac{\beta}{\alpha}\Big)\Big)}{D_{-\frac{\lambda}{\alpha}-1}\Big(\mathrm{sign}(x_0-S)\frac{\sqrt{2\alpha}}{\sigma}\Big(S+\frac{\beta}{\alpha}\Big)\Big)}, & \alpha>0, \end{cases} \tag{41}
$$

where $\mathrm{sign}(z)$ denotes the sign function that returns -1 if $z<0$, $+1$ if $z>0$ and 0 otherwise. Moreover, by setting $\lambda=0$ in (41) and recalling (40), for $x_0\neq S$ one has

$$
P_Y(S|x_0) = \int_0^{+\infty} g_Y(S,t|x_0)\,dt = \begin{cases} 1, & \alpha<0,\\ \dfrac{\mathrm{Erfc}\Big(\frac{\sqrt{\alpha}}{\sigma}\Big(x_0+\frac{\beta}{\alpha}\Big)\Big)}{\mathrm{Erfc}\Big(\frac{\sqrt{\alpha}}{\sigma}\Big(S+\frac{\beta}{\alpha}\Big)\Big)}, & \alpha>0, \end{cases} \tag{42}
$$

so that the first passage through the state S is a sure event for $\alpha<0$.

The inverse LT of $g_\lambda^{(Y)}(S|x_0)$ can be obtained in closed form only if $S=-\beta/\alpha$.

Proposition 6. *For the TH-OU process, the FPT density through the boundary $S=-\beta/\alpha$ is*

$$
g_Y\Big(-\frac{\beta}{\alpha},t\Big|x_0\Big) = \frac{2\,e^{\alpha t}|x_0+\beta/\alpha|}{\sigma\sqrt{\pi}}\Big[\frac{\alpha}{e^{2\alpha t}-1}\Big]^{3/2}\exp\Big\{-\frac{\alpha\,e^{2\alpha t}\,(x_0+\beta/\alpha)^2}{\sigma^2(e^{2\alpha t}-1)}\Big\},\quad x_0\neq-\beta/\alpha, \tag{43}
$$

and the ultimate FPT probability is

$$
P_Y\Big(-\frac{\beta}{\alpha}\Big|x_0\Big) = \int_0^{+\infty} g_Y\Big(-\frac{\beta}{\alpha},t\Big|x_0\Big)\,dt = \begin{cases} 1, & \alpha<0,\\ \mathrm{Erfc}\Big(\frac{\sqrt{\alpha}}{\sigma}\Big(x_0+\frac{\beta}{\alpha}\Big)\Big), & \alpha>0. \end{cases} \tag{44}
$$

Proof. Because

$$
D_\nu(0) = \frac{2^{\nu/2}\sqrt{\pi}}{\Gamma\Big(\frac{1-\nu}{2}\Big)},
$$

from (41) for $\alpha\neq0$ and $x_0\neq-\beta/\alpha$ one has

$$
g_\lambda^{(Y)}\Big(-\frac{\beta}{\alpha}\Big|x_0\Big) = \begin{cases} \exp\Big\{\frac{|\alpha|}{2\sigma^2}\Big(x_0+\frac{\beta}{\alpha}\Big)^2\Big\}\,\frac{2^{\frac{\lambda}{2|\alpha|}}}{\sqrt{\pi}}\,\Gamma\big(\tfrac{1}{2}+\tfrac{\lambda}{2|\alpha|}\big)\,D_{-\frac{\lambda}{|\alpha|}}\Big(\frac{\sqrt{2|\alpha|}}{\sigma}\Big|x_0+\frac{\beta}{\alpha}\Big|\Big), & \alpha<0,\\[2ex] \exp\Big\{-\frac{\alpha}{2\sigma^2}\Big(x_0+\frac{\beta}{\alpha}\Big)^2\Big\}\,\frac{2^{\frac{\lambda}{2\alpha}+\frac{1}{2}}}{\sqrt{\pi}}\,\Gamma\big(1+\tfrac{\lambda}{2\alpha}\big)\,D_{-\frac{\lambda}{\alpha}-1}\Big(\frac{\sqrt{2\alpha}}{\sigma}\Big|x_0+\frac{\beta}{\alpha}\Big|\Big), & \alpha>0. \end{cases} \tag{45}
$$

Equation (43) follows by taking the inverse LT of (45) and making use of the following result (cf. Erdèlyi et al. [36], p. 290, no. 9):

$$
\int_0^{+\infty} e^{-pt}\Big[\frac{e^{t/2}}{(e^t-1)^{\nu+1/2}}\exp\Big\{-\frac{\gamma^2}{4(e^t-1)}\Big\}D_{2\nu}\Big(\frac{\gamma}{\sqrt{1-e^{-t}}}\Big)\Big]\,dt = 2^{p+\nu}\Gamma(p+\nu)\,D_{-2p}(\gamma),\qquad \mathrm{Re}\,p>0.
$$

Moreover, by setting $\lambda=0$ in (45) and recalling (40), one obtains (44). □

When $\alpha < 0$, the FPT moments through S starting from x_0 can be evaluated by making use of Siegert Formulas (10) and (11) with $r_1 = -\infty$ and $r_2 = +\infty$. In particular, for $n = 1$ and $\alpha < 0$ one has

$$t_1^{(Y)}(S|x_0) = \frac{1}{|\alpha|}\left\{\frac{\pi}{2}\left[\mathrm{Erfi}\left(\frac{\sqrt{|\alpha|}}{\sigma}\left(x_0 \vee S + \frac{\beta}{\alpha}\right)\right) - \mathrm{Erfi}\left(\frac{\sqrt{|\alpha|}}{\sigma}\left(x_0 \wedge S + \frac{\beta}{\alpha}\right)\right)\right]\right.$$
$$\left. + \psi_1\left(\frac{\sqrt{|\alpha|}}{\sigma}\left(S + \frac{\beta}{\alpha}\right)\right) - \psi_1\left(\frac{\sqrt{|\alpha|}}{\sigma}\left(x_0 + \frac{\beta}{\alpha}\right)\right)\right\}, \qquad x_0 \neq S,,$$

where

$$\mathrm{Erfi}(z) = \frac{2}{\sqrt{\pi}}\int_0^z e^{u^2}\,du = \frac{2}{\sqrt{\pi}}\sum_{k=0}^{+\infty}\frac{z^{2k+1}}{(2k+1)\,k!}, \qquad \psi_1(z) = \sum_{k=0}^{+\infty}\frac{2^k z^{2k+2}}{(k+1)\,(2k+1)!!}.$$

Furthermore, for $\alpha < 0$ from (10) and (11) one obtains (cf. Ricciardi et al. [12])

$$\lim_{S\to+\infty}\frac{t_n^{(Y)}(S|x_0)}{[t_1^{(Y)}(S|x_0)]^n} = n! \quad (x_0 < S)$$

for $n = 1, 2, \ldots$ so that for $\alpha < 0$ the FPT density of the Ornstein–Uhlenbeck process exhibits an exponential asymptotic behavior as the boundary moves away from the starting point.

In Tables 3 and 4, the mean $t_1^{(Y)}(S|x_0)$, the variance $\mathrm{Var}^{(Y)}(S|x_0)$, the coefficient of variation $\mathrm{Cv}^{(Y)}(S|x_0)$, and the skewness $\Sigma^{(Y)}(S|x_0)$ of the FPT, obtained by using (10) and (11), are listed for $x_0 = 4$, $\alpha = -0.02$, $\sigma = 1$ and some choices of β and S.

Table 3. For the TH-OU process, with $B_1(x) = -0.02\,x + \beta$ and $B_2(x) = 1$, the mean, the variance, the coefficient of variation, and the skewness of FPT are listed for $x_0 = 4$, $\beta = -0.1, 0, 0.1$ and for increasing values the boundary $S > x_0$.

	S	$t_1^{(Y)}(S\|x_0)$	$\mathrm{Var}^{(Y)}(S\|x_0)$	$\mathrm{Cv}^{(Y)}(S\|x_0)$	$\Sigma^{(Y)}(S\|x_0)$
$\beta = -0.1$	5	1.491996×10^2	1.265053×10^5	2.383893	3.759894
	10	3.97436×10^3	1.842427×10^7	1.080010	2.020738
	15	1.005474×10^5	1.017251×10^{10}	1.003097	1.999075
	20	7.036678×10^6	4.951899×10^{13}	1.000012	1.986165
	25	1.413375×10^9	1.997625×10^{18}	0.9999992	1.970391
$\beta = 0$	5	3.077237×10^1	5.966963×10^3	2.510243	4.446831
	10	4.475225×10^2	2.545499×10^5	1.127383	2.098001
	15	4.272683×10^3	1.855231×10^7	1.008088	2.000674
	20	1.008457×10^5	1.017263×10^{10}	1.000136	1.993045
	25	7.036975×10^6	4.951901×10^{13}	1.000000	1.980451
$\beta = 0.1$	5	1.161050×10^1	7.316673×10^2	2.329732	4.818854
	10	1.162808×10^2	1.533674×10^4	1.065021	2.181767
	15	5.330310×10^2	2.639197×10^5	0.9637923	2.002620
	20	4.358192×10^3	1.856168×10^7	0.9885583	1.997674
	25	1.009312×10^5	1.017264×10^{10}	0.9992894	1.998622

From Table 3, we note that for the TH-OU process $Y(t)$ the coefficient of variation approaches the value 1 and the skewness approaches the value 2 for large boundaries. Hence, when $\alpha < 0$ the FPT density of the TH-OU process exhibits an exponential behavior for large boundaries S, such that $S > x_0$.

Table 4. As in Table 3 with $x_0 = 4$, $\alpha = 0.02$, $\sigma = 1$, $\beta = -0.1, 0, 0.1$ and for decreasing values the boundary $S \in [0, x_0)$.

	S	$t_1^{(Y)}(S\|x_0)$	$\mathrm{Var}^{(Y)}(S\|x_0)$	$\mathrm{Cv}^{(Y)}(S\|x_0)$	$\Sigma^{(Y)}(S\|x_0)$
$\beta = -0.1$	3.5	2.321819	38.35014	2.667198	6.633648
	3.0	4.740922	81.06132	1.899083	4.708105
	2.5	7.264863	128.7704	1.561998	3.861617
	2.0	9.902021	182.2269	1.363272	3.362382
	1.5	12.66172	242.3166	1.229416	3.026467
	1.0	15.55435	310.0904	1.132119	2.782932
	0.5	18.59156	386.8000	1.057858	2.597858
	0.0	21.7864	473.9432	0.9992582	2.452724
$\beta = 0$	3.5	3.763743	130.8265	3.03898	6.851246
	3.0	7.756523	281.9104	2.164654	4.855690
	2.5	12.00269	457.1913	1.781437	3.978597
	2.0	16.53010	661.5233	1.555955	3.462116
	1.5	21.37079	900.9213	1.404503	3.115650
	1.0	26.56169	1182.882	1.294836	2.865643
	0.5	32.14558	1516.804	1.211556	2.676884
	0.0	38.17219	1914.549	1.146268	2.530113
$\beta = 0.1$	3.5	7.745813	697.7332	3.410183	6.676430
	3.0	16.23615	1550.603	2.425310	4.719137
	2.5	25.58427	2600.685	1.993291	3.859455
	2.0	35.92567	3903.424	1.739074	3.354977
	1.5	47.42327	5532.504	1.568445	3.018737
	1.0	60.27406	7586.697	1.445093	2.778458
	0.5	74.71751	10199.60	1.351666	2.599416
	0.0	91.04643	13553.66	1.278690	2.462534

From Table 4, we note that for the TH-OU process $Y(t)$ the coefficient of variation and the skewness decreases as S decreases.

Moreover, taking the Laplace transform in (3) one has

$$a_\lambda^{(Y)}(x|x_0) = f_\lambda^{(Y)}(x|x_0) - g_\lambda^{(Y)}(0|x_0)\, f_\lambda^{(Y)}(x|0), \qquad x_0 > 0, x > 0, \tag{46}$$

so that, recalling (38) and (41), one can obtain the LT of $a_Y(x,t|x_0)$ for the TH-OU process in $(0, +\infty)$ with 0 absorbing boundary.

Proposition 7. *Let $Y(t)$ be a TH-OU process.*

- *For $\alpha < 0$, one has*

$$L_Y(x|x_0) = \int_0^{+\infty} a_Y(x,t|x_0)\, dt = \frac{1}{\sigma}\sqrt{\frac{\pi}{|\alpha|}} \exp\Big\{-\frac{|\alpha|}{\sigma^2}\Big(x+\frac{\beta}{\alpha}\Big)^2\Big\} \times\Big[\mathrm{Erfi}\Big(\frac{\sqrt{|\alpha|}}{\sigma}\Big(x_0\wedge x+\frac{\beta}{\alpha}\Big)\Big) - \mathrm{Erfi}\Big(\frac{\sqrt{|\alpha|}}{\sigma}\frac{\beta}{\alpha}\Big)\Big], \qquad x_0 > 0, x > 0. \tag{47}$$

- *For $\alpha > 0$, it results in*

$$L_Y(x|x_0) = \int_0^{+\infty} a_Y(x,t|x_0)\, dt = \frac{1}{\sigma}\sqrt{\frac{\pi}{\alpha}} \exp\Big\{\frac{\alpha}{\sigma^2}\Big(x+\frac{\beta}{\alpha}\Big)^2\Big\} \frac{\mathrm{Erfc}\Big(\frac{\sqrt{\alpha}}{\sigma}\Big(x_0\vee x+\frac{\beta}{\alpha}\Big)\Big)}{\mathrm{Erfc}\Big(\frac{\sqrt{\alpha}}{\sigma}\frac{\beta}{\alpha}\Big)} \times\Big[\mathrm{Erf}\Big(\frac{\sqrt{\alpha}}{\sigma}\Big(x_0\wedge x+\frac{\beta}{\alpha}\Big)\Big) - \mathrm{Erf}\Big(\frac{\sqrt{\alpha}}{\sigma}\frac{\beta}{\alpha}\Big)\Big], \qquad x_0 > 0, x > 0. \tag{48}$$

Proof. Because $+\infty$ is a nonattracting boundary for $\alpha < 0$ and attracting for $\alpha > 0$, Equations (47) and (48) follow from (12) making use of (37) and (42). □

From (47) and (48), for $\alpha \neq 0$, $\beta \in \mathbb{R}$ and $\sigma > 0$ one obtains $\lim_{x\downarrow 0} L_Y(x|x_0) = 0$ and $\lim_{x\uparrow +\infty} L_Y(x|x_0) = 0$.

In Figure 2, the asymptotic average of the local time for the TH-OU process $Y(t)$ is plotted for $x_0 = 4, \sigma = 1$ and some choices of α and β.

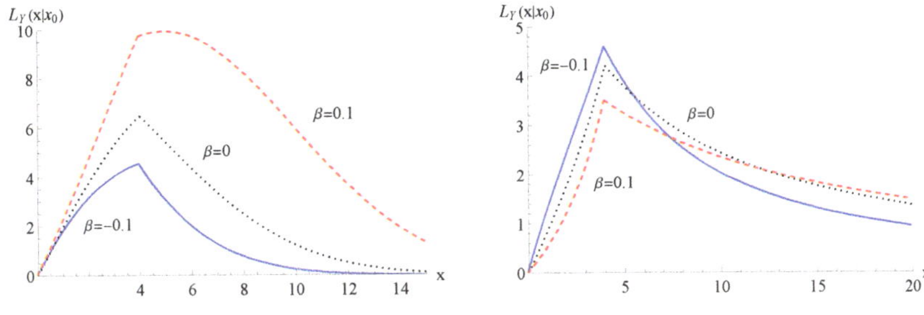

Figure 2. $L_Y(x|x_0)$, given in Proposition 7, with $x_0 = 4, \sigma = 1$, and some choices of β. In (**a**) $\alpha = -0.02$ and in (**b**) $\alpha = 0.02$.

4. Feller-Type Diffusion Process

Let $\{Z(t), t \geq t_0\}, t_0 \geq 0$ be a TNH-F process having infinitesimal drift and infinitesimal variance

$$C_1(x,t) = \alpha(t)\, x + \beta(t), \qquad C_2(x,t) = 2\, r(t)\, x, \tag{49}$$

with state space $(0, +\infty)$, where $\alpha(t) \in \mathbb{R}, \beta(t) \in \mathbb{R}, r(t) > 0$ continuous functions.

We point out that the processes (27) and (49) have identical infinitesimal drifts; instead, the infinitesimal variances are different in terms of the involved noise intensity functions.

The TNH-F process is used to describe the growth of a population (cf. Feller [37], Giorno and Nobile [38]) and the number of customers in queueing models (cf. Di Crescenzo and Nobile [39]). This process is also applied in mathematical finance to study stochastic volatility and interest rates (see Tian and Zhang [40], Cox et al. [41], Di Nardo and D'Onofrio [42]) and in neurobiology to model the input–output behavior of single neurons (see Ditlevsen and Lánský [43], D'Onofrio et al. [44]).

We consider the TNH-F process $Z(t)$, having infinitesimal moments (49), with an absorbing condition placed in the zero state and we denote with $a_Z(x,t|x_0,t_0)$ its PDF. We assume that $\alpha(t) \in \mathbb{R}, \beta(t) \in \mathbb{R}, r(t) > 0, \beta(t) \leq \xi\, r(t)$, with $0 \leq \xi < 1$. For the TNH-F process $Z(t)$ with an absorbing boundary in zero, we consider two cases: the proportional case in which $\beta(t) = \xi\, r(t)$, with $0 \leq \xi < 1$ and $r(t) > 0$, and the time-homogeneous case.

4.1. Proportional Case for the Feller Process

We assume that $\alpha(t) \in \mathbb{R}$, $r(t) > 0$ and $\beta(t) = \xi\, r(t)$, with $0 \leq \xi < 1$, in (49). As proven in Giorno and Nobile [45] one has

$$a_Z(x,t|x_0,t_0) = \begin{cases} \dfrac{e^{-A(t|t_0)}}{\Gamma(2-\xi)} \left[\dfrac{1}{R(t|t_0)}\right]^{2-\xi} x_0^{1-\xi} \exp\Big\{-\dfrac{x_0}{R(t|t_0)}\Big\}, & x = 0, \\[2ex] \dfrac{e^{-A(t|t_0)}}{R(t|t_0)} \left(\dfrac{x_0}{x}\right)^{(1-\xi)/2} \exp\Big\{-\dfrac{x_0 + x\, e^{-A(t|t_0)}}{R(t|t_0)}\Big\} & \\ \qquad \times \exp\Big\{\dfrac{1-\xi}{2} A(t|t_0)\Big\}\, I_{1-\xi}\left[\dfrac{2\sqrt{x\, x_0\, e^{-A(t|t_0)}}}{R(t|t_0)}\right], & x > 0, \end{cases} \tag{50}$$

with $A(t|t_0)$ given in (30), $I_\nu(z)$ defined in (24) and

$$R(t|t_0) = \int_{t_0}^{t} r(\theta)\, e^{-A(\theta|t_0)}\, d\theta. \tag{51}$$

Proposition 8. *Let $\alpha(t) \in \mathbb{R}$, $r(t) > 0$ and $\beta(t) = \xi\, r(t)$, with $0 \le \xi < 1$, in (49). For the TNH-F process $Z(t)$ one has*

$$g_Z(0,t|x_0,t_0) = \frac{1}{\Gamma(1-\xi)}\,\frac{r(t)\,e^{-A(t|t_0)}}{R(t|t_0)}\,\Big[\frac{x_0}{R(t|t_0)}\Big]^{1-\xi}\exp\Big\{-\frac{x_0}{R(t|t_0)}\Big\}, \qquad x_0 > 0, \quad (52)$$

with $R(t|t_0)$ given in (51). Moreover, it results in

$$P_Z(0|x_0,t_0) = \int_{t_0}^{+\infty} g_Z(0,t|x_0,t_0)\,dt = \begin{cases} 1, & \lim_{t\to+\infty} R(t|t_0) = +\infty, \\ \dfrac{\Gamma\big(1-\xi, \frac{x_0}{c}\big)}{\Gamma\big(1-\xi\big)}, & \lim_{t\to+\infty} R(t|t_0) = c. \end{cases} \quad (53)$$

Proof. From (50), one has (cf. Erdèlyi et al. [36], p. 197, no. 19)

$$\int_0^{+\infty} a_Z(x,t|x_0,t_0)\,dx = \frac{1}{\Gamma(1-\xi)}\,\gamma\Big(1-\xi, \frac{x_0}{R(t|t_0)}\Big), \qquad 0 \le \xi < 1, \quad (54)$$

where $\Gamma(\nu)$ is the Euler gamma function and $\gamma(\nu,z) = \int_0^z y^{\nu-1} e^{-y}\,dy$, with $\nu > 0$, is the incomplete gamma function. Hence, due to (5) and recalling (54), Equation (52) follows. Finally, Equation (53) is obtained, making use of (54) in (6). □

We point out that the general TNH-F process with an absorbing boundary in zero is considered in Giorno and Nobile [45], Masoliver and Perelló [46], Masoliver [47] and Lavigne and Roques [48].

4.2. Time-Homogeneous Case for the Feller Process

Let $Z(t)$ be the TH-F process, obtained by setting $\alpha(t) = \alpha$, $\beta(t) = \beta$ and $r(t) = r$ in (49). From (9), the scale function and the speed density of the TH-F process $Z(t)$ are

$$h_Z(x) = x^{-\beta/r}\exp\Big\{-\frac{\alpha x}{r}\Big\}, \qquad s_Z(x) = \frac{x^{\beta/r-1}}{r}\exp\Big\{\frac{\alpha x}{r}\Big\}, \quad (55)$$

respectively. As proven by Feller, the state 0 is an exit boundary for $\beta \le 0$, regular for $0 < \beta < r$ and entrance for $\beta \ge 0$. Furthermore, the end point $+\infty$ is a nonattracting natural boundary for $\alpha \le 0$ and an attracting natural boundary for $\alpha > 0$. In the sequel, we assume that $\alpha \in \mathbb{R}$, $\beta \in \mathbb{R}$, $r > 0$, with $\beta < r$, and an absorbing condition is set in the zero-state.

As proven in Giorno and Nobile [45], for a TH-F process $Z(t)$ having $\beta \in \mathbb{R}$, $r > 0$, with $\beta < r$, one has

- If $\alpha = 0$ one has

$$a_Z(x,t|x_0) = \begin{cases} \frac{1}{\Gamma(2-\beta/r)}\Big(\frac{1}{rt}\Big)^{2-\beta/r} x_0^{1-\beta/r}\exp\Big\{-\frac{x_0}{rt}\Big\}, & x = 0, \\ \frac{1}{rt}\Big(\frac{x_0}{x}\Big)^{(1-\beta/r)/2}\exp\Big\{-\frac{x+x_0}{rt}\Big\} I_{1-\beta/r}\Big[\frac{2\sqrt{x\,x_0}}{rt}\Big], & x > 0. \end{cases} \quad (56)$$

- If $\alpha \neq 0$ one obtains

$$a_Z(x,t|x_0) = \begin{cases} \frac{e^{-\alpha t}}{\Gamma(2-\beta/r)}\Big[\frac{\alpha\,e^{\alpha t}}{r(e^{\alpha t}-1)}\Big]^{2-\beta/r} x_0^{1-\beta/r}\exp\Big\{-\frac{\alpha\,x_0\,e^{\alpha t}}{r(e^{\alpha t}-1)}\Big\}, & x = 0, \\ \frac{\alpha\,e^{\alpha(1-\beta/r)t/2}}{r(e^{\alpha t}-1)}\Big(\frac{x_0}{x}\Big)^{(1-\beta/r)/2}\exp\Big\{-\frac{\alpha(x+x_0\,e^{\alpha t})}{r(e^{\alpha t}-1)}\Big\} I_{1-\beta/r}\Big[\frac{2\alpha\sqrt{x\,x_0\,e^{\alpha t}}}{r(e^{\alpha t}-1)}\Big], & x > 0. \end{cases} \quad (57)$$

Proposition 9. *Let $\alpha \in \mathbb{R}$, $\beta \in \mathbb{R}$, $r > 0$, with $\beta < r$. For the TH-F process $Z(t)$, with $x_0 > 0$, one has*

$$g_Z(0,t|x_0) = \begin{cases} \frac{1}{t\Gamma(1-\beta/r)} \left(\frac{x_0}{rt}\right)^{1-\beta/r} \exp\left\{-\frac{x_0}{rt}\right\}, & \alpha = 0, \\ \frac{1}{\Gamma(1-\beta/r)} \frac{\alpha}{e^{\alpha t}-1} \left[\frac{\alpha\, x_0\, e^{\alpha t}}{r(e^{\alpha t}-1)}\right]^{1-\beta/r} \exp\left\{-\frac{\alpha\, x_0\, e^{\alpha t}}{r(e^{\alpha t}-1)}\right\}, & \alpha \neq 0 \end{cases} \tag{58}$$

and

$$P_Z(0|x_0) = \int_0^{+\infty} g_Z(0,t|x_0)\, dt = \begin{cases} 1, & \alpha \leq 0, \\ \frac{\Gamma\left(1-\frac{\beta}{r}, \frac{\alpha x_0}{r}\right)}{\Gamma\left(1-\beta/r\right)}, & \alpha > 0. \end{cases} \tag{59}$$

Proof. From (56) and (57), one obtains (cf. Erdèlyi et al. [36], p. 197, no. 19)

$$\int_0^{+\infty} a_Z(x,t|x_0)\, dx = \begin{cases} \frac{1}{\Gamma(1-\beta/r)} \gamma\left(1-\frac{\beta}{r}, \frac{x_0}{rt}\right), & \alpha = 0, \\ \frac{1}{\Gamma(1-\beta/r)} \gamma\left(1-\frac{\beta}{r}, \frac{\alpha\, x_0\, e^{\alpha t}}{r(e^{\alpha t}-1)}\right), & \alpha \neq 0. \end{cases} \tag{60}$$

Making use of (60) in (5), Equation (58) follows. Finally, by virtue of (6) and (60), we obtain the FPT probability (59). □

By applying the Siegert Formula (11) with $r_2 = +\infty$ and recalling (55), for $\alpha = 0$ and $\beta < r$ one has that the FPT mean $t_1^{(Z)}(0|x_0)$ diverges, whereas for $\alpha < 0$ and $\beta < r$ one obtains

$$t_1^{(Z)}(S|x_0) = \frac{1}{|\alpha|} \int_{|\alpha|S/r}^{|\alpha|x_0/r} x^{-\beta/r} e^x\, \Gamma\left(\frac{\beta}{r}, x\right) dx, \qquad x_0 > S \geq 0.$$

In Table 5, the mean $t_1^{(Z)}(S|x_0)$, the variance $\mathrm{Var}^{(Z)}(S|x_0)$, the coefficient of variation $\mathrm{Cv}^{(Z)}(S|x_0)$, and the skewness $\Sigma^{(Z)}(S|x_0)$ of the FPT, obtained by using the Siegert Formula (11), are listed for $x_0 = 4$ and some choices of S, with $\alpha = -0.02$, $\beta = -0.1, 0, 0.1$ and $r = 0.5$.

As shown in Table 5, for the TH-F process $Z(t)$ the mean and the variance of the FPT increases as S decreases; instead, the coefficient of variation and the skewness decrease as S decreases.

Proposition 10. *Let $Z(t)$ be a TH-F process having $\beta \in \mathbb{R}$, $r > 0$, with $\beta < r$.*

- *If $\alpha \leq 0$, for $x_0 > 0$ and $x > 0$ one has*

$$\begin{aligned} L_Z(x|x_0) &= \int_0^{+\infty} a_Z(x,t|x_0)\, dt \\ &= \begin{cases} \frac{1}{r} \left(\frac{r}{|\alpha| x}\right)^{1-\beta/r} e^{-|\alpha| x/r} \int_0^{|\alpha|(x_0 \wedge x)/r} y^{-\beta/r} e^y\, dy, & \alpha < 0 \\ \frac{1}{r} \frac{1}{1-\beta/r} \left(\frac{x_0 \wedge x}{x}\right)^{1-\beta/r}, & \alpha = 0. \end{cases} \end{aligned} \tag{61}$$

- *If $\alpha > 0$, for $x_0 > 0$ and $x > 0$ one obtains*

$$\begin{aligned} L_Z(x|x_0) &= \int_0^{+\infty} a_Z(x,t|x_0)\, dt \\ &= \frac{1}{r} \left(\frac{r}{\alpha x}\right)^{1-\beta/r} e^{\alpha x/r} \gamma\left(1-\frac{\beta}{r}, \frac{\alpha(x_0 \wedge x)}{r}\right) \frac{\Gamma\left(1-\frac{\beta}{r}, \frac{\alpha(x_0 \vee x)}{r}\right)}{\Gamma\left(1-\frac{\beta}{r}\right)}. \end{aligned} \tag{62}$$

Proof. Because $+\infty$ is a nonattracting boundary for $\alpha \leq 0$ and attracting for $\alpha > 0$, Equations (61) and (62) follow from (12), making use of (55) and (59). □

Table 5. For the TH-F process, with $C_1(x) = -0.02\,x + \beta$ and $C_2(x) = x$, the mean, the variance, the coefficient of variation, and the skewness of FPT are listed for $x_0 = 4$, $\beta = -0.1, 0, 0.1$ and for decreasing values the boundary $S \in [0, x_0)$.

	S	$t_1^{(Z)}(S\|x_0)$	$\mathrm{Var}^{(Z)}(S\|x_0)$	$\mathrm{Cv}^{(Z)}(S\|x_0)$	$\Sigma^{(Z)}(S\|x_0)$
$\beta = -0.1$	3.5	1.392620	43.55136	4.738799	13.55745
	3.0	2.859008	89.01922	3.300096	9.501619
	2.5	4.412074	136.5521	2.648539	7.684201
	2.0	6.069704	186.2977	2.248724	6.585614
	1.5	7.858374	238.3733	1.964699	5.823477
	1.0	9.821571	292.7690	1.742134	5.249299
	0.5	12.04587	348.9465	1.550747	4.792382
	0.0	14.86611	401.3413	1.347596	4.429951
$\beta = 0$	3.5	1.702121	69.83433	4.909578	13.0398
	3.0	3.512494	144.3878	3.420973	9.119134
	2.5	5.452850	224.4206	2.747311	7.356073
	2.0	7.554206	310.9409	2.334265	6.285290
	1.5	9.864220	405.3553	2.041057	5.536382
	1.0	12.46511	509.782	1.811324	4.964136
	0.5	15.53300	627.8781	1.613178	4.494585
	0.0	19.91651	768.9171	1.392280	4.068225
$\beta = 0.1$	3.5	2.126455	115.1039	5.045321	12.41484
	3.0	4.415487	240.9926	3.515793	8.659792
	2.5	6.904328	380.0698	2.823646	6.964637
	2.0	9.647790	535.8021	2.399244	5.929435
	1.5	12.73337	713.5752	2.097861	5.199456
	1.0	16.31983	922.7997	1.861394	4.631900
	0.5	20.77405	1184.008	1.656366	4.155107
	0.0	28.39302	1607.070	1.411906	3.602509

From (61) and (62), for $\alpha \in \mathbb{R}$, $\beta \in \mathbb{R}$, $r > 0$, with $\beta < r$, one has $\lim_{x\uparrow+\infty} L_Z(x|x_0) = 0$ and

$$\lim_{x\downarrow 0} L_Z(x|x_0) = \begin{cases} \dfrac{1}{r}\,\dfrac{1}{1-\beta/r} & \alpha \leq 0, \\[2ex] \dfrac{1}{r}\,\dfrac{1}{1-\beta/r}\,\dfrac{\Gamma\left(1-\frac{\beta}{r}, \frac{\alpha\, x_0}{r}\right)}{\Gamma\left(1-\frac{\beta}{r}\right)}, & \alpha > 0. \end{cases}$$

Therefore, for the TH-F process the asymptotic average of local time tend to zero as x increases, whereas it is positive for $x \downarrow 0$.

In Figure 3, the asymptotic average of the local time for the TH-F process $Z(t)$ is plotted for $x_0 = 4$, $r = 0.5$ and some choices of α and β.

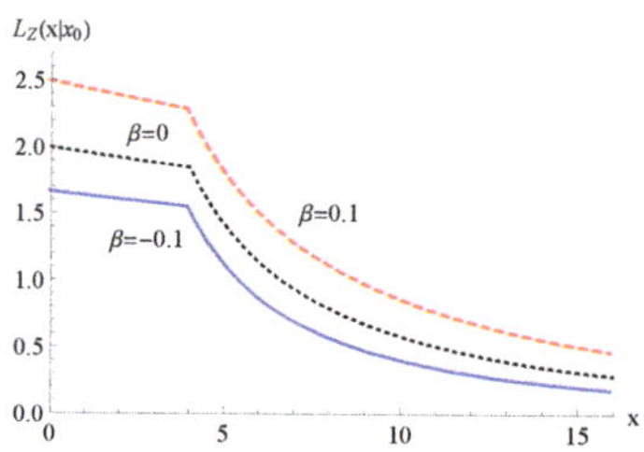

(a)

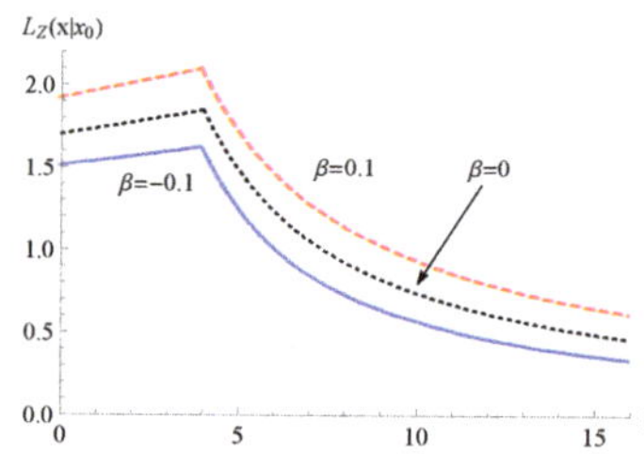

(b)

Figure 3. *Cont.*

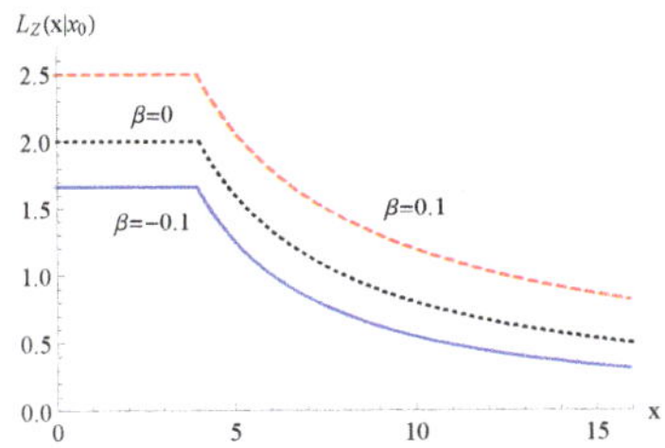

(c)

Figure 3. $L_Z(x|x_0)$, given in Proposition 12, with $x_0 = 4$, $r = 0.5$, and some choices of β. In **(a)** $\alpha = -0.02$, in **(b)** $\alpha = 0.02$ and in **(c)** $\alpha = 0$.

5. Relationships and Asymptotic Results

In this section, for $\beta(t) = r(t)/2$ some relationships between the PDF in the presence of an absorbing boundary in the zero state and between the FPT densities through zero for Wiener, Ornstein–Uhlenbeck and Feller processes are proven; moreover, for $\beta(t) = \xi\, r(t)$ $(0 < \xi < 1)$ some asymptotic results for large times between the FPT densities are provided.

5.1. Relations between the Transition Densities with an Absorbing Boundary in the Zero State

We consider the TNH-F process (49) with $\beta(t) = r(t)/2$ in the presence of an absorbing boundary in the zero state, and we show that its PDF can be related to the PDF of the Wiener and of the Ornstein–Uhlenbeck processes with an absorbing boundary in the zero state.

Proposition 11. *Let $Z(t)$ be a TNH-F process with $C_1(t) = r(t)/2$ and $C_2(x,t) = 2\,r(t)\,x$, where $r(t) > 0$, and let $X(t)$ be a TNH-W process with $A_1 = 0$ and $A_2(t) = r(t)/2$. One has*

$$a_Z(x,t|x_0,t_0) = \frac{1}{2\sqrt{x}}\, a_X(\sqrt{x},t|\sqrt{x_0},t_0), \qquad x_0 > 0, x > 0, \tag{63}$$

$$g_Z(0,t|x_0,t_0) = g_X(0,t|\sqrt{x_0},t_0), \qquad x_0 > 0. \tag{64}$$

Proof. For the TNH-F process $Z(t)$, by setting $\alpha(t) = 0$ and $\beta(t) = r(t)/2$ in (50) and in Proposition 9, recalling that

$$I_{1/2}(x) = \sqrt{\frac{2}{\pi}}\,\frac{\sinh(x)}{\sqrt{x}}, \qquad \gamma\Big(\frac{1}{2},x\Big) = \sqrt{\pi}\,\mathrm{Erf}(\sqrt{x}), \tag{65}$$

one has

$$a_Z(x,t|x_0,t_0) = \begin{cases} 2\sqrt{\frac{x_0}{\pi\,[\widetilde{R}(t|t_0)]^3}}\,\exp\Big\{-\frac{x_0}{\widetilde{R}(t|t_0)}\Big\}, & x = 0, \\ \frac{1}{2\sqrt{\pi\,\widetilde{R}(t|t_0)\,x}}\left[\exp\Big\{-\frac{(\sqrt{x}-\sqrt{x_0})^2}{\widetilde{R}(t|t_0)}\Big\} - \exp\Big\{-\frac{(\sqrt{x}+\sqrt{x_0})^2}{\widetilde{R}(t|t_0)}\Big\}\right], & x > 0, \end{cases} \tag{66}$$

and

$$g_Z(0,t|x_0,t_0) = r(t)\sqrt{\frac{x_0}{\pi\,[\widetilde{R}(t|t_0)]^3}}\,\exp\Big\{-\frac{x_0}{\widetilde{R}(t|t_0)}\Big\}, \qquad x_0 > 0, \tag{67}$$

where $\widetilde{R}(t|t_0) = \int_{t_0}^{t} r(\theta)\,d\theta$. Furthermore, for the TNH-W process $X(t)$ with $\beta(t) = 0$ and $\sigma^2(t) = r(t)/2$, one has $V_X(t|t_0) = \widetilde{R}(t|t_0)/2$. Then, (63) and (64) follow by comparing (66) and (67) with (17) and (18), respectively. □

Under the assumptions of the Proposition 13, one has $L_Z(x|x_0,t_0) = L_X(\sqrt{x}\,|\sqrt{x_0},t_0)/(2\sqrt{x})$ for $x > 0, x_0 > 0$ and, if $\lim_{t\to+\infty}\widetilde{R}(t|t_0) = +\infty$, one obtains $t_n^{(Z)}(0|x_0,t_0) = t_n^{(X)}(0|\sqrt{x_0},t_0)$ for $n = 1,2,\ldots$ with $x_0 > 0$.

Proposition 12. *Let $Z(t)$ be a TNH-F process with $C_1(x,t) = \alpha(t)\,x + r(t)/2$ and $C_2(x,t) = 2\,r(t)\,x$, where $\alpha(t)$ is not always zero and $r(t) > 0$, and let $Y(t)$ be a TNH-OU process with $B_1(x,t) = \alpha(t)\,x/2$ and $B_2(t) = r(t)/2$. One has*

$$a_Z(x,t|x_0,t_0) = \frac{1}{2\sqrt{x}}\,a_Y(\sqrt{x},t|\sqrt{x_0},t_0), \qquad x_0 > 0, x > 0, \tag{68}$$

$$g_Z(0,t|x_0,t_0) = g_Y(0,t|\sqrt{x_0},t_0), \qquad x_0 > 0. \tag{69}$$

Proof. For the TNH-F process $Z(t)$, by setting $\beta(t) = r(t)/2$ in (50) and in Proposition 9, recalling (65), one obtains

$$a_Z(x,t|x_0,t_0) = \begin{cases} 2\,e^{-A(t|t_0)}\sqrt{\frac{x_0}{\pi\,[R(t|t_0)]^3}}\exp\Big\{-\frac{x_0}{R(t|t_0)}\Big\}, & x = 0,\\ \frac{e^{-A(t|t_0)/2}}{2\sqrt{\pi\,R(t|t_0)x}}\Big[\exp\Big\{-\frac{\big(\sqrt{x\,e^{-A(t|t_0)}}-\sqrt{x_0}\big)^2}{R(t|t_0)}\Big\} \\ \qquad -\exp\Big\{-\frac{\big(\sqrt{x\,e^{-A(t|t_0)}}+\sqrt{x_0}\big)^2}{R(t|t_0)}\Big\}\Big], & x > 0, \end{cases} \tag{70}$$

and

$$g_Z(0,t|x_0,t_0) = r(t)\,e^{-A(t|t_0)}\sqrt{\frac{x_0}{\pi\,[R(t|t_0)]^3}}\exp\Big\{-\frac{x_0}{R(t|t_0)}\Big\}, \qquad x_0 > 0, \tag{71}$$

with $A(t|t_0)$ and $R(t|t_0)$ given in (30) and (51), respectively. Moreover, in the TNH-OU process $Y(t)$ we set $\beta(t) = 0$, $\sigma^2(t) = r(t)/2$ and we change $\alpha(t)$ into $\alpha(t)/2$, so that, by virtue of (29) and (51), one has $V_Y(t|t_0) = R(t|t_0)e^{A(t|t_0)}/2$. Then, (68) and (69) follow by comparing (70) and (71) with (33) and (34), respectively. □

Under the assumptions of Proposition 14, one has $L_Z(x|x_0,t_0) = L_Y(\sqrt{x}\,|\sqrt{x_0},t_0)/(2\sqrt{x})$ for $x > 0, x_0 > 0$ and, if $\lim_{t\to+\infty} R(t|t_0) = +\infty$, one obtains $t_n^{(Z)}(0|x_0,t_0) = t_n^{(Y)}(0|\sqrt{x_0},t_0)$ for $n = 1,2,\ldots$ with $x_0 > 0$.

5.2. Asymptotic Behaviors between the FPT Densities

In this section, for $\beta(t) = \xi\,r(t)$ $(0 < \xi < 1)$ some asymptotic results for large times between the FPT densities of TNH-W, TNH-OU and TNH-F processes are shown.

Proposition 13. *Let $Z(t)$ be a TNH-F process with $C_1(t) = \xi\,r(t)$ and $C_2(x,t) = 2\,r(t)\,x$, where $r(t) > 0$, $0 < \xi < 1$, and let $X(t)$ be a TNH-W process with $A_1 = 0$ and $A_2(t) = \xi\,r(t)$. If $\lim_{t\to+\infty}\widetilde{R}(t|t_0) = +\infty$, and one has*

$$\lim_{t\to+\infty}\frac{g_Z(0,t|x_0,t_0)\,[\widetilde{R}(t|t_0)]^{1/2-\xi}}{g_X(0,t|x_0^{1-\xi},t_0)} = \frac{\sqrt{2\,\pi\,\xi}}{\Gamma(1-\xi)}, \qquad x_0 > 0. \tag{72}$$

Proof. Recalling (18) and (67) and noting that $V_X(t|t_0) = \xi\,\widetilde{R}(t|t_0)$, one has

$$\frac{g_Z(0,t|x_0,t_0)\,[\widetilde{R}(t|t_0)]^{1/2-\xi}}{g_X(0,t|x_0^{1-\xi},t_0)} = \frac{\sqrt{2\,\pi\,\xi}}{\Gamma(1-\xi)}\exp\Big\{-\frac{x_0}{\widetilde{R}(t|t_0)} + \frac{x_0^{2(1-\xi)}}{2\,\xi\,\widetilde{R}(t|t_0)}\Big\},$$

from which, under the assumption $\lim_{t\to+\infty}\widetilde{R}(t|t_0) = +\infty$, Equation (72) follows. □

Proposition 14. *Let $Z(t)$ be a TNH-F process having $C_1(x,t) = \alpha(t)\,x + \xi\,r(t)$ and $C_2(x,t) = 2\,r(t)\,x$, with $\alpha(t)$ not always zero, $r(t) > 0$, $0 < \xi < 1$, and let $Y(t)$ be a TNH-OU process with $B_1(x,t) = \alpha(t)\,x/2$ and $B_2(t) = \xi\,r(t)$. If $\lim_{t\to+\infty} R(t|t_0) = +\infty$, and one has*

$$\lim_{t\to+\infty}\frac{g_Z(0,t|x_0,t_0)\,[R(t|t_0)]^{1/2-\xi}}{g_Y(0,t|x_0^{1-\xi},t_0)} = \frac{\sqrt{2\,\pi\,\xi}}{\Gamma(1-\xi)}, \qquad x_0 > 0. \tag{73}$$

Proof. Making use of (34) and (52) and noting that $V_Y(t|t_0) = \xi\, R(t|t_0)\, e^{A(t|t_0)}$, one obtains

$$\frac{g_Z(0,t|x_0,t_0)\,[R(t|t_0)]^{1/2-\xi}}{g_Y(0,t|x_0^{1-\xi},t_0)} = \frac{\sqrt{2\pi\xi}}{\Gamma(1-\xi)}\,\exp\Big\{-\frac{x_0}{R(t|t_0)} + \frac{x_0^{2(1-\xi)}}{2\,\xi\,R(t|t_0)}\Big\},$$

from which, recalling that $\lim_{t\to+\infty} R(t|t_0) = +\infty$, Equation (73) follows. □

6. Conclusions

For the Wiener, Ornstein–Uhlenbeck, and Feller processes, we analyze the transition densities in the presence of an absorbing boundary in the zero state and the FPT problem to the zero state. Particular attention is dedicated to the proportional cases and to the time-homogeneous cases, by achieving the FPT densities through the zero state. Extensive computation are performed with MATHEMATICA to obtain the mean, the variance, the coefficient of variation and the skewness of FPT for TH-W, TH-OU and TH-F processes. Moreover, for these processes, a detailed study of the asymptotic average of local time with an absorbing boundary in the zero-state is carried out.

In Table 6, a summary containing the conditions and the most important equations numbering in Sections 2–4 concerning the absorbing problem for Wiener, Ornstein–Uhlenbeck and Feller diffusion processes is given.

Table 6. Summary containing conditions and the most important equations numbering in Sections 2–4 for Wiener, Ornstein–Uhlenbeck and Feller diffusion processes.

	Conditions	Results—Equations Numbering
Wiener process $A_1(t) = \beta(t)$ $A_2(t) = \sigma^2(t)$ $(\beta(t) \in \mathbb{R}, \sigma(t) > 0)$	$\beta(t) = \gamma\,\sigma^2(t)$ $(\gamma \in \mathbb{R}, \sigma(t) > 0)$	$a_X(x,t|x_0,t_0)$ – (17) $g_X(0,t|x_0,t_0)$ – (18) $P_X(0|x_0,t_0)$ – (19)
	$\beta(t) = \beta, \sigma^2(t) = \sigma^2$ $(\beta \in \mathbb{R}, \sigma > 0)$	$g_X(S,t|x_0)$ – (22) $P_X(S|x_0)$ – (23) $a_X(x,t|x_0)$ – (25) $L_X(x|x_0)$ – (26)
Ornstein–Uhlenbeck process $B_1(x,t) = \alpha(t)\,x + \beta(t)$ $B_2(t) = \sigma^2(t)$ $(\alpha(t) \in \mathbb{R}, \beta(t) \in \mathbb{R}, \sigma(t) > 0)$	$\beta(t) = \gamma\,\sigma^2(t)\,e^{-A(t|0)}$ $(\gamma \in \mathbb{R}, \alpha(t) \in \mathbb{R}, \sigma(t) > 0)$	$a_Y(x,t|x_0,t_0)$ – (33) $g_Y(0,t|x_0,t_0)$ – (34) $P_Y(0|x_0,t_0)$ – (35)
	$\alpha(t) = \alpha, \beta(t) = \beta, \sigma^2(t) = \sigma^2$ $(\alpha \neq 0, \beta \in \mathbb{R}, \sigma > 0)$	$g_\lambda^{(Y)}(S|x_0)$ – (41) $P_Y(S|x_0)$ – (42) $a_\lambda^{(Y)}(x|x_0)$ – (46) $L_Y(x|x_0)$ – (47), (48)
Feller process $C_1(x,t) = \alpha(t)\,x + \beta(t)$ $C_2(x,t) = 2\,r(t)\,x$ $(\alpha(t) \in \mathbb{R}, \beta(t) \in \mathbb{R}, r(t) > 0,$ $(\beta(t) \leq \xi\,r(t), 0 \leq \xi < 1)$	$\beta(t) = \xi\,r(t)$ $(0 \leq \xi < 1, r(t) > 0)$	$a_Z(x,t|x_0,t_0)$ – (50) $g_Z(0,t|x_0,t_0)$ – (52) $P_Z(0|x_0,t_0)$ – (53)
	$\alpha(t) = \alpha, \beta(t) = \beta, r(t) = r$ $(\alpha \in \mathbb{R}, \beta \in \mathbb{R}, r > 0, \beta < r)$	$a_Z(x,t|x_0)$ – (56), (57) $g_Z(0,t|x_0)$ – (58) $P_Z(0|x_0)$ – (59) $L_Z(x|x_0)$ – (61), (62)

As shown in Table 6, by setting $\beta(t) = 0$ in TNH-W, TNH-OU and TNH-F processes, the PDF in the presence of an absorbing boundary in the zero state and the FPT density through zero are given in closed form. Moreover, in TH-W, TH-F processes, the previous densities are obtainable, whereas for the TH-OU process only the LT is available.

The knowledge of the PDF in the presence of an absorbing boundary in the zero state is of interest in the context of biological systems because it allows us to evaluate the survival probabilities (20), (36) and (54) for Wiener, Ornstein–Uhlenbeck and Feller processes, respectively. Moreover, such PDF allows one to get information on the FPT density through zero (extinction density) (18), (34), and (52) and on the probability of extinction (19), (35) and (53) of the considered processes. Furthermore, the asymptotic average of the local time for TH-W, TH-OU, and TH-F processes provides information on the average of the sojourn time in the various states before the absorption occurs in the zero state.

The results of Section 5 show that the same FPT density through the zero-state (extinction density) may correspond to different diffusion processes with modified initial states.

Author Contributions: Conceptualization, V.G. and A.G.N.; methodology, V.G. and A.G.N.; validation, V.G. and A.G.N.; formal analysis, V.G. and A.G.N.; investigation, V.G. and A.G.N.; supervision, V.G. and A.G.N. All authors have read and agreed to the published version of the manuscript.

Funding: This research is partially supported by MIUR - PRIN 2017, Project "Stochastic Models for Complex Systems".

Institutional Review Board Statement: Not applicable.

Informed Consent Statement: Not applicable.

Data Availability Statement: Not applicable.

Acknowledgments: The authors are members of the research group GNCS of INdAM.

Conflicts of Interest: The authors declare no conflict of interest.

Abbreviations

The following abbreviations are used in this manuscript:

PDF	Transition Probability Density Function
FPT	First Passage Time
TNH-D	Time Inhomogeneous Diffusion
TNH-W	Time Inhomogeneous Wiener
TNH-OU	Time Inhomogeneous Ornstein-Uhlenbeck
TNH-F	Time Inhomogeneous Feller
TH-D	Time Homogeneous Diffusion
TH-W	Time Homogeneous Wiener
TH-OU	Time Homogeneous Ornstein-Uhlenbeck
TH-F	Time Homogeneous Feller

References

1. Bailey, N.T.J. *The Elements of Stochastic Processes with Applications to the Natural Sciences*; John Wiley & Sons, Inc.: New York, NY, USA, 1964.
2. Ricciardi, L.M. *Diffusion Processes and Related Topics in Biology*; Springer: Berlin/Heidelberg, Germany, 1977.
3. Gardiner, C.W. *Handbook of Stochastic Methods: For Physics, Chemistry and the Natural Sciences*; Springer: Berlin/Heidelberg, Germany, 1985.
4. Stirzaker, D. *Stochastic Processes and Models*; Oxford University Press: Oxford, UK, 2005.
5. Janssen, J.; Manca, O.; Manca, R. *Applied Diffusion Processes from Engineering to Finance*; Wiley-ISTE: London, UK, 2013.
6. Pavliotis, G.A. *Stochastic Processes and Applications. Diffusion Processes, the Fokker-Planck and Langevin Equations*; Springer: New York, NY, USA, 2014.
7. Dynkin, E.B. Kolmogorov and the theory of Markov processes. *Ann. Probab.* **1989**, *17*, 822–832. [CrossRef]
8. Cox, D.R.; Miller, H.D. *The Theory of Stochastic Processes*; Chapman & Hall/CRC: Boca Raton, FL, USA, 1996.
9. Allen, L.J.S. *An Introduction to Stochastic Processes with Applications to Biology*, 2nd ed.; Chapman and Hall/CRC: Lubbock, TX, USA, 2010.
10. Allen, L.J.S. *Stochastic Population and Epidemic Models Persistence and Extinction*; Springer: Cham, Switzerland, 2015.
11. Blake, I.F.; Lindsey, W.C. Level-Crossing Problems for Random Processes. *IEEE Trans. Inf. Theory* **1973**, *19*, 295–315. [CrossRef]
12. Ricciardi, L.M.; Di Crescenzo, A.; Giorno, V.; Nobile, A.G. An outline of theoretical and algorithmic approaches to first passage time problems with applications to biological modeling. *Math. Jpn.* **1999**, *50*, 247–322.
13. Ding, M.; Rangarajan, G. First Passage Time Problem: A Fokker-Planck Approach. In *New Directions in Statistical Physics*; Wille, L.T., Ed.; Springer: Berlin/Heidelberg, Germany, 2004; pp. 31–46.
14. Molini, A.; Talkner, P.; Katul, G.G.; Porporato, A. First passage time statistics of Brownian motion with purely time dependent drift and diffusion. *Physica A* **2011**, *390*, 1841–1852. [CrossRef]
15. Giorno, V.; Nobile, A.G. On the first-passage time problem for a Feller-type diffusion process. *Mathematics* **2021**, *9*, 2470. [CrossRef]
16. Masoliver, J. *Random Processes: First-Passage and Escape*; World Scientific: Singapore, 2018.

7. Linetsky, V. Computing hitting time densities for CIR and OU diffusions. Applications to mean-reverting models. *J. Comput. Finance* **2004**, *7*, 1–22. [CrossRef]
8. Guo, G.; Chen, B.; Zhao, X.; Zhao, F.; Wang, Q. First passage time distribution of a modified fractional diffusion equation in the semi-infinite interval. *Physica A* **2015**, *433*, 279–290. [CrossRef]
9. Wiese, K.J. First passage in an interval for fractional Brownian motion. *Phys. Rev. E* **2019**, *99*, 032106. [CrossRef]
0. Abundo, M. An inverse first-passage problem revisited: The case of fractional Brownian motion, and time-changed Brownian motion. *Stoch. Anal. Appl.* **2019**, *37*, 708–716. [CrossRef]
1. Leonenko, N.; Pirozzi, E. First passage times for some classes of fractional time-changed diffusions. *Stoch. Anal. Appl.* **2022**, *40*, 735–763. [CrossRef]
2. Siegert, A.J.F. On the first passage time probability problem. *Phys. Rev.* **1951**, *81*, 617–623. [CrossRef]
3. Karlin, S.; Taylor, H.W. *A Second Course in Stochastic Processes*; Academic Press: New York, NY, USA, 1981.
4. Aït-Sahalia, Y.; Park, J.Y. Stationarity-based specification tests for diffusions when the process is nonstationary. *J. Econom.* **2012**, *169*, 279–292. [CrossRef]
5. Feller, W. Two singular diffusion problems. *Ann. Math.* **1951**, *5*, 173–182. [CrossRef]
6. Feller, W. The parabolic differential equations and the associated semi-groups of transformations. *Ann. Math.* **1952**, *55*, 468–518. [CrossRef]
7. Giorno, V.; Nobile, A.G. Asymptotic average of the local time for one-dimensional diffusion processes. *Lect. Notes Semin. Interdiscip. Mat.* **2015**, *12*, 139–160.
8. Knight, F.B. *Essentials of Brownian Motion and Diffusions*; American Mathematical Society: Providence, RI, USA, 1981.
9. Abramowitz, I.A.; Stegun, M. *Handbook of Mathematical Functions*; Dover Publications: Mineola, NY, USA, 1972.
0. Uhlenbeck, G.E.; Ornstein, L.S. On the theory of the Brownian motion. *Phys. Rev.* **1930**, *36*, 823–841. [CrossRef]
1. Ricciardi, L.M.; Sacerdote, L. The Ornstein-Uhlenbeck process as a model for neuronal activity. I. Mean and variance of the firing time. *Biol. Cybern.* **1979**, *35*, 1–9. [CrossRef]
2. Lánský, P.; Ditlevsen, S. A review of the methods for signal estimation in stochastic diffusion leaky integrate-and-fire neuronal models. *Biol. Cybern.* **2008**, *99*, 253–262. [CrossRef]
33. Vasicek, O. An equilibrium characterization of the term structure. *J. Financ. Econ.* **1977**, *5*, 177–188. [CrossRef]
34. Hull, J.; White, A. Pricing interest-rate-derivative securities. *Rev. Financ. Stud.* **1990**, *3*, 573–592. [CrossRef]
35. Gradshteyn, I.S.; Ryzhik, I.M. *Table of Integrals, Series and Products*; Academic Press Inc.: New York, NY, USA, 2014.
36. Erdélyi, A.; Magnus, W.; Oberthettinger, F.; Tricomi, F.G. *Tables of Integral Transforms*; Mc Graw-Hill: New York, NY, USA, 1954; Volume 1.
37. Feller, W. Diffusion processes in genetics. In Proceedings of the Second Berkeley Symposium on Mathematical Statistics and Probability, Berkeley, CA, USA, 31 July–12 August 1950; pp. 227–246.
38. Giorno, V.; Nobile, A.G. Time-inhomogeneous Feller-type diffusion process in population dynamics. *Mathematics* **2021**, *9*, 1879. [CrossRef]
39. Di Crescenzo, A.; Nobile, A.G. Diffusion approximation to a queueing system with time-dependent arrival and service rates. *Queueing Syst.* **1995**, *19*, 41–62. [CrossRef]
40. Tian, Y.; Zhang, H. Skew CIR process, conditional characteristic function, moments and bond pricing. *Appl. Math. Comput.* **2018**, *329*, 230–238. [CrossRef]
41. Cox, J.C.; Ingersoll, J.E., Jr.; Ross, S.A. A theory of the term structure of interest rates. *Econometrica* **1985**, *53*, 385–407. [CrossRef]
42. Di Nardo, E.; D'Onofrio, G. A cumulant approach for the first-passage-time problem of the Feller square-root process. *Appl. Math. Comput.* **2021**, *391*, 125707. [CrossRef]
43. Ditlevsen, S.; Lánský, P. Estimation of the input parameters in the Feller neuronal model. *Phys. Rev. E* **2006**, *73*, 061910. [CrossRef]
44. D'Onofrio, G.; Lánský, P.; Pirozzi, E. On two diffusion neuronal models with multiplicative noise: The mean first-passage time properties. *Chaos* **2018**, *28*, 043103. [CrossRef]
45. Giorno, V.; Nobile A.G. Time-inhomogeneous Feller-type diffusion process with absorbing boundary condition. *J. Stat. Phys.* **2021**, *183*, 1–27. [CrossRef]
46. Masoliver, J.; Perelló, J. First-passage and escape problems in the Feller process. *Phys. Rev. E* **2012**, *86*, 041116. [CrossRef]
47. Masoliver, J. Extreme values and the level-crossing problem: An application to the Feller process. *Phys. Rev. E* **2014**, *89*, 042106. [CrossRef]
48. Lavigne, F.; Roques, L. Extinction times of an inhomogeneous Feller diffusion process: A PDF approach. *Expo. Math.* **2021**, *39*, 137–142. [CrossRef]

fractal and fractional

MDPI

Article

Ergodic Stationary Distribution and Threshold Dynamics of a Stochastic Nonautonomous SIAM Epidemic Model with Media Coverage and Markov Chain

Chao Liu [1,2,*], Peng Chen [1,2] and Lora Cheung [3]

1 Institute of Systems Science, Northeastern University, Shenyang 110169, China
2 School of Mathematics and Statistics, Northeastern University at Qinhuangdao, Qinhuangdao 066004, China
3 Department of Mathematics and Statistics, York University, Toronto, ON M3J 1P3, Canada
* Correspondence: liuchao@mail.neu.edu.cn

Abstract: A stochastic nonautonomous SIAM (Susceptible individual–Infected individual–Aware individual–Media coverage) epidemic model with Markov chain and nonlinear noise perturbations has been constructed, which is used to research the hybrid dynamic impacts of media coverage and Lévy jumps on infectious disease transmission. The uniform upper bound and lower bound of the positive solution are studied. Based on defining suitable random Lyapunov functions, we researched the existence of a nontrival positive T-periodic solution. Sufficient conditions are derived to discuss the exponential ergodicity based on verifying a Foster–Lyapunov condition. Furthermore, the persistence in the average sense and extinction of infectious disease are investigated using stochastic analysis techniques. Finally, numerical simulations are utilized to provide evidence for the dynamical properties of the stochastic nonautonomous SIAM.

Keywords: media coverage; Lévy jumps; nontrival positive T-periodic solution; exponential ergodicity; persistence in mean; extinction

Citation: Liu, C.; Chen, P.; Cheung, L. Ergodic Stationary Distribution and Threshold Dynamics of a Stochastic Nonautonomous SIAM Epidemic Model with Media Coverage and Markov Chain. *Fractal Fract.* **2022**, *6*, 699. https://doi.org/10.3390/fractalfract6120699

Academic Editors: Mario Abundo and Vassili Kolokoltsov

Received: 16 October 2022
Accepted: 23 November 2022
Published: 26 November 2022

1. Introduction

Recent studies have shown that public health alerts via social media exert a positive influence on usefully informing people of the prevalence about infectious disease [1]. Therefore, media coverage has effectively reduced the prevalence and shortened the duration of disease [2]. The influence of media message reminders on local behavioral response and public awareness response was studied in [3], and pharmaceutical interventions and the response of infected people to information have also been successful in controlling of the epidemic.

As the mass media has directed people's attention, it is often focused on infectious disease; thus, relying on the mass media to publicize the law of infectious disease transmission is extremely constructive for the effective treatment of the epidemic [4,5]. Assuming that the implementation of a public health alert program is proportional to the infected population, recent studies have shown that progress has been made in the social cost–benefit analysis of media campaigns for vaccination against infectious disease [6–9].

There is also a series of studies that specifically discussed the increased vaccination coverage of people due to social media advertising and television programs [10–12], which includes the example of discussing the function of media alerts to reduce the number of infected people. Particularly, in [12] a SIAM (Susceptible individual—Infected individual—Aware individual—Media coverage) epidemic model with media coverage and public health alerts was established as follows, and stability analysis around the endemic equilibrium was studied.

$$
\begin{cases}
\frac{dS(t)}{dt} = \Lambda - \beta S(t)I(t) - \lambda S(t)\frac{M(t)}{M(t)+p} + \nu I(t) + \lambda_0 A(t) - hS(t),\\
\frac{dI(t)}{dt} = \beta S(t)I(t) - (\nu + \alpha + h)I(t),\\
\frac{dA(t)}{dt} = \lambda S(t)\frac{M(t)}{M(t)+p} - (\lambda_0 + h)A(t),\\
\frac{dM(t)}{dt} = r\left(1 - \theta\frac{A(t)}{w+A(t)}\right)I(t) - r_0(M(t) - M_0),
\end{cases}
\tag{1}
$$

with the initial value for model (1) takes the following from:

$$S(0) \geq 0, I(0) \geq 0, A(0) \geq 0, M(0) \geq M_0,$$

$S(t), I(t), A(t)$ denotes the number of susceptible individuals, infected individuals and aware individuals, respectively. $M(t)$ is the cumulative number of TV programs and social media. Λ denotes the increase in the number of people who are susceptible. β stands for the rate of contact between susceptible individuals and infected individuals. ν, α and h denote the rate of recovery, disease-induced death and natural death, respectively. Furthermore, λ represents the rate of awareness among the susceptible, and λ_0 is the transfer rate of aware individuals to susceptible individuals. r is the growth rate in media coverage, and r_0 represents the diminution rate of advertisements [12].

It is well-known that dynamical effects of a periodically varying situation are different from those in a relatively stable situation [13]. Some parameters describing seasonal effects are affected by disturbances in time and usually exhibit [14]. Therefore, it is more accurate to assume periodicity of the surrounding situation and introduce time-varying periodic function parameters into the epidemic models, which can be found in [15–17] and the references therein.

Recently, there have been studies concentrated on discussing the spread dynamics of infectious disease using a stochastic mathematical model with Brownian motion [14–17]. Recently, some scholars found that, compared with Gaussian white noise and Brownian motion, Lévy jumps can more accurately describe the unexpected violent disturbances in the real situation [18,19]. Furthermore, Markov chain [20] is usually used to describe the vital transient transitions of important rates between two or more infectious states [21,22].

Taking the above mentioned content into account, media coverage, random perturbations and time-varying periodic function parameters are important disciplines in the modeling and dynamical analysis of infectious disease transmission. In this work, a random nonautonomous SIAM infectious disease model with Markov chain and nonlinear noise perturbations has been established as follows:

$$
\begin{cases}
dS(t) = \left[\Lambda(t) - \beta(t)S(t)I(t) - \lambda(t)S(t)\frac{M(t)}{M(t)+p(t)} + \nu(t)I(t) + \lambda_0(t)A(t) - h(t)S(t)\right]dt\\
\qquad + [\sigma_{11}(\gamma(t)) + \sigma_{12}(\gamma(t))S(t)]S(t)dB_1(t) + \int_{\mathbb{Y}} c_1(u)S(t-)\widetilde{X}(du, dt),\\
dI(t) = [\beta(t)S(t)I(t) - (\nu(t) + \alpha(t) + h(t))I(t)]dt\\
\qquad + [\sigma_{21}(\gamma(t)) + \sigma_{22}(\gamma(t))I(t)]I(t)dB_2(t) + \int_{\mathbb{Y}} c_2(u)I(t-)\widetilde{X}(du, dt),\\
dA(t) = \left[\lambda(t)S(t)\frac{M(t)}{M(t)+p(t)} - (\lambda_0(t) + h(t))A(t)\right]dt\\
\qquad + [\sigma_{31}(\gamma(t)) + \sigma_{32}(\gamma(t))A(t)]A(t)dB_3(t) + \int_{\mathbb{Y}} c_3(u)A(t-)\widetilde{X}(du, dt),\\
dM(t) = \left[r(t)\left(1 - \theta(t)\frac{A(t)}{w(t)+A(t)}\right)I(t) - r_0(t)(M(t) - M_0(t))\right]dt\\
\qquad + [\sigma_{41}(\gamma(t)) + \sigma_{42}(\gamma(t))M(t)]M(t)dB_4(t) + \int_{\mathbb{Y}} c_4(u)M(t-)\widetilde{X}(du, dt).
\end{cases}
\tag{2}
$$

where $\Lambda(t), \beta(t), \lambda(t), p(t), \nu(t), \lambda_0(t), h(t), \alpha(t), r(t), \theta(t), w(t), r_0(t)$ are continuous T-periodic functions. $\sigma_{ij}^2(\cdot) > 0 (i = 1, 2, 3, 4, j = 1, 2)$ represent white noises. $\gamma(t)$ denotes a irreducible and continuous Markov chain, which is defined in $\mathbb{N} = \{1, 2, 3, \ldots, K\}$. $\gamma(t)$ is supposed to be generated by the following transition rate matrix $\Gamma = (\mu_{nj})_{K\times K}$,

$$\mathbb{P}\{\gamma(\tau + \triangle\tau) = j | \gamma(\tau) = n\} = \begin{cases} \mu_{nj}\triangle\tau + o(\triangle\tau), & n \neq j,\\ 1 + \mu_{nn}\triangle\tau + o(\triangle\tau), & n = j, \end{cases}$$

where $\mu_{nj} > 0$ is the transition rate from state n to state j, and $\mu_{nn} = -\Sigma_{n\neq j,n=1}^{K}\mu_{nj}$ holds for $n \neq j$.

Due primary to $\gamma(t)$ being an irreducible Markov procedure, it exists as a unique stationary probability distribution $\phi = (\phi_1, \phi_2, \cdots, \phi_K) \in \mathbb{R}^{1\times K}$ subject to $\sum_{n=1}^{K}\phi_n = 1$ and $\phi_n > 0$ hold for any $n \in \mathbb{N}$. $S(t-), I(t-), A(t-), M(t-)$ denotes left limit of $S(t), I(t), A(t), M(t)$, respectively. $\mathbb{Y}$ represents for a measurable subset of $\mathbb{R}_+$, X depicts an independent Poisson counting measure with Lévy measure ρ on $\mathbb{Y}$ with $\rho(\mathbb{Y}) < \infty$ such that $\widetilde{X}(dt, du) = X(dt, du) - \rho(du)dt$. It is supposed that $c_i(u) > -1$, and there are four constants $\kappa_i > 0 (i = 1, 2, 3, 4)$ are constructed as below,

$$\max\{\int_{\mathbb{Y}}(\ln(1 + c_i(u))\rho du, \int_{\mathbb{Y}}(\ln(1 + c_i(u))^2\rho du\} \leq \kappa_i. \tag{3}$$

Based on the properties of Markov chain, we can regard system (2) as the subsystems defined as below:

$$\begin{cases} dS(t) = \Big[\Lambda(t) - \beta(t)S(t)I(t) - \lambda(t)S(t)\frac{M(t)}{M(t)+p(t)} + \nu(t)I(t) + \lambda_0(t)A(t) - h(t)S(t)\Big]dt \\ \qquad +[\sigma_{11}(n) + \sigma_{12}(n)S(t)]S(t)dB_1(t) + \int_{\mathbb{Y}} c_1(u)S(t-)\widetilde{X}(du, dt), \\ dI(t) = [\beta(t)S(t)I(t) - (\nu(t) + \alpha(t) + h(t))I(t)]dt \\ \qquad +[\sigma_{21}(n) + \sigma_{22}(n)I(t)]I(t)dB_2(t) + \int_{\mathbb{Y}} c_2(u)I(t-)\widetilde{X}(du, dt), \\ dA(t) = \Big[\lambda(t)S(t)\frac{M(t)}{M(t)+p(t)} - (\lambda_0(t) + h(t))A(t)\Big]dt \\ \qquad +[\sigma_{31}(n) + \sigma_{32}(n)A(t)]A(t)dB_3(t) + \int_{\mathbb{Y}} c_3(u)A(t-)\widetilde{X}(du, dt), \\ dM(t) = \Big[r(t)\Big(1 - \theta(t)\frac{A(t)}{w(t)+A(t)}\Big)I(t) - r_0(t)(M(t) - M_0(t))\Big]dt \\ \qquad +[\sigma_{41}(n) + \sigma_{42}(n)M(t)]M(t)dB_4(t) + \int_{\mathbb{Y}} c_4(u)M(t-)\widetilde{X}(du, dt). \end{cases} \tag{4}$$

Remark 1. *In recent related work, stochastic perturbations are usually represented by linear form perturbation of white noise, and the influences of linear noises perturbations on nonautonomous epidemic models were studied in [13–17]. However, in order to accurately depict some stochastic phenomena arising from infectious disease transmission in the real world, it is more constructive to introduce nonlinear noise perturbations into a nonautonomous epidemic model. Furthermore, some stochastic models have been established to discuss the prevalence mechanism of infectious diseases [23–31] without Lévy jumps.*

A SIS infectious disease system with regime-switching driven by Lévy jumps was investigated in [32], while the random dynamics for infectious disease system with hybrid dynamic impacts of Lévy jumps and media coverage are rarely reported. Taking the media coverage and random disturbance into dynamic impacts on threshold dynamics of random infectious disease model were investigated in [33–36], while Lévy jumps and periodic function parameters were not considered in [33–36].

The dynamic behavior of infectious disease systems in [37–39] were investigated under nonlinear noise perturbations and Lévy jumps, while all parameters were assumed to be constant values in [37–39], periodicity factors during transmission within the infectious disease regimes were not considered.

Although the stochastic infectious disease model and its dynamic analysis have attracted wide attention, as far as the authors know, the hybrid dynamic impacts of Lévy jumps and media coverage on random dynamics of the nonautonomous SIAM epidemic model with Markov chain and nonlinear noise perturbations have not been reported in previous related studies.

By incorporating Lévy jumps, nonlinear noise perturbations and periodic function parameters into the the epidemic system, we aim to study the hybrid dynamic impacts of media coverage on infectious disease transmission driven by Lévy jumps. For the rest of this work, we will make some arrangements as below: In the next section, the uniform upper bound and lower bound of the solution for stochastic nonautonomous system will be investigated. Based on constructing certain appropriate stochastic Lyapunov functions, sufficient conditions for existence of a nontrival positive T-periodic solution will be discussed.

Based on verifying a Foster–Lyapunov criterion, sufficient conditions for the exponential ergodicity are discussed. Furthermore, some sufficient conditions are derived to discuss the persistence in the mean and extinction of the infectious disease. In the third section, numerical simulations are used to prove the accuracy of the theoretical derivation. Lastly, section four is the conclusion of this paper.

2. Qualitative Analysis

For the sake of the narrative, we define the following mathematical symbols,

$$g^u = \sup_{t\in\mathbb{R}_+} g(t), \quad g^l = \inf_{t\in\mathbb{R}_+} g(t), \quad \langle g(t)\rangle = \frac{1}{t}\int_0^t g(t)\mathrm{d}t.$$

Lemma 1. *For any initial value $(S(0), I(0), A(0), M(0), n) \in \mathbb{R}_+^4 \times \mathbb{N}$, when the sufficient condition* (3) *holds, then there exists a uniform upper bound and a uniform lower bound for the solution of system* (4).

Proof. Let $y_1(t) = S^\eta(t), \eta \in (0,1)$. Utilizing the Itô's formula to $e^t y_1(t)$ and, integrating both sides from 0 to t, the following results can be obtained.

$$\begin{aligned}
&\mathbb{E}(e^t y_1(t)) \\
= \ & \mathbb{E}\int_0^t e^s\left[1+\eta\left(\frac{\Lambda(t)}{S(t)} - \beta(t)I(t) - \frac{\lambda(t)M(t)}{M(t)+p(t)} + \frac{v(t)I(t)}{S(t)} + \frac{\lambda_0(t)A(t)}{S(t)} - h(t)\right)\right]y_1(t)\mathrm{d}s \\
&+y_1(0) + \mathbb{E}\int_0^t e^s\left[\int_{\mathbb{Y}}((1+c_1(u))^\eta - 1)\widetilde{X}(\mathrm{d}u,\mathrm{d}t)\right]y_1(t)\mathrm{d}s \\
&-\mathbb{E}\int_0^t e^s\left[\frac{\eta(1-\eta)}{2}(\sigma_{11}(n) + \sigma_{12}(n)S(t))^2\right]y_1(t)\mathrm{d}s \\
\leq \ & \mathbb{E}\int_0^t e^s\left[1+\eta\left(\frac{\Lambda(t)}{S(t)} - \beta(t)I(t) - \frac{\lambda(t)M(t)}{M(t)+p(t)} + \frac{v(t)I(t)}{S(t)} + \frac{\lambda_0(t)A(t)}{S(t)} - h(t)\right)\right]y_1(t)\mathrm{d}s \\
&+y_1(0) - \mathbb{E}\int_0^t e^s\left[\frac{\eta(1-\eta)}{2}(\sigma_{11}(n) + \sigma_{12}(n)S(t))^2\right]y_1(t)\mathrm{d}s \\
&+\mathbb{E}\int_0^t e^s\left[\int_{\mathbb{Y}}((1+c_1(u))^\eta - 1 - \eta c_1(u))\rho\mathrm{d}u\right]y_1(t)\mathrm{d}s.
\end{aligned}$$

When $S(t) \geq 0$ and $0 < \eta < 1$, based on the inequality $S^\eta(t) \leq 1 + \eta(S(t) - 1)$, if sufficient condition (3) holds, we can obtain the following results

$$\begin{aligned}
& y_1(t)\left[1+\eta\left(\frac{\Lambda(t)}{S(t)} + \frac{v(t)I(t)}{S(t)} + \frac{\lambda_0(t)A(t)}{S(t)} - \beta(t)I(t) - \frac{\lambda(t)M(t)}{M(t)+p(t)} - h(t)\right)\right] \\
& -y_1(t)\left[\frac{\eta(1-\eta)}{2}(\sigma_{11}(n)+\sigma_{12}(n)S(t))^2 - \int_{\mathbb{Y}}((1+c_1(u))^\eta - 1 - \eta c_1(u))\rho\mathrm{d}u\right] \\
\leq \ & \left[1+\eta\left(\frac{\Lambda(t)}{S(t)} - \beta(t)I(t) - \frac{\lambda(t)M(t)}{M(t)+p(t)} + \frac{v(t)I(t)}{S(t)} + \frac{\lambda_0(t)A(t)}{S(t)} - h(t)\right)\right]S^\eta(t) \\
\leq \ & G_1(\eta),
\end{aligned}$$

where $G_1(\eta)$ is a positive definite function associated with η.

Hence, we can reach the following conclusion

$$\mathbb{E}(e^t y_1(t)) \leq y_1(0) + \mathbb{E}\int_0^t e^s G_1(\eta)\mathrm{d}s,$$

which reveals that $\limsup_{t\to\infty} \mathbb{E}(S(t)^\eta) \leq G_1(\eta)$.

Basing on utilizing the similar arguments, one can find that

$$\begin{cases} \limsup_{t\to\infty} \mathbb{E}(I(t)^{\eta}) \leq G_2(\eta), \\ \limsup_{t\to\infty} \mathbb{E}(A(t)^{\eta}) \leq G_3(\eta), \\ \limsup_{t\to\infty} \mathbb{E}(M(t)^{\eta}) \leq G_4(\eta). \end{cases}$$

Let $(\chi(t), n) = (S(t), I(t), A(t), M(t), n) \in \mathbb{R}^4_+ \times \mathbb{N}$, it yields that

$$2^{(1-\frac{\eta}{2})\wedge 0} \mid \chi(t) \mid^{\eta} \leq S^{\eta}(t) + I^{\eta}(t) + A^{\eta}(t) + M^{\eta}(t),$$

which follows that

$$\begin{aligned} \limsup_{t\to\infty} \mathbb{E} \mid \chi(t) \mid^{\eta} &\leq 0.5^{(1-\frac{\eta}{2})\wedge 0} \limsup_{t\to\infty} \mathbb{E}[S^{\eta}(t) + I^{\eta}(t) + A^{\eta}(t) + M^{\eta}(t)] \\ &\leq 0.5^{(1-\frac{\eta}{2})\wedge 0}[G_1(\eta) + G_2(\eta) + G_3(\eta) + G_4(\eta)] := G(\eta). \end{aligned}$$

For any $\eta \in (0,1)$, let $G(\varepsilon) = \left(\frac{G(\eta)}{\varepsilon}\right)^{\frac{1}{\eta}}$, by applying the Chebyshev's inequality, we will obtain the following results

$$\begin{cases} \mathbb{P}[\chi(t) < G(\varepsilon)] \leq G(\varepsilon)^{\eta} \mathbb{P}[\chi(t)^{-\eta}(t)], \\ \\ \liminf_{t\to\infty}[\chi(t) \leq G(\varepsilon)] \geq 1 - \varepsilon. \end{cases}$$

Nextly, based on using Chebyshev's inequality and similar arguments, one can find a constant $Q(\varepsilon) > 0$ subject to

$$\liminf_{t\to\infty}[\chi(t) \geq Q(\varepsilon)] \geq 1 - \varepsilon.$$

Taking the above mentioned discussions into consideration, one can draw a conclusion that there exists a uniform upper bound and a uniform lower bound for the solution of system (4) with any initial value $(S(0), I(0), A(0), M(0), n)$. □

Lemma 2. *For every initial value* $(S(0), I(0), A(0), M(0), n) \in \mathbb{R}^4_+ \times \mathbb{N}$. *When sufficient condition* (3) *holds, then system* (4) *exists a unique positive solution* $(S(t), I(t), A(t), M(t), n)$ *that remains in* $\mathbb{R}^4_+ \times \mathbb{N}$ *with probability one.*

Proof. First, based on some standard arguments and analysis, it is not difficult to show that system (4) meets the local Lipschitz conditions. Thus, system (4) exists with a unique local positive solution on $t \in [0, \tau_e)$ most likely for any initial value $(S(0), I(0), A(0), M(0), n)$, where τ_e represents the explosion time. For the sake of proving the positive solution is global, next, we will show that $\tau_e = \infty$.

Secondly, it is assumed that there exists a sufficiently large integer $N_0^* \geq 0$ subject to $(S(0), I(0), A(0), M(0))$ all on the interval $\left[\frac{1}{N_0^*}, N_0^*\right]$. For any positive integer $n \geq N_0^*$, we can construct the stopping time as below,

$$\tau_s = \inf\left\{ t \in [0, \tau_e) \middle| \begin{array}{c} \min\{S(t), I(t), A(t), M(t)\} \leq \frac{1}{n}, \text{or} \\ \max\{S(t), I(t), A(t), M(t)\} \geq n \end{array} \right\}.$$

According to the mathematical properties of τ_s, it is clear that τ_e increases as $n \to \infty$. Let $\tau_\infty = \lim_{n\to\infty} \tau_s$, and then we can obtain that $\tau_\infty \leq \tau_e$ most likely. If $\tau_\infty = \infty$ holds most likely, it can be obtained that $\tau_e = \infty$ most likely holds, which obtains that $(S(t), I(t), A(t), M(t), n) \in \mathbb{R}^4_+ \times \mathbb{N}$ holds for all $t \geq 0$.

If $\tau_\infty = \infty$ most likely does not hold, then we can find two positive constants $\tilde{N}_0^* > 0$ and $\epsilon < 1$ subject to $\mathbb{P}\{\tau_\infty \leq \tilde{N}_0^*\} \geq \epsilon$. Therefore, we can find a positive integer $N_1 > \tilde{N}_0^*$ subject to $\mathbb{P}\{\tau_s \leq \tilde{N}_0^*\} \geq \epsilon$ holds for any $n > N_1$.

By utilizing a C^4- function $V : \mathbb{R}^4_+ \to \mathbb{R}_+ \cup \{0\}$ as below,

$$
\begin{aligned}
V(S(t),I(t),A(t),M(t)) &= S(t) - a_1 - a_1 \ln \frac{S(t)}{a_1} + I(t) - 1 - \ln I(t) \\
&\quad + A(t) - 1 + \ln A(t) + a_1(M(t) - 1 - \ln M(t)),
\end{aligned}
$$

where $a_1 = \frac{\alpha^l + h^l}{r^u + \beta^u}$.

Based primary on utilizing Itô's formula, we can find the following results

$$
\begin{aligned}
& dV(S(t),I(t),A(t),M(t)) \\
= & \left(1 - \frac{a_1}{S(t)}\right)\left[\Lambda(t) - \beta(t)S(t)I(t) - \frac{\lambda(t)S(t)M(t)}{M(t)+p(t)}\right]dt \\
& + \left(1 - \frac{a_1}{S(t)}\right)[\nu(t)I(t) + \lambda_0(t)A(t) - h(t)S(t)]dt \\
& + \left(1 - \frac{1}{I(t)}\right)[(\beta(t)S(t)I(t) - (\nu(t) + \alpha(t) + h(t))I(t))]dt \\
& + \left(1 - \frac{1}{A(t)}\right)\left[\lambda(t)S(t)\frac{M(t)}{M(t)+p(t)} - (\lambda_0(t) + h(t))A(t)\right]dt \\
& + \left(a_1 - \frac{a_1}{M(t)}\right)\left[r(t)\left(1 - \theta\frac{A(t)}{w(t)+A(t)}\right)I(t) - r_0(t)(M(t) - M_0)\right]dt \\
& + \left[\frac{a_1}{2}(\sigma_{11}(n)S(t) + \sigma_{12}(n))^2 + \frac{1}{2}(\sigma_{21}(n)I(t) + \sigma_{22}(n))^2\right]dt \\
& + \left[\frac{1}{2}(\sigma_{31}(n)A(t) + \sigma_{32}(n))^2 + \frac{a_1}{2}(\sigma_{41}(n)M(t) + \sigma_{42}(n))^2\right]dt \\
& + \left[a_1\int_{\mathbb{Y}}(c_1(u) - \ln(1 + c_1(u)))\rho du + \int_{\mathbb{Y}}(c_2(u) - \ln(1 + c_2(u)))\rho du\right]dt \\
& + \left[\int_{\mathbb{Y}}(c_3(u) - \ln(1 + c_3(u)))\rho du + a_1\int_{\mathbb{Y}}(c_4(u) - \ln(1 + c_4(u)))\rho du\right]dt \\
& + (\sigma_{11}(n)S(t) + \sigma_{12}(n))(S(t) - a_1)dB_1(t) + (\sigma_{21}(n)I(t) + \sigma_{22}(n))(I(t) - 1)dB_2(t) \\
& + (\sigma_{31}(n)A(t) + \sigma_{32}(n))(A(t) - 1)dB_3(t) + a_1(\sigma_{41}(n)M(t) + \sigma_{42}(n))(M(t) - 1)dB_4(t) \\
& + a_1\int_{\mathbb{Y}}[c_1(u)S(t) - \ln(1 + c_1(u))]\widetilde{X}(du,dt) + \int_{\mathbb{Y}}[c_2(u)I(t) - \ln(1 + c_2(u))]\widetilde{X}(du,dt) \\
& + \int_{\mathbb{Y}}[c_3(u)A(t) - \ln(1 + c_3(u))]\widetilde{X}(du,dt) + a_1\int_{\mathbb{Y}}[c_4(u)M(t) - \ln(1 + c_4(u))]\widetilde{X}(du,dt)
\end{aligned}
$$

Furthermore, it follows from simple computations that

$$
\begin{aligned}
& dV(S(t),I(t),A(t),M(t)) \\
= & \mathcal{L}V dt + (\sigma_{11}(n)S(t) + \sigma_{12}(n))(S(t) - a_1)dB_1(t) + (\sigma_{21}(n)I(t) + \sigma_{22}(n))(I(t) - 1)dB_2(t) \\
& + (\sigma_{31}(n)A(t) + \sigma_{32}(n))(A(t) - 1)dB_3(t) + a_1(\sigma_{41}(n)S(t) + \sigma_{42}(n))(M(t) - 1)dB_4(t) \\
& + a_1\int_{\mathbb{Y}}[c_1(u)S(t) - \ln(1 + c_1(u))]\widetilde{X}(du,dt) + \int_{\mathbb{Y}}[c_2(u)I(t) - \ln(1 + c_2(u))]\widetilde{X}(du,dt) \\
& + \int_{\mathbb{Y}}[c_3(u)A(t) - \ln(1 + c_3(u))]\widetilde{X}(du,dt) + a_1\int_{\mathbb{Y}}[c_4(u)M(t) - \ln(1 + c_4(u))]\widetilde{X}(du,dt),
\end{aligned}
$$

where $\mathcal{L}V$ is defined as follows,

$$\begin{aligned}
\mathcal{L}V &= \Lambda(t) - h(t)S(t) - (\alpha(t) + h(t))I(t) - h(t)A(t) + a_1 r(t)\left(1 - \theta(t)\frac{A(t)}{\omega(t) + A(t)}\right)I(t) \\
&+a_1\left[r_0(t)(M_0(t) - M(t)) - \frac{\Lambda(t)}{S(t)} + \beta(t)I(t)\right] + (a_1 - S(t))\frac{\lambda(t)M(t)}{M(t) + p(t)} \\
&-a_1\left[\frac{\nu(t)I(t)}{S(t)} + a_1\frac{\lambda_0(t)A(t)}{S(t)} - a_1 h(t)\right] - \beta(t)S(t) + \nu(t) + \alpha(t) + h(t) \\
&-a_1\left[\frac{r(t)I(t)}{M(t)} - \frac{\theta(t)r(t)I(t)}{M(t)}\frac{A(t)}{\omega(t) + A(t)} - r_0(t) + r_0(t)\frac{M_0(t)}{M(t)}\right] \\
&+\frac{a_1}{2}(\sigma_{11}(n)S(t) + \sigma_{12}(n))^2 + \frac{1}{2}(\sigma_{21}(n)I(t) + \sigma_{22}(n))^2 \\
&+\frac{1}{2}(\sigma_{31}(n)A(t) + \sigma_{32}(n))^2 + \frac{a_1}{2}(\sigma_{41}(n)M(t) + \sigma_{42}(n))^2 \\
&+\int_{\mathbb{Y}}(c_1(u) - \ln(1 + c_1(u)))\rho du + a_1\int_{\mathbb{Y}}(c_2(u) - \ln(1 + c_2(u)))\rho du \\
&+\int_{\mathbb{Y}}(c_3(u) - \ln(1 + c_3(u)))\rho du + a_1\int_{\mathbb{Y}}(c_4(u) - \ln(1 + c_4(u)))\rho du.
\end{aligned}$$

When the condition (3) are met, we can obtain the following results based on simple computations

$$\begin{aligned}
\mathcal{L}V &\leq \Lambda(t) + [a_1(r(t) + \beta(t)) - (\alpha(t) + h(t))]I(t) + a_1 r_0(t)M_0(t) + a_1\lambda(t) \\
&+a_1(h(t) + r_0(t)) + \nu(t) + \alpha(t) + 2h(t) + \lambda_0(t) + a_1\frac{r(t)\theta(t)I(t)A(t)}{M(t)(\omega(t) + A(t))} \\
&+\frac{a_1}{2}(\sigma_{11}(n)S(t) + \sigma_{12}(n))^2 + \frac{1}{2}(\sigma_{21}(n)I(t) + \sigma_{22}(n))^2 + \frac{1}{2}(\sigma_{31}(n)A(t) + \sigma_{32}(n))^2 \\
&+\frac{a_1}{2}(\sigma_{41}(n)M(t) + \sigma_{42}(n))^2 + a_1(\kappa_1 + \kappa_4) + \kappa_2 + \kappa_3.
\end{aligned}$$

Based on the properties of parametric function and Lemma 1 of this paper, one can find that

$$\begin{aligned}
\mathcal{L}V &\leq \Lambda^u + [a_1(r^u + \beta^u) - (\alpha^l + h^l)]I(t) + a_1(r_0^u M_0^u + \lambda^u + h^u) + \nu^u + \alpha^u + h^u \\
&+(\lambda_0 + h)^u + a_1 r_0^u + a_1\frac{r^u\theta^u G^2(\varepsilon)}{Q(\varepsilon)(\omega^l + Q(\varepsilon))} + \frac{a_1}{2}(\sigma_{11}G(\varepsilon) + \sigma_{12})^2 + \frac{1}{2}(\sigma_{21}G(\varepsilon) + \sigma_{22})^2 \\
&+\frac{1}{2}(\sigma_{31}G(\varepsilon) + \sigma_{32})^2 + \frac{a_1}{2}(\sigma_{41}G(\varepsilon) + \sigma_{42})^2 + a_1(\kappa_1 + \kappa_4) + \kappa_2 + \kappa_3,
\end{aligned}$$

where $a_1 = \frac{\alpha^l + h^l}{r^u + \beta^u}$.

The rest of the discussions resemble those in [16,20]; thus, we omitted them. One can find that $\tau_\infty = \infty$, which means that the solution of (4) will not explosion in a finite time most likely. □

Lemma 3. *If a sufficient condition* (3) *holds, the following properties holds for the positive solution of* (4) *with every initial value* $(S(0), I(0), A(0), M(0), n) \in \mathbb{R}_+^4 \times \mathbb{N}$,

$$\begin{cases}
\limsup_{t\to\infty}\langle S(t)\rangle \leq \frac{\Lambda^u}{h^l}, & \limsup_{t\to\infty}\langle I(t)\rangle \leq \frac{\Lambda^u}{\alpha^l + h^l}, \\
\limsup_{t\to\infty}\langle A(t)\rangle \leq \frac{\Lambda^u}{h^l}, & \limsup_{t\to\infty}\langle M(t)\rangle \leq \frac{r^u\Lambda^u}{r_0^l\alpha^l} + M_0^u.
\end{cases}$$

Proof. Based on the first three formulas of system (4), we can find the results as below

$$\begin{aligned}
&\frac{S(t)-S(0)}{t}+\frac{I(t)-I(0)}{t}+\frac{A(t)-A(0)}{t} \\
\leq\ & \Lambda^u - h^l\langle S(t)\rangle - (\alpha^l + h^l)\langle I(t)\rangle - h^l\langle A(t)\rangle \\
&+\frac{1}{t}\sum_{i=1}^{3}\left[\int_0^t (\sigma_{i1}(n)+\sigma_{i2}(n)\chi_i(t))\chi_i(t)\mathrm{d}B_i(t) + \int_0^t\int_{\mathbb{Y}} c_i(u)\chi_i(t-)\widetilde{X}(\mathrm{d}u,\mathrm{d}s)\right],
\end{aligned}$$

which reveals that

$$\begin{aligned}
h^l\langle S(t)\rangle \ \leq\ & \Lambda^u + \frac{S(0)+I(0)+A(0)-S(t)-I(t)-A(t)}{t} \\
&+\frac{1}{t}\sum_{i=1}^{3}\left[\int_0^t (\sigma_{i1}(n)+\sigma_{i2}(n)\chi_i(t))\chi_i(t)\mathrm{d}B_i(t) + \int_0^t\int_{\mathbb{Y}} c_i(u)\chi_i(t-)\widetilde{X}(\mathrm{d}u,\mathrm{d}s)\right].
\end{aligned}$$

Hence, it is not difficult to show that

$$\begin{aligned}
\langle S(t)\rangle \ \leq\ & \frac{\Lambda^u}{h^l} + \frac{S(0)+I(0)+A(0)-S(t)-I(t)-A(t)}{h^l t} \\
&+\frac{1}{h^l t}\sum_{i=1}^{3}\left[\int_0^t (\sigma_{i1}(n)+\sigma_{i2}(n)\chi_i(t))\chi_i(t)\mathrm{d}B_i(t) + \int_0^t\int_{\mathbb{Y}} c_i(u)\chi_i(t-)\widetilde{X}(\mathrm{d}u,\mathrm{d}s)\right],
\end{aligned}$$

which follows that

$$\langle S(t)\rangle \leq \frac{\Lambda^u}{h^l} + \frac{S(0)+I(0)+A(0)}{h^l t} + \sum_{i=1}^{3}\left(\frac{\psi_{1i}}{h^l t} + \frac{\psi_{2i}}{h^l t}\right),$$

where ψ_{1i} and ψ_{2i} $(i=1,2,3)$ will be defined as follows,

$$\begin{cases}
\psi_{1i} = \int_0^t (\sigma_{i1}(n)+\sigma_{i2}(n)\chi_i(t))\chi_i(t)\mathrm{d}B_i(t), \\
\psi_{2i} = \int_0^t \int_{\mathbb{Y}} c_i(u)\chi_i(t-)\widetilde{X}(\mathrm{d}u,\mathrm{d}s).
\end{cases}$$

By using Lemma 1, Lemma 2 and exponential martingale inequalities, it can be obtained that

$$\langle \psi_{1i}, \psi_{1i}\rangle = \int_0^t (\sigma_{i1}(n)+\sigma_{i2}(n)\chi_i(t))^2\chi_i^2(t)\mathrm{d}s,$$

and

$$\begin{aligned}
\limsup_{t\to\infty}\frac{\langle \psi_{1i}, \psi_{1i}\rangle}{t} \ =\ & \limsup_{t\to\infty}\frac{1}{t}\int_0^t (\sigma_{i1}(n)+\sigma_{i2}(n)\chi_i(t))^2\chi_i^2(t)\mathrm{d}s \\
\leq\ & (\sigma_{i1}^u + \sigma_{i2}^u G(\varepsilon))^2 G^2(\varepsilon) \\
<\ & \infty,
\end{aligned}$$

which follows that

$$\limsup_{t\to\infty}\frac{\psi_{1i}}{t} = 0, \tag{5}$$

holds for $i = 1,2,3$.

Further computations show that

$$\mathbb{P}\left\{\sup_{0\leq i\leq j}\left[\psi_{2i} - \frac{1}{2}\langle\psi_{2i},\psi_{2i}\rangle\right] > 2\ln j\right\} \leq \frac{1}{j^2}. \tag{6}$$

It is easy to show that we can find a random integer $j_0 = j_0(\omega)$ holds with almost all $\omega \in \Omega$. Hence, it can be concluded that

$$\sup_{0\le i\le j}\left[\psi_{2i} - \frac{1}{2}\langle\psi_{2i}, \psi_{2i}\rangle\right] \le 2\ln j_0 \tag{7}$$

holds for $\omega \in \Omega$ most likely, which reveals that

$$\psi_{2i} \le 2\ln j_0 + \frac{1}{2}\langle\psi_{2i}, \psi_{2i}\rangle \tag{8}$$

holds for $i = 1, 2, 3$ and all $0 \le t \le j_0$.

Consequently, we have

$$\begin{aligned}\langle S(t)\rangle &\le \frac{\Lambda^u}{h^l} + \frac{S(0)+I(0)+A(0)}{h^l t} + \frac{2\ln j_0}{h^l t} + \sum_{i=1}^{3}\frac{\psi_{1i}}{h^l t}\\ &\le \frac{\Lambda^u}{h^l} + \frac{S(0)+I(0)+A(0)}{h^l t} + \frac{2\ln j_0}{h^l (j_0-1)} + \sum_{i=1}^{3}\frac{\psi_{1i}}{h^l t}.\end{aligned}$$

By taking the superior limit of $\langle S(t)\rangle$, for all $0 \le t \le j_0$, it yields

$$\begin{aligned}\limsup_{t\to\infty}\langle S(t)\rangle &\le \limsup_{t\to\infty}\frac{\Lambda^u}{h^l} + \limsup_{t\to\infty}\frac{S(0)+I(0)+A(0)}{h^l t}\\ &\quad + \limsup_{t\to\infty}\frac{2\ln j_0}{h^l t} + \limsup_{t\to\infty}\sum_{i=1}^{3}\frac{\psi_{1i}}{h^l t}\\ &\le \limsup_{t\to\infty}\frac{\Lambda^u}{h^l} + \limsup_{t\to\infty}\frac{S(0)+I(0)+A(0)}{h^l t}\\ &\quad + \limsup_{t\to\infty}\frac{2\ln j_0}{h^l (j_0-1)} + \limsup_{t\to\infty}\sum_{i=1}^{3}\frac{\psi_{1i}}{h^l t}\\ &\le \frac{\Lambda^u}{h^l}.\end{aligned}$$

Based on using the similar arguments and discussions mentioned above, one can find

$$\limsup_{t\to\infty}\langle I(t)\rangle < \frac{\Lambda^u}{\alpha^l + h^l}, \quad \limsup_{t\to\infty}\langle A(t)\rangle < \frac{\Lambda^u}{h^l}, \quad \limsup_{t\to\infty}\langle M(t)\rangle < \frac{r^u\Lambda^u}{r_0^l\alpha^l} + M_0^u,$$

and the proofs are omitted here. Hence, we can draw the next conclusions

$$\begin{cases}\limsup_{t\to\infty}\langle S(t)\rangle \le \frac{\Lambda^u}{h^l}, & \limsup_{t\to\infty}\langle I(t)\rangle \le \frac{\Lambda^u}{\alpha^l + h^l},\\ \limsup_{t\to\infty}\langle A(t)\rangle \le \frac{\Lambda^u}{h^l}, & \limsup_{t\to\infty}\langle M(t)\rangle \le \frac{r^u\Lambda^u}{r_0^l\alpha^l} + M_0^u.\end{cases}$$

This proof is ending. □

Theorem 1. *When $R_1 > 0$ and $R_2 > 0$ holds, there exists a nontrival positive T-periodic solution of system (4), where R_i $(i = 1, 2)$ will be constructed as below:*

$$\begin{cases} R_1 = \left\langle \frac{\beta(t)\Lambda(t)}{\nu(t)+\alpha(t)+h(t)+\zeta_2(u)+\frac{1}{2}\sum_{n=1}^{K}\phi_n\sigma_{21}^2(n)} - \frac{2(\lambda(t)+h(t)+l_2(t)\Lambda(t)+\zeta_1(u))+\sum_{n=1}^{K}\phi_n\sigma_{11}^2(n)}{2}\right\rangle,\\ R_2 = h(t) - (\sigma_{12}^2(n) \vee \sigma_{22}^2(n) \vee \sigma_{32}^2(n)).\end{cases} \tag{9}$$

and $\zeta_i(u), l_i(t)$ $(i = 1, 2)$ will be constructed as below,

$$\begin{cases} \zeta_i(u) = \int_{\mathbb{Y}}(c_i(u) - \ln(1 + c_i(u)))\rho du, \\ l_1(t) = \frac{\beta(t)\Lambda(t)}{[\nu(t)+\alpha(t)+h(t)+\zeta_1(u)+\frac{1}{2}\sum_{n=1}^{K}\phi_n\sigma_{21}^2(n)]^2}, \\ l_2(t) = \frac{\max\{\sigma_{11}(n)\sigma_{12}(n), l_1(t)\sigma_{21}(n)\sigma_{22}(n)\}}{h(t)}. \end{cases} \tag{10}$$

Furthermore, $\vec{\omega}(n)$ is assumed to be a twice continuously differentiable function that characterizes a Markov process and its Itô's derivative is defined as follows:

$$\mathcal{L}\vec{\omega}(n) = \sum_{n=1,j=1}^{K} \mu_{nj}\vec{\omega}(n). \tag{11}$$

Proof. First, we define $U_1(t)$ as follows

$$U_1(t) = -\ln S(t) - l_1(t)\ln I(t) + l_2(t)(S(t) + I(t) + A(t)) + \vec{\omega}(n).$$

Based primary on utilizing Itô's formula, we can find the results as below

$$\begin{aligned} \mathcal{L}U_1(t) &= -\frac{\Lambda(t) + \nu(t)I(t) + \lambda_0(t)A(t)}{S(t)} + \beta(t)I(t) + \frac{\lambda(t)M(t)}{M(t) + p(t)} \\ &+h(t) + l_1(t)[-\beta(t)S(t) + \nu(t) + \alpha(t) + h(t)] \\ &+\frac{1}{2}[\sigma_{11}(n) + \sigma_{12}(n)S(t)]^2 + \frac{l_1(t)}{2}[\sigma_{21}(n) + \sigma_{22}(n)I(t)]^2 \\ &+ \int_{\mathbb{Y}}(c_1(u) - \ln(1 + c_1(u))\rho du + l_1(t)\int_{\mathbb{Y}}(c_2(u) - \ln(1 + c_2(u)))\rho du \\ &+ \sum_{n=1,j=1}^{K} \mu_{nj}\vec{\omega}(n) + l_2(t)\Lambda(t) - l_2(t)[h(t)(S(t) + I(t) + A(t)) - \alpha(t)I(t)] \\ &\leq -2\sqrt{l_1(t)\beta(t)\Lambda(t)} + l_1(t)\left(\nu(t) + \alpha(t) + h(t) + \frac{1}{2}\sigma_{21}^2(n) + \zeta_2(u)\right) \\ &+\lambda(t) + h(t) + \frac{1}{2}\sigma_{11}^2(n) + \zeta_1(u) + l_2(t)[\Lambda(t) - h(t)(S(t) + I(t))] \\ &+ \max\{\sigma_{11}(n)\sigma_{12}(n), l_1(t)\sigma_{21}(n)\sigma_{22}(n)\}(S(t) + I(t)) + \sum_{n=1,j=1}^{K} \mu_{nj}\vec{\omega}(n) \\ &+\beta(t)I(t) + \max\{\sigma_{12}^2(n), l_1(t)\sigma_{22}^2(n)\}(S^2(t) + I^2(t)). \end{aligned} \tag{12}$$

According to irreducibility property of n, for $(\sigma_{11}^2(1), \sigma_{11}^2(2), \ldots, \sigma_{11}^2(K))$, there exists a functional vector $\vec{\omega}(n) = (\omega(1), \omega(2), \ldots, \omega(K))$, and $\vec{\omega}(n)$ has been mentioned in (11),

$$\frac{1}{2}\sigma_{11}^2(n) + \sum_{n=1,j=1}^{K} \mu_{nj}\vec{\omega}(n) = \frac{1}{2}\sum_{n=1}^{K} \phi_n\sigma_{11}^2(n).$$

By using similar arguments, for $(\sigma_{21}^2(1), \sigma_{21}^2(2), \ldots, \sigma_{21}^2(K))$, we have

$$\frac{1}{2}\sigma_{21}^2(n) + \sum_{n=1,j=1}^{K} \mu_{nj}\vec{\omega}(n) = \frac{1}{2}\sum_{n=1}^{K} \phi_n\sigma_{21}^2(n),$$

where $n \in \mathbb{N}$ and $\mu_{nj} > 0$ depicts the rate that switch from state n to state j.

Based on the above analysis, $R^s(t)$ is constructed as below,

$$\begin{aligned} R^s(t) \;:=\; & \frac{\beta(t)\Lambda(t)}{\nu(t)+\alpha(t)+h(t)+\zeta_2(u)+\frac{1}{2}\sum_{n=1}^{K}\phi_n\sigma_{21}^2(n)} \\ & -\lambda(t)-h(t)-\zeta_1(u)-l_2(t)\Lambda(t)-\frac{1}{2}\sum_{n=1}^{K}\phi_n\sigma_{11}^2(n), \end{aligned}$$

one can obtain the following results

$$\begin{aligned} \mathcal{L}U_1(t) \;\le\; & -R^s(t)+\max\{\sigma_{12}^2(n),l_1(t)\sigma_{22}^2(n)\}(S^2(t)+I^2(t)) \\ & +\max\{\sigma_{11}(n)\sigma_{12}(n),l_1(t)\sigma_{21}(n)\sigma_{22}(n)+\beta(t)\}(S(t)+I(t)). \end{aligned} \tag{13}$$

In the following part, we construct a T-periodic function as follows,

$$\Phi(t)=-\int_0^t (R_1-R^s(\tau))\mathrm{d}\tau,$$

where $R_1=\langle R^s\rangle_T$ is construct as below

$$\begin{aligned} R_1=\langle R^s\rangle_T \;=\; & \Bigg\langle \frac{\beta(t)\Lambda(t)}{\nu(t)+\alpha(t)+h(t)+\zeta_2(u)+\frac{1}{2}\sum_{n=1}^{K}\phi_n\sigma_{21}^2(n)} \\ & -\lambda(t)-h(t)-\zeta_1(u)-l_2(t)\Lambda(t)-\frac{1}{2}\sum_{n=1}^{K}\phi_n\sigma_{11}^2(n)\Bigg\rangle. \end{aligned}$$

Based on some simple computations, we can find the following results

$$\begin{aligned} \mathcal{L}(U_1(t)+\Phi(t)) \;\le\; & -R_1+\max\{\sigma_{12}^2(n),l_1(t)\sigma_{22}^2(n)\}(S^2(t)+I^2(t)) \\ & +\max\{\sigma_{11}(n)\sigma_{12}(n),l_1(t)\sigma_{21}(n)\sigma_{22}(n)+\beta(t)\}(S(t)+I(t)). \end{aligned} \tag{14}$$

Secondly, we define $U_2(t)$ as follows

$$U_2(t)=S(t)+I(t)+A(t)+\frac{1}{S(t)+I(t)+A(t)}.$$

By using Itô formula, it yields that

$$\begin{aligned} \mathcal{L}U_2(t) \;=\; & \Lambda(t)-h(t)U_1(t)-\alpha I(t)-\frac{\Lambda(t)-h(t)(S(t)+I(t)+A(t))-\alpha I(t)}{(S(t)+I(t)+A(t))^2} \\ & +\frac{S^2(t)(\sigma_{11}(n)+\sigma_{12}(n)S(t))^2+I^2(t)(\sigma_{21}(n)+\sigma_{22}(n)I(t))^2}{(S(t)+I(t)+A(t))^3} \\ & +\frac{A^2(t)(\sigma_{31}(n)+\sigma_{32}(n)A(t))^2}{(S(t)+I(t)+A(t))^3}+\frac{\int_{\mathbb{Y}}\left(\frac{1}{c_{\min}(u)}-1+c_{\max}(u)\right)\rho du}{S(t)+I(t)+A(t)} \\ \le\; & -\Big[h(t)-(\sigma_{12}^2(n)\vee\sigma_{22}^2(n)\vee\sigma_{32}^2(n))\Big]\left[S(t)+I(t)+A(t)+\frac{1}{S(t)+I(t)+A(t)}\right] \\ & +\Lambda(t)-\frac{\Lambda(t)}{(S(t)+I(t)+A(t))^2}+\frac{2h(t)+\alpha(t)+\int_{\mathbb{Y}}\left(\frac{1}{c_{\min}(u)}-1+c_{\max}(u)\right)\rho du}{S(t)+I(t)+A(t)} \\ & +(\sigma_{11}^2(n)\vee\sigma_{21}^2(n)\vee\sigma_{31}^2(n))\frac{1}{S(t)+I(t)+A(t)} \\ & +2[\sigma_{11}(n)\sigma_{12}(n)\vee\sigma_{21}(n)\sigma_{22}(n)\vee\sigma_{31}(n)\sigma_{32}(n)]. \end{aligned}$$

Hence, one can find the following results

$$\mathcal{L}U_2(t) \leq -R_2\left[S(t)+I(t)+A(t)+\frac{1}{S(t)+I(t)+A(t)}\right]+W_1(t), \tag{15}$$

where R_2 is defined in (9) and

$$\begin{cases} c_{\min}(u) = \min\{c_1(u), c_2(u), c_3(u)\}, \\ c_{\max}(u) = \max\{c_1(u), c_2(u), c_3(u)\}, \end{cases} \tag{16}$$

and

$$\begin{aligned} W_1(t) &= \Lambda(t)+\frac{\left[2h(t)+\alpha(t)+\int_{\mathbb{Y}}\left(\frac{1}{c_{\min}(u)}-1+c_{\max}(u)\right)\rho du\right]^2}{2\Lambda(t)} \\ &\quad +2(\sigma_{11}(n)\sigma_{12}(n)\vee\sigma_{21}(n)\sigma_{22}(n)\vee\sigma_{31}(n)\sigma_{32}(n)) \\ &\quad +\frac{[(\sigma_{11}^2(n)\vee\sigma_{21}^2(n)\vee\sigma_{31}^2(n))-(\sigma_{12}^2(n)\vee\sigma_{22}^2(n)\vee\sigma_{32}^2(n))]^2}{2\Lambda(t)}. \end{aligned}$$

Based on the boundedness of the parametric functions, the following results can be obtained that

$$\begin{aligned} W_1(t) \leq \overline{W}_1 &= \Lambda^u+2(\sigma_{11}^u\sigma_{12}^u\vee\sigma_{21}^u\sigma_{22}^u\vee\sigma_{31}^u\sigma_{32}^u) \\ &\quad +\frac{\left[2h^u+\alpha^u+\int_{\mathbb{Y}}\left(\frac{1}{c_{\min}^l(u)}-1+c_{\max}^u(u)\right)\rho du\right]^2}{2\Lambda^l} \\ &\quad +\frac{\left[((\sigma_{11}^2)^u\vee(\sigma_{21}^2)^u\vee(\sigma_{31}^2)^u)-((\sigma_{12}^2)^l\vee(\sigma_{22}^2)^l\vee(\sigma_{32}^2)^l)\right]^2}{2\Lambda^l}, \end{aligned} \tag{17}$$

where $\overline{W}_1$ represents the supreme of $W_1(t)$.

Thirdly, for any constant $\xi \in (0,1)$, $U_3(t)$ is defined as below,

$$U_3(t) = \Sigma_{i=1}^4(1+\chi_i(t))^{\xi}+l_3(t)(\Sigma_{i=1}^3\chi_i(t)),$$

For the convenience of proof, $F_i(t)$ $(i=1,\cdots,4)$ are constructed as below,

$$F_i(t) = -\frac{\xi(1-\xi)(1+\chi_i(t))^{\xi}\chi_i^2(t)}{2}(\sigma_{i1}(n)\wedge\sigma_{i2}(n))^2-\int_{\mathbb{Y}}(1+\chi_i(t))^{\xi}\rho du.$$

where $\chi(t) = (\chi_1(t), \chi_2(t), \chi_3(t), \chi_4(t)) = (S(t), I(t), A(t), M(t))$.

Based on utilizing the simple computations, one can be yield the results as below

$$\begin{aligned} \mathcal{L}U_3(t) &\leq [\max\{\xi\nu^u, \xi\Lambda_0^u+\xi\Lambda^u, \xi r^u\}-l_3(t)h(t)](\Sigma_{i=1}^3\chi_i(t)) \\ &\quad +W_2(t)+\Sigma_{i=1}^4F_i(t), \end{aligned} \tag{18}$$

where $l_3(t)$ and $W_2(t)$ are defined as follows,

$$l_3(t) = \frac{\max\{\xi\nu^u, \xi\Lambda_0^u+\xi\Lambda^u, \xi r^u\}}{h(t)},$$

$$\begin{aligned} W_2(t) &= \xi\Lambda^u + l_3(t)\Lambda^u - \xi(1+S(t))^{\xi-1}\left[\beta^l S(t)I(t) + \lambda^l \frac{S(t)M(t)}{M(t)+p^u} + h^l S(t)\right]. \\ &\quad -\xi(1+I(t))^{\xi-1}(\nu^l + \alpha^l + h^l)I(t) - \xi(1+A(t))^{\xi-1}(\lambda_0^l + h^l)A(t) \\ &\quad +\xi\beta^u S(t)I(t) - \xi(1+M(t))^{\xi-1}r_0^l M(t) + \xi r_0^u M_0^u \\ &\quad +\sum_{i=1}^{4}\int_{\mathbb{Y}}\left[(1+c_i(u)\chi_i(t)+\chi_i(t))^{\xi} - \xi(1+\chi_i(t))^{\xi-1}c_i(u)\chi_i(t)\right]\rho du. \end{aligned}$$

It is not difficult to show $W_2(t)$ is continuous in $(0,+\infty)$ and it follows from Lemma 1 that

$$\begin{aligned} W_2(t) \le \overline{W}_2 &= (\xi + l_3^u)\Lambda^u - \xi(1+G(\varepsilon))^{\xi-1}\left[\beta^l Q(\varepsilon) + \lambda^l \frac{Q(\varepsilon)}{G(\varepsilon)+p^u} + h^l Q(\varepsilon)\right] \\ &\quad -\xi(1+G(\varepsilon))^{\xi-1}(\nu^l + \alpha^l + \lambda_0^l + r_0^l + 2h^l)Q(\varepsilon) + \xi(\beta^u G(\varepsilon) + r_0^u M_0^u) \\ &\quad +\sum_{i=1}^{4}\int_{\mathbb{Y}}\left[(1+c_i(u)G(\varepsilon)+G(\varepsilon))^{\xi} - \xi(1+Q(\varepsilon))^{\xi-1}c_i(u)Q(\varepsilon)\right]\rho du, \end{aligned} \tag{19}$$

where $\overline{W}_2$ represents the supreme of $W_2(t)$.

Finally, we define $U(t)$ as follows,

$$U(t) = \Theta(U_1(t) + \Phi(t)) + U_2(t) + U_3(t),$$

where Θ is a sufficient large positive constant such that for $\chi_i(t) \to 0^+$ $(i = 1, \cdots, 4)$

$$-\Theta R_1 + \overline{W}_1 + \overline{W}_2 + \sup\sum_{i=1}^{4} F_i(t) < -2. \tag{20}$$

A continuous function $U(t)$ will be defined as below, and there exists a minimum $U(S_0, I_0, A_0, M_0, n_0)$ around $(S_0, I_0, A_0, M_0, n_0)$ when $U(t)$ tends to ∞.

Hence, we formulate a non-negative function as follows,

$$\tilde{U}(t) = U(S(t), I(t), A(t), M(t), n) - U_0(S_0, I_0, A_0, M_0, n_0).$$

By using (14), (15) and (18), one can be yielded that

$$\begin{aligned} \mathcal{L}\tilde{U}(t) &= \mathcal{L}U(S(t), I(t), A(t), M(t), n) - \mathcal{L}U(S_0, I_0, A_0, M_0, n_0) \\ &\le -\Theta R_1 + \Theta\max\{\sigma_{12}^2(n), l_1(t)\sigma_{22}^2(n)\}(S^2(t) + I^2(t)) \\ &\quad +\Theta\max\{\sigma_{11}(n)\sigma_{12}(n), l_1(t)\sigma_{21}(n)\sigma_{22}(n) + \beta(t)\}(S(t) + I(t)) \\ &\quad -\frac{R_2[1 + (\Sigma_{i=1}^3\chi_i(t))^2]}{\Sigma_{i=1}^3\chi_i(t)} + \overline{W}_1 + \overline{W}_2 + \Sigma_{i=1}^4 F_i(t), \end{aligned}$$

where $\overline{W}_1$ and $\overline{W}_2$ have been defined in (17) and (19).

When $\chi_i(t) \to 0$ or $\chi_i(t) \to \infty$, if $R_2 > 0$, one can find that

$$-\frac{R_2[1 + (\Sigma_{i=1}^3\chi_i(t))^2]}{\Sigma_{i=1}^3\chi_i(t)} \to -\infty. \tag{21}$$

Based on Lemma 1, when $\chi_i(t) \to 0$, it yields that

$$\begin{aligned} &\Theta\max\{\sigma_{12}^2(n), l_1(t)\sigma_{22}^2(n)\}(S^2(t) + I^2(t)) \\ &+\Theta\max\{\sigma_{11}(n)\sigma_{12}(n), l_1(t)\sigma_{21}(n)\sigma_{22}(n) + \beta(t)\}(S(t) + I(t)) \\ \le\ & 2\Theta\max\{\sigma_{12}^2(n), l_1(t)\sigma_{22}^2(n)\}G^2(\varepsilon) \\ &+2\Theta\max\{\sigma_{11}(n)\sigma_{12}(n), l_1(t)\sigma_{21}(n)\sigma_{22}(n) + \beta(t)\}G(\varepsilon). \end{aligned} \tag{22}$$

By using (20)–(22), if $R_1 > 0$ hold, when $\chi_i(t) \to 0$, it gives that

$$\begin{aligned} \mathcal{L}\widetilde{U}(t) &\leq -\frac{R_2[1 + (\Sigma_{i=1}^3 \chi_i(t))^2]}{\Sigma_{i=1}^3 \chi_i(t)} - \Theta R_1 + \overline{W}_1 + \overline{W}_2 \\ &\quad + \Theta \max\{\sigma_{12}^2(n), l_1(t)\sigma_{22}^2(n)\}(S^2(t) + I^2(t)) + \sup \Sigma_{i=1}^4 F_i(t) \\ &\quad + \Theta \max\{\sigma_{11}(n)\sigma_{12}(n), l_1(t)\sigma_{21}(n)\sigma_{22}(n) + \beta(t)\}(S(t) + I(t)) \\ &\leq -1. \end{aligned} \tag{23}$$

When $\chi_i(t) \to \infty$ $(i = 1, \cdots, 4)$, it is easy to show that

$$\widetilde{\Theta} \to -\infty, \tag{24}$$

where $\widetilde{\Theta}$ is constructed as follows,

$$\begin{aligned} \widetilde{\Theta} &= \Theta \max\{\sigma_{11}(n)\sigma_{12}(n), l_1(t)\sigma_{21}(n)\sigma_{22}(n) + \beta(t)\}(S(t) + I(t)) \\ &\quad + \Theta \max\{\sigma_{12}^2(n), l_1(t)\sigma_{22}^2(n)\}(S^2(t) + I^2(t)) - \int_{\mathbb{Y}} (1 + \chi_i(t))^{\xi} \rho du \\ &\quad - \frac{\xi(1-\xi)(1 + \chi_i(t))^{\xi} \chi_i^2(t)}{2} (\sigma_{i1}(n) \wedge \sigma_{i2}(n))^2. \end{aligned} \tag{25}$$

Furthermore, if $R_1 > 0$ and $R_2 > 0$ hold, it follows from (21) and (24), it yields that

$$\begin{aligned} \mathcal{L}\widetilde{U}(t) &\leq -\frac{R_2[1 + (\Sigma_{i=1}^3 \chi_i(t))^2]}{\Sigma_{i=1}^3 \chi_i(t)} - \Theta R_1 + \overline{W}_1 + \overline{W}_2 \\ &\quad - \frac{\xi(1-\xi)(1 + \chi_i(t))^{\xi} \chi_i^2(t)}{2} (\sigma_{i1}(n) \wedge \sigma_{i2}(n))^2 - \int_{\mathbb{Y}} (1 + \chi_i(t))^{\xi} \rho du \\ &\quad + \Theta \max\{\sigma_{11}(n)\sigma_{12}(n), l_1(t)\sigma_{21}(n)\sigma_{22}(n) + \beta(t)\}(S(t) + I(t)) \\ &\quad + \Theta \max\{\sigma_{12}^2(n), l_1(t)\sigma_{22}^2(n)\}(S^2(t) + I^2(t)) \\ &\leq -\infty - \Theta R_1 + \overline{W}_1 + \overline{W}_2 - \infty \\ &< -1, \end{aligned} \tag{26}$$

where $\overline{W}_1$ and $\overline{W}_2$ have been defined in (17) and (19).

According to Lemma 1 and (26), we can find following results

(i) system (4) exists a unique global solution;
(ii) we can find a T-periodic function $\widetilde{U}(t) \in C^1 \times \mathbb{N}$ and $\mathcal{L}\widetilde{U}(t) < -1$ on the outside of some compact set.

Hence, sufficient condition (i) and condition (ii) in Theorem 3.8 [40] all hold, which means that system (4) exists a nontrival positive T-periodic solution.

The proof is ending. □

Theorem 2. *When $\widetilde{R}^s > 0$ holds, the solution of system* (4) *is f-exponentially ergodic, where $\widetilde{R}^s = \sum_{n=1}^K \phi_n \widetilde{R}_n$, and $\widetilde{R}_n$ $(n = 1, 2, \cdots, K)$ are defined as follows,*

$$\begin{aligned} \widetilde{R}_n &= \beta^l Q(\varepsilon) - (\nu^u + \alpha^u + h^u) - \frac{(\vartheta + 1)(\sigma_{21}^u(n) + \sigma_{22}^u(n) G(\varepsilon))^2}{2} \\ &\quad - \int_{\mathbb{Y}} \left[\frac{(1 + c_2(u))^{-\vartheta} - 1}{\vartheta} - c_2(u) \right] \rho du, \quad (0 < \vartheta < 1). \end{aligned} \tag{27}$$

Proof. For the diffusion matrix form of system (4), we have

$$\begin{cases} D_{\min} \parallel \chi(t) \parallel^2 \leq \Sigma_{i=1}^4 \left[\sigma_{i1}(n)\chi_i(t) + \sigma_{i2}(n)\chi_i^2(t)\right]^2 \chi_i^2(t), \\ \Sigma_{i=1}^4 \left[\sigma_{i1}(n)\chi_i(t) + \sigma_{i2}(n)\chi_i^2(t)\right]^2 \chi_i^2(t) \leq D_{\max} \parallel \chi(t) \parallel^2, \end{cases} \tag{28}$$

where $D_{\min}$ and $D_{\max}$ are defined as follows,

$$\begin{cases} D_{\min} = \min\Big\{\Sigma_{i=1}^{4}\big[\sigma_{i1}(n)\chi_i(t) + \sigma_{i2}(n)\chi_i^2(t)\big]^2\chi_i^2(t)\Big\}, \\ D_{\max} = \max\Big\{\Sigma_{i=1}^{4}\big[\sigma_{i1}(n)\chi_i(t) + \sigma_{i2}(n)\chi_i^2(t)\big]^2\chi_i^2(t)\Big\}. \end{cases}$$

and $\chi(t) = (\chi_1(t), \cdots, \chi_4(t)) = (S(t), I(t), A(t), M(t))$.

It follows from (28) that uniform elliptic conditions hold for the diffusion matrix of system (4). Furthermore, the diffusion of initial value $(S(t), I(0), A(0), M(0), n)$ transition probability exists a positive smooth density on $\mathbb{R}^4 \times \mathbb{R}^4 \times \mathbb{R}^4 \times \mathbb{R}^4 \times \mathbb{N}$.

Furthermore, according to Lemma 2 in [41], for the following linear equation,

$$(\widetilde{R}_1, \cdots, \widetilde{R}_K)^T - \Gamma\chi(t) = (\widetilde{R}^s, \cdots, \widetilde{R}^s)^T, \tag{29}$$

where $\widetilde{R}^s = \Sigma_{n=1}^{K}\phi_n\widetilde{R}_n$. It follows from simple computations, we can find a unique positive solution $(\varpi_1, \cdots, \varpi_K)^T$ of Equation (29).

If $\widetilde{R}_n > 0\ (n = 1, 2, \cdots, K)$ hold, it is easy to show that $\widetilde{R}^s = \Sigma_{n=1}^{K}\phi_n\widetilde{R}_n > 0$, where $\Sigma_{n=1}^{K}\phi_n = 1$ and $\phi_n > 0$ hold for any $n \in \mathbb{N}$.

Based on the above analysis, we define $U_4(t)$ and $U_5(t)$ as follows,

$$\begin{cases} U_4(t) = \Sigma_{i=1}^{3}\chi_i(t) + \frac{\alpha^l}{r^u}M(t), \\ U_5(t) = (1 + \vartheta\varpi_n)I(t)^{-\vartheta}. \end{cases}$$

Based primary on utilizing the Itô formula to system (4), it yields that

$$\begin{aligned} \mathcal{L}U_4(t) &= \Lambda^u - h^l(\Sigma_{i=1}^{3}\chi_i(t)) - \alpha^l I(t) \\ &\quad + \frac{\alpha^l}{r^u}\left[r^u I(t)\left(1 - \theta^l\frac{A(t)}{\omega^u + A(t)}\right) - r_0^l M(t) + r_0^u M_0^u\right]. \end{aligned} \tag{30}$$

and

$$\begin{aligned} \mathcal{L}U_5(t) &= \vartheta(1 + \vartheta\varpi_n)I(t)^{-\vartheta}[\nu(t) + \alpha(t) + h(t) - \beta(t)S(t)] \\ &\quad + \frac{\vartheta(1 + \vartheta\varpi_n)I(t)^{-\vartheta}(\vartheta + 1)(\sigma_{21}(n) + \sigma_{22}(n)I(t))^2}{2} \\ &\quad + I(t)^{-\vartheta}\left\{\vartheta\sum_{n=1,j=1}\mu_{nj}\varpi_n + (1 + \vartheta\varpi_n)\int_{\mathbb{Y}}[(1 + c_2(u))^{-\vartheta} - 1 + \vartheta c_2(u)]\rho du\right\} \\ &= \vartheta I(t)^{-\vartheta}(1 + \vartheta\varpi_n)[-\beta(t)S(t) + \nu(t) + \alpha(t) + h(t)] \\ &\quad + \vartheta I(t)^{-\vartheta}(1 + \vartheta\varpi_n)\frac{(\vartheta + 1)(\sigma_{21}(n) + \sigma_{22}(n)I(t))^2}{2} \\ &\quad + \vartheta I(t)^{-\vartheta}\sum_{n=1,j=1}\mu_{nj}\varpi_n + \int_{\mathbb{Y}}\left[\frac{(1 + c_2(u))^{-\vartheta} - 1}{\vartheta} + c_2(u)\right]\rho du \\ &= \vartheta I(t)^{-\vartheta}\sum_{n=1,j=1}\mu_{nj}\varpi_n - (1 + \vartheta\varpi_n)[\beta(t)S(t) - (\nu(t) + \alpha(t) + h(t)] \\ &\quad - \vartheta I(t)^{-\vartheta}\left\{\frac{\vartheta + 1}{2}[\sigma_{21}(n) + \sigma_{22}(n)I(t)]^2 + \int_{\mathbb{Y}}\left[\frac{(1 + c_2(u))^{-\vartheta} - 1}{\vartheta} + c_2(u)\right]\rho du\right\}. \end{aligned}$$

If (29) holds, then it is easy to show that $\sum_{n=1,j=1}\mu_{nj}\varpi_n = \widetilde{R_n} - \widetilde{R^s}$, which yields that

$$
\begin{aligned}
\mathcal{L}U_5(t) &= \vartheta I(t)^{-\vartheta}\left\{\sum_{n=1,j=1}\mu_{nj}\varpi_n-(1+\vartheta\varpi_n)[\beta(t)S(t)-(\nu(t)+\alpha(t)+h(t))]\right\}\\
&\quad -\vartheta I(t)^{-\vartheta}\left\{\frac{\vartheta+1}{2}[\sigma_{21}(n)+\sigma_{22}(n)I(t)]^2+\int_{\mathbb{Y}}\left[\frac{(1+c_2(u))^{-\vartheta}-1}{\vartheta}+c_2(u)\right]\rho du\right\}\\
&\le \vartheta I(t)^{-\vartheta}\left\{\widetilde{R_n}-\widetilde{R^s}-(1+\vartheta\varpi_n)\left[\beta^l Q(\varepsilon)-(\nu^u+\alpha^u+h^u)\right]\right\}\\
&\quad +\frac{(\vartheta+1)\vartheta I(t)^{-\vartheta}(1+\vartheta\varpi_n)}{2}[\sigma_{21}(n)+\sigma_{22}(n)G(\varepsilon)]^2\\
&\quad +\vartheta I(t)^{-\vartheta}(1+\vartheta\varpi_n)\int_{\mathbb{Y}}\left[\frac{(1+c_2(u))^{-\vartheta}-1}{\vartheta}+c_2(u)\right]\rho du\\
&\le \vartheta I(t)^{-\vartheta}\left[\widetilde{R_n}-\widetilde{R^s}-(1+\vartheta\varpi_n)\widetilde{R_n}\right]\\
&= \vartheta I(t)^{-\vartheta}[-\vartheta\varpi_n\widetilde{R_n}-\widetilde{R^s}]. \qquad (31)
\end{aligned}
$$

By using (30) and (31), it can be obtained that

$$
\begin{aligned}
\mathcal{L}[U_4(t)+U_5(t)] &= \mathcal{L}\left[\Sigma_{i=1}^3\chi_i(t)+\frac{\alpha^l}{r^u}M(t)+(1+\vartheta\varpi_n)I(t)^{-\vartheta}\right]\\
&\le -\widetilde{M}\left[\Sigma_{i=1}^3\chi_i(t)+\frac{\alpha^l}{r^u}M(t)+(1+\vartheta\varpi_n)I(t)^{-\vartheta}\right]+\Lambda^u+\frac{r_0^u\alpha^l}{r^u}\\
&= -\widetilde{M}[U_4(t)+U_5(t)]+\Lambda^u+\frac{r_0^u\alpha^l}{r^u},
\end{aligned}
$$

where $\widetilde{M}=\min\left\{h^l, r_0^l, \frac{K(K\varpi_n\widetilde{R}_n+\widetilde{R^s})}{1+K\varpi_n}\right\}$ $(n=1,2,\cdots,K)$.

According to Theorem 6.1 in [42] and Theorem 6.3 in [43], all the sufficient conditions for existence of exponential ergodicity hold.

Hence, based on the above analysis, if $\widetilde{R}^s>0$, the positive solution of system (4) is f-exponentially ergodic.

The proof is ending. □

Remark 2. *Let $\mathbb{P}(t,(\chi(t),n),\cdot)$ depict the transition probability of $(\chi(t),n)$. According to Theorem 2 of this paper, for some positive constant $\delta\in(0,1)$, it can be found that $(\chi(t),n)$ is considered to be f-exponentially ergodic if there exists a probability measure $\pi(\cdot)$ and a finite-valued function $v(\chi(t),n)$ such that*

$$\|\mathbb{P}(t,(\chi(t),n),\cdot)-\pi(\cdot)\|\le v(\chi(t),n)\delta^t,$$

holds for all $t\ge 0$ and $(\chi(t),n)\in\mathbb{R}_+^4\times\mathbb{N}$.

In the next part, we will concentrate on hybrid dynamic impacts of random perturbations and media coverage on the variations of epidemic transmission.

Theorem 3. *For the infected individual $I(t)$ of system* (4),

(i) if $R_I<1$ and R_I is defined in (32),

$$R_I=\frac{\beta^u\Lambda^u}{h^l(\nu^l+\alpha^l+h^l+\zeta_2^l)}+\frac{\sigma_2^u\Lambda^u}{(\alpha^l+h^l)(\nu^l+\alpha^l+h^l+\zeta_2^l)}, \qquad (32)$$

then the number of infected individual $I(t)$ of system (4) *satisfies*

$$\lim_{t\to\infty}I(t)=0,$$

which means infected individual tends to zero exponentially;

(ii) *if $R_E > 0$ and R_E is defined in (33),*

$$R_E = \frac{\Lambda^u(\beta^l + h^l)}{\lambda^u + h^u + \zeta_1^u + \frac{1}{2}\sum_{n=1}^{K}\phi_n(\sigma_{11}^u + \sigma_{12}^u G(\varepsilon))^2} - \left[r_0^u M_0^u + \nu^u + \alpha^u + h^u + \frac{1}{2}(\sigma_{21}^u + \sigma_{22}^u G(\varepsilon))^2 + \zeta_2^u\right], \tag{33}$$

then the number of infected individual $I(t)$ of system (4) meets

$$\liminf_{t\to\infty}\frac{1}{t}\int_0^t I(s)ds > 0,$$

which means infected individual will be persistent in the average sense.

Proof. (i) Based on applying Itô's formula to system (4), we can obtain the results as below

$$\begin{aligned} d\ln I(t) = {} & [\beta(t)S(t) - (\nu(t) + \alpha(t) + h(t))]dt - \frac{(\sigma_{21}(n) + \sigma_{22}(n)I(t))^2}{2}dt \\ & + \left[\int_{\mathbb{Y}}[\ln(1 + c_2(u)) - c_2(u)]\rho du\right]dt + (\sigma_{21}(n) + \sigma_{22}(n)I(t))dB_2(t) \\ & + \int_{\mathbb{Y}}\ln(1 + c_2(u))\widetilde{X}(dt, du). \end{aligned}$$

Based primary on integrating from 0 to t among both sides of the above equation, the following results can be yielded

$$\begin{aligned} & \ln I(t) - \ln I(0) \\ = {} & \int_0^t[\beta(s)S(s) - (\nu(s) + \alpha(s) + h(s))]ds \\ & - \int_0^t \frac{(\sigma_{21}(n) + \sigma_{22}(n)I(s))^2}{2}ds + \int_0^t\left[\int_{\mathbb{Y}}(\ln(1 + c_2(u)) - c_2(u))\rho du\right]ds \\ & + \int_0^t(\sigma_{21}(n) + \sigma_{22}(n)I(s))dB_2(t) + \int_0^t\int_{\mathbb{Y}}\ln(1 + c_2(u))\widetilde{X}(du, ds). \end{aligned}$$

Further computations show

$$\begin{aligned} & \frac{\ln I(t) - \ln I(0)}{t} \\ \leq {} & \beta^u\langle S(t)\rangle - (\nu^l + \alpha^l + h^l) - \sigma_{21}(n)\sigma_{22}(n)\langle I(t)\rangle - \zeta_2^l \\ & + \frac{\int_0^t(\sigma_{21}(n) + \sigma_{22}(n)I(t))dB_2(t)}{t} + \frac{\int_0^t\int_{\mathbb{Y}}\ln(1 + c_2(u))\widetilde{X}(du, ds)}{t} \\ \leq {} & \beta^u\frac{\Lambda^u}{h^l} + \widetilde{\sigma}^*\frac{\Lambda^u}{(\alpha^l + h^l)} - (\nu^l + \alpha^l + h^l) - \zeta_2^l \\ & + \frac{\int_0^t(\sigma_{21}(n) + \sigma_{22}(n)I(t))dB_2(t)}{t} + \frac{\int_0^t\int_{\mathbb{Y}}\ln(1 + c_2(u))\widetilde{X}(du, ds)}{t}, \end{aligned} \tag{34}$$

where $\widetilde{\sigma}^* = \max\{-\sigma_{21}(n)\sigma_{22}(n)\}$.

By using the mathematical properties of white noise, it is not difficult to show that $\widetilde{\sigma}^* \geq 0$. For $\int_0^t (\sigma_{21}(n) + \sigma_{22}(n)I(t))\mathrm{d}B_2(t)$, based on Lemma 1 (the boundedness of $I(t)$) and exponential martingale inequality from Lemma 3 that

$$\begin{aligned}
&\limsup_{t\to\infty} \frac{\left\langle \int_0^t (\sigma_{21}(n) + \sigma_{22}(n)I(t))\mathrm{d}B_2(t), \int_0^t (\sigma_{21}(n) + \sigma_{22}(n)I(t))\mathrm{d}B_2(t) \right\rangle}{t} \\
= &\limsup_{t\to\infty} \frac{1}{t}\int_0^t [(\sigma_{21}(n) + \sigma_{22}(n)I(t))I(t)]^2 \mathrm{d}s \\
< &(\sigma_{21}^u + \sigma_{22}^u G(\varepsilon))^2 G(\varepsilon)^2 \\
< &\infty.
\end{aligned} \tag{35}$$

Hence, it can be concluded that

$$\lim_{t\to\infty} \frac{\int_0^t (\sigma_{21}(n) + \sigma_{22}(n)I(t))\mathrm{d}B_2(t)}{t} = 0. \tag{36}$$

Let $\psi_3(t) = \int_0^t \int_{\mathbb{Y}} \ln(1 + c_2(u))\widetilde{X}(\mathrm{d}u, \mathrm{d}s)$, by applying the exponential martingales inequality, it follows from similar arguments in Lemma 3 that

$$\mathbb{P}\left\{ \sup_{0\leq i\leq j} \left[\psi_3 - \frac{1}{2}\langle \psi_3, \psi_3 \rangle \right] > 2\ln j \right\} \leq \frac{1}{j^2}. \tag{37}$$

We can easily to find a random integer $j_0^* = j_0^*(\omega)$ holds for the almost whole $\omega \in \Omega$, and it can be obtained that

$$\sup_{0\leq i\leq j_0^*} \left[\psi_3 - \frac{1}{2}\langle \psi_3, \psi_3 \rangle \right] \leq 2\ln j_0^* \tag{38}$$

holds for $\omega \in \Omega$ most likely.

Hence, it can be obtained that

$$\psi_3 \leq 2\ln j_0^* + \frac{1}{2}\langle \psi_3, \psi_3 \rangle \tag{39}$$

holds for all $0 \leq t \leq j_0^*$.

By taking the superior limit for (34), if $R_I < 1$ holds, then it yields that

$$\begin{aligned}
\limsup_{t\to\infty} \frac{\ln I(t) - \ln I(0)}{t} \leq\ & \beta^u \frac{\Lambda^u}{h^l} + \widetilde{\sigma}^* \frac{\Lambda^u}{(\alpha^l + h^l)} - (\nu^l + \alpha^l + h^l + \zeta_2^l) \\
&+ \limsup_{t\to\infty} \frac{\int_0^t \int_{\mathbb{Y}} \ln(1 + c_2(u))\widetilde{X}(\mathrm{d}u, \mathrm{d}s)}{t} \\
&+ \limsup_{t\to\infty} \frac{\int_0^t (\sigma_{21}(n) + \sigma_{22}(n)I(t))\mathrm{d}B_2(t)}{t} \\
\leq\ & \beta^u \frac{\Lambda^u}{h^l} + \widetilde{\sigma}^* \frac{\Lambda^u}{(\alpha^l + h^l)} - (\nu^l + \alpha^l + h^l + \zeta_2^l) + \limsup_{t\to\infty} \frac{2\ln j_0^*}{j_0^* - 1} \\
&+ \limsup_{t\to\infty} \frac{\int_0^t (\sigma_{21} + \sigma_{22}(n)I(t))\mathrm{d}B_2(t)}{t}.
\end{aligned} \tag{40}$$

holds for all $t \leq j_0^*$. By utilizing the above equation, we can find the following results

$$\limsup_{t\to\infty} \frac{\ln I(t) - \ln I(0)}{t} < 0,$$

which derives that $\lim_{t\to\infty} I(t) = 0$.

The proof of (i) of Theorem 3 is ending.

(ii) First, we construct $U_6(t)$ as follows,

$$U_6(t) = S(t) - l_4 - l_4 \ln \frac{S(t)}{l_4} + I(t) - 1 - \ln I(t) + A(t) + M(t) + \vec{\omega}(n),$$

where l_4 is defined as follows: $l_4 = \frac{\Lambda^u(\beta^l + d^l)}{\left[\lambda^u + h^u + \zeta_1^u + \sum_{n=1}^{K} \frac{\phi_n}{2}(\sigma_{11}^u + \sigma_{12}^u G(\varepsilon))^2\right]^2}$.

Based primary on utilizing Itô's formula and simple computations, one can find that

$$\begin{aligned}
\mathcal{L}U_6(t) \leq\ & \Lambda^u - h^l S(t) + r^u I(t) + r_0^u M_0^u - l_4\left[\frac{\Lambda^l}{S(t)} - \beta^u - \lambda^u - h^u\right] \\
& -\beta^l S(t) + \frac{1}{2}\left[l_4(\sigma_{11}^u + \sigma_{21}^u S(t))^2 + (\sigma_{21}^u + \sigma_{22}^u I(t))^2\right] \\
& +\nu^u + \alpha^u + h^u + l_4\zeta_1^u + \zeta_2^u + \sum_{n=1,j=1}^{K} \mu_{nj}\vec{\omega}(n) \\
\leq\ & -2\sqrt{l_4\Lambda^u(\beta^l + d^l)} + l_4\left[\lambda^u + h^u + \frac{1}{2}(\sigma_{11}^u + \sigma_{12}^u G(\varepsilon))^2 + \zeta_1^u\right] \\
& +r_0^u M_0^u + \nu^u + \alpha^u + h^u + \frac{1}{2}[\sigma_{21}^u + \sigma_{22}^u G(\varepsilon)]^2 \\
& +(l_4\beta^u + r^u)I(t) + \zeta_2^u + \sum_{n=1,j=1}^{K} \mu_{nj}\vec{\omega}(n),
\end{aligned}$$

which derives that

$$\begin{aligned}
dU_6(t) \leq\ & -\frac{\Lambda^u(\beta^l + d^l)}{\lambda^u + h^u + \zeta_1^u + \frac{1}{2}\sum_{n=1}^{K}\phi_n[\sigma_{11}^u + \sigma_{12}^u G(\varepsilon)]^2} \\
& +r_0^u M_0^u + (\nu^u + \alpha^u + h^u) + \frac{1}{2}(\sigma_{21}^u + \sigma_{22}^u G(\varepsilon))^2 + \zeta_2^u + (l_1\beta^u + r^u)I(t) \\
& +\sum_{i=1}^{4}\left[(\sigma_{i1}(n) + \sigma_{i2}(n)\chi_i(t))\chi_i(t)dB_i(t) + \int_{\mathbb{Y}} c_i(u)\chi_i(t-)\widetilde{X}(du, dt)\right] \\
& +l_4[\sigma_{11}(n) + \sigma_{12}(n)S(t)]dB_1(t) + [\sigma_{21}(n) + \sigma_{22}(n)I(t)]dB_2(t) \\
& +l_4\int_{\mathbb{Y}} \ln(1 + c_1(u))\widetilde{X}(du, dt) + \int_{\mathbb{Y}} \ln(1 + c_2(u))\widetilde{X}(du, dt),
\end{aligned}$$

holds for $i = 1, 2, 3, 4$.

By integrating both sides of above equation from 0 to t and dividing by t, one can yields that

$$\begin{aligned}
\frac{U_6(t) - U_6(0)}{t} \leq\ & -R_E + \frac{l_4\beta^u + r^u}{t}\int_0^t I(s)ds \\
& +\sum_{i=1}^{4}\left[\frac{1}{t}\int_0^t (\sigma_{i1}(n) + \sigma_{i2}(n)\chi_i(t))\chi_i(t)dB_i + \frac{1}{t}\int_0^t\int_{\mathbb{Y}} c_i(u)\chi_i(t-)\widetilde{X}(du, ds)\right] \\
& +\frac{l_4}{t}\int_0^t (\sigma_{11}(n) + \sigma_{12}(n)S(t))dB_1 + \frac{1}{t}\int_0^t (\sigma_{21}(n) + \sigma_{22}(n)I(t))dB_2 \\
& +\frac{l_4}{t}\int_0^t\int_{\mathbb{Y}} \ln(1 + c_1(u))\widetilde{X}(du, ds) + \frac{1}{t}\int_0^t\int_{\mathbb{Y}} \ln(1 + c_2(u))\widetilde{X}(du, ds),
\end{aligned} \tag{41}$$

where R_E has been defined in (33).

Based on similar arguments utilized in Lemma 3 of this paper, it gives that

$$\begin{cases} \lim_{t\to\infty}\frac{1}{t}\int_0^t(\sigma_{i1}(n)+\sigma_{i2}(n)\chi_i(t))\chi_i(t)\mathrm{d}B_i(t)=0,\\ \lim_{t\to\infty}\frac{1}{t}\int_0^t\int_{\mathbb{Y}}c_i(u)\chi_i(t-)\widetilde{X}(du,\mathrm{d}s)=0,\\ \lim_{t\to\infty}\frac{l_4}{t}\int_0^t\int_{\mathbb{Y}}\ln(1+c_1(u))\widetilde{X}(du,\mathrm{d}s)=0,\\ \lim_{t\to\infty}\frac{l_4}{t}\int_0^t(\sigma_{11}(n)+\sigma_{12}(n)S(t))\mathrm{d}B_1(t)=0,\\ \lim_{t\to\infty}\frac{1}{t}\int_0^t(\sigma_{21}(n)+\sigma_{22}(n)I(t))\mathrm{d}B_2(t)=0,\\ \lim_{t\to\infty}\frac{1}{t}\int_0^t\int_{\mathbb{Y}}\ln(1+c_2(u))\widetilde{X}(du,\mathrm{d}s)=0, \end{cases}$$

holds for $i=1,2,3,4$.

Based on taking the inferior limit on the both sides of (41), if $R_E>0$ holds, then one can be obtained that

$$\liminf_{t\to\infty}\frac{1}{t}\int_0^t I(s)\mathrm{d}s\geq\frac{R_E}{l_4\beta^u+r^u}>0,$$

which means infected individual is persistent in mean.

This proof of (ii) of Theorem 3 is ending. □

3. Numerical Simulations

In this chapter, we will prove the results obtained above through numerical simulation, which are utilized to show hybrid dynamic impacts of media coverage and nonlinear perturbations on random dynamics of system (4). The parameter functions utilized in this section are as follows,

$$\begin{cases} \Lambda(t)=5+0.5\sin t, & \beta(t)=3+0.3\sin t\\ \lambda(t)=0.12+0.05\sin t, & \lambda_0(t)=0.08+0.01\sin t,\\ \nu(t)=0.02+0.01\sin t, & \alpha(t)=1\times10^{-3}+5\times10^{-4}\sin t,\\ h(t)=4\times10^{-3}+8\times10^{-4}\sin t, & r(t)=0.006+0.003\sin t,\\ r_0(t)=0.05+0.01\sin t, & \theta(t)=5\times10^{-3}+1\times10^{-3}\sin t,\\ \omega(t)=0.06+0.01\sin t, & p(t)=1.2+0.6\sin t,\\ M_0(t)=5+0.02\times\sin t. \end{cases}$$

It is assumed that $n\in\mathbb{N}=\{1,2,3,4\}$, $\mathbb{Y}=\{1,2,3,4,5\}$ and the transition matrix is given as follows:

$$\Gamma=\begin{pmatrix} -1 & 1 & 0 & 0\\ 2 & -2 & 0 & 0\\ 0 & 0 & -3 & 3\\ 0 & 0 & 4 & -4 \end{pmatrix}.$$

Hence, it follows from simple algebraic computations that $(\phi_1,\phi_2,\phi_3,\phi_4)=\frac{1}{\sqrt{30}}(2,1,4,3)$.

3.1. Numerical Simulation I

It is assumed that $\sigma_{i1}(n)=0.04+0.01\sin n,\sigma_{i2}(n)=\sqrt{0.1+0.05\sin n}$ $(i=1,2,3,4)$, if $c_1(u)=0.5,c_2(u)=0.03,c_3(u)=0.02,c_4(u)=0.05,\vartheta=0.5$, then it is easy to show that $\widetilde{R}_1=0.0349,\widetilde{R}_2=-0.0252,\widetilde{R}_3=0.0445,\widetilde{R}_4=-0.0273$, which follows that $\widetilde{R}_s=0.0257>0$. Based on Theorem 2, one can be concluded that $(S(t),I(t),A(t),M(t),n)$ of system (4) is f-exponentially ergodic. The dynamical responses $S(t),I(t),A(t),M(t)$ of system (4) with initial value $(0.4,0.1,0.05,5)$ are plotted in Figure 1a, Figure 1b, Figure 1c, Figure 1d, respectively, which indicates an exponential convergence.

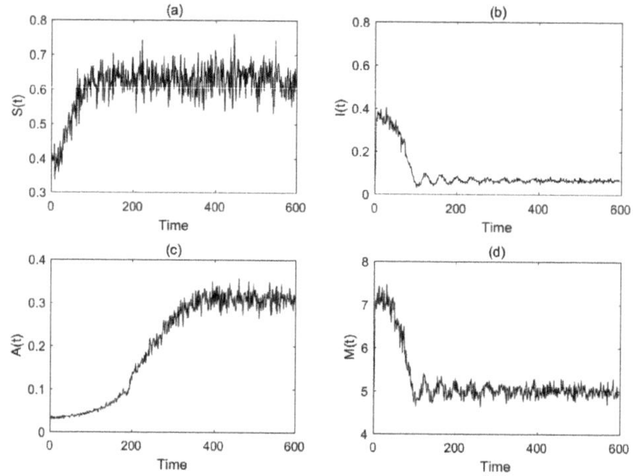

Figure 1. If $\sigma_{i1}(n) = 0.04 + 0.01 \sin n, \sigma_{i2}(n) = \sqrt{0.1 + 0.05 \sin n}$ $(i = 1,2,3,4)$, $c_1(u) = 0.5, c_2(u) = 0.03, c_3(u) = 0.02, c_4(u) = 0.05, \vartheta = 0.5$, then it is easy to show that $\widetilde{R}_1 = 0.0349, \widetilde{R}_2 = -0.0252, \widetilde{R}_3 = 0.0445, \widetilde{R}_4 = -0.0273$, which follows that $\widetilde{R}_s = 0.0257 > 0$. The dynamical responses $S(t), I(t), A(t), M(t)$ of system (4) with initial value $(0.4, 0.1, 0.05, 5)$ are plotted in (**a–d**), respectively, which indicates an exponential convergence.

If $c_1(u) = 0.4, c_2(u) = 0.03, c_3(u) = 0.02, c_4(u) = 0.05, \vartheta = 0.5$, then it is easy to show that $\widetilde{R}_1 = 0.0426, \widetilde{R}_2 = -0.0849, \widetilde{R}_3 = 0.0551, \widetilde{R}_4 = -0.0415$, which follows that $\widetilde{R}_s = 0.0176 > 0$. Based on Theorem 2, one can be concluded that $(S(t), I(t), A(t), M(t), n)$ of system (4) is f-exponentially ergodic. The dynamical responses $S(t), I(t), A(t), M(t)$ of system (4) with initial value $(0.35, 0.15, 0.04, 5)$ are plotted in Figure 2a, Figure 2b, Figure 2c, Figure 2d, respectively, which indicates an exponential convergence.

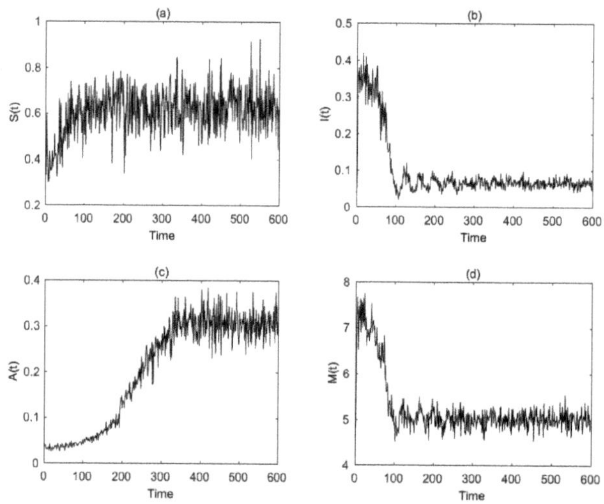

Figure 2. If $\sigma_{i1}(n) = 0.04 + 0.01 \sin n, \sigma_{i2}(n) = \sqrt{0.1 + 0.05 \sin n}$ $(i = 1,2,3,4)$, $c_1(u) = 0.4, c_2(u) = 0.03, c_3(u) = 0.02, c_4(u) = 0.05, \vartheta = 0.5$, then it is easy to show that $\widetilde{R}_1 = 0.0426, \widetilde{R}_2 = -0.0849, \widetilde{R}_3 = 0.0551, \widetilde{R}_4 = -0.0415$, which follows that $\widetilde{R}_s = 0.0176 > 0$. The dynamical responses $S(t), I(t), A(t), M(t)$ of system (4) with initial value $(0.35, 0.15, 0.04, 5)$ are plotted in (**a–d**), respectively, which indicates an exponential convergence.

In order to show the dynamic effects of Lévy jumps, values of $\widetilde{R}_s$ are given in Table 1 under four different $c_i(u)$ $(i = 1, 2, 3, 4)$, and the detailed values can be found in Table 1. Furthermore, total variation norms $\|\mathbb{P}(t, (\chi(t), n), \cdot) - \pi(\cdot)\|$ are plotted in Figure 3 due to variations of c_i $(i = 1, 2, 3, 4)$ under four different cases.

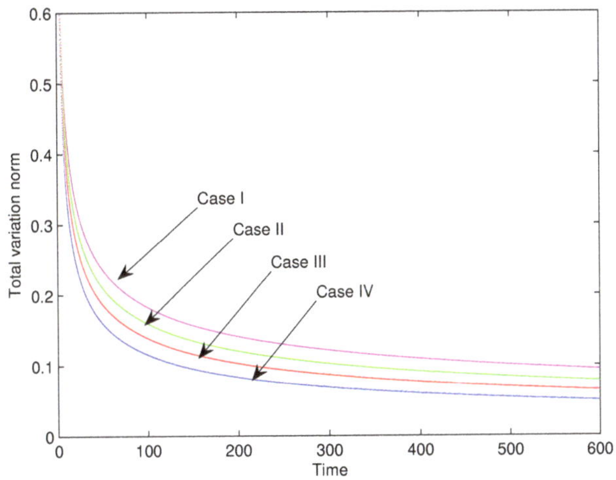

Figure 3. Total variation norms $\|\mathbb{P}(t, (\chi(t), n), \cdot) - \pi(\cdot)\|$ are plotted due to variations of $c_i(u)$ $(i = 1, 2, 3, 4)$ under four different cases corresponding to Table 1.

Table 1. When $\sigma_{i1}(n) = 0.04 + 0.01 \sin n, \sigma_{i2}(n) = \sqrt{0.1 + 0.05 \sin n}$ $(i = 1, 2, 3, 4)$, values of $\widetilde{R}_s$ are given under four different values of c_i $(i = 1, 2, 3, 4)$.

	Values of $c_i(u)$ $(i = 1, 2, 3, 4)$	$\widetilde{R}_s$
Case I	$c_1(u) = 0.5, c_2(u) = 0.03, c_3(u) = 0.02, c_4(u) = 0.05$	0.0257
Case II	$c_1(u) = 0.4, c_2(u) = 0.03, c_3(u) = 0.02, c_4(u) = 0.05$	0.0176
Case III	$c_1(u) = 0.3, c_2(u) = 0.02, c_3(u) = 0.01, c_4(u) = 0.03$	0.0159
Case IV	$c_1(u) = 0.2, c_2(u) = 0.02, c_3(u) = 0.01, c_4(u) = 0.02$	0.0113

3.2. Numerical Simulation II

It is assumed that $\sigma_{i1}(n) = 0.04 + 0.01 \sin n, \sigma_{i2}(n) = \sqrt{0.1 + 0.05 \sin n}$ $(i = 1, 2, 3, 4)$. If $c_1(u) = 0.5, c_2(u) = 0.03, c_3(u) = 0.02, c_4(u) = 0.05$ and initial value $(0.4, 0.1, 0.05, 5)$, then it can be obtained that $R_I = 0.7143 < 1$, which follows from Theorem 3-(i) that the infected individual $I(t)$ of system (4) tends to zero exponentially. On the other hand, if $c_1(u) = 0.5, c_2(u) = 0.02, c_3(u) = 0.01, c_4(u) = 0.03$, then it can be obtained that $R_E = 0.2683 > 0$, which follows from Theorem 3-(ii) that the number of infected individual $I(t)$ of system (4) is persistent in average sense. Dynamical responses of the number of infected individual $I(t)$ are shown in Figure 4a and Figure 4b, respectively.

It is assumed that $\sigma_{i1}(n) = 0.09 + 0.04 \sin n, \sigma_{i2}(n) = \sqrt{0.2 + 0.09 \sin n}$ $(i = 1, 2, 3, 4)$. If $c_1(u) = 0.04, c_2(u) = 0.3, c_3(u) = 0.06, c_4(u) = 0.03$ and initial value $(0.35, 0.15, 0.04, 5)$, then it can be obtained that $R_I = 0.8926 < 1$, which follows from Theorem 3-(i) that the number of infected individual $I(t)$ of system (4) tends to zero exponentially. On the other hand, if $c_1(u) = 0.4, c_2(u) = 0.03, c_3(u) = 0.02, c_4(u) = 0.05$, then it can be obtained that $R_E = 0.1972 > 0$, which follows from Theorem 3-(ii) that the number of infected individual $I(t)$ of system (4) is persistent in average sense. Dynamical responses of the infected individual $I(t)$ are shown in Figure 5a and Figure 5b, respectively.

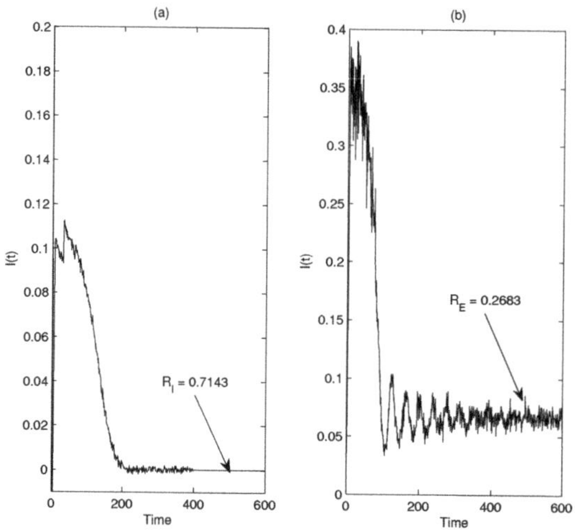

Figure 4. If $\sigma_{i1}(n) = 0.04 + 0.01 \sin n, \sigma_{i2}(n) = \sqrt{0.1 + 0.05 \sin n}$ $(i = 1,2,3,4)$, $c_1(u) = 0.5, c_2(u) = 0.03, c_3(u) = 0.02, c_4(u) = 0.05$ and initial value $(0.4, 0.1, 0.05, 5)$, dynamical responses of the infected individual $I(t)$ are shown in (**a**,**b**), respectively. It can be obtained that $R_I = 0.7143 < 1$, which follows that infectious disease becomes extinct exponentially. In other words, if $c_1(u) = 0.5, c_2(u) = 0.02, c_3(u) = 0.01, c_4(u) = 0.03$, then it can be obtained that $R_E = 0.2683 > 0$, which follows that infectious disease persists in mean.

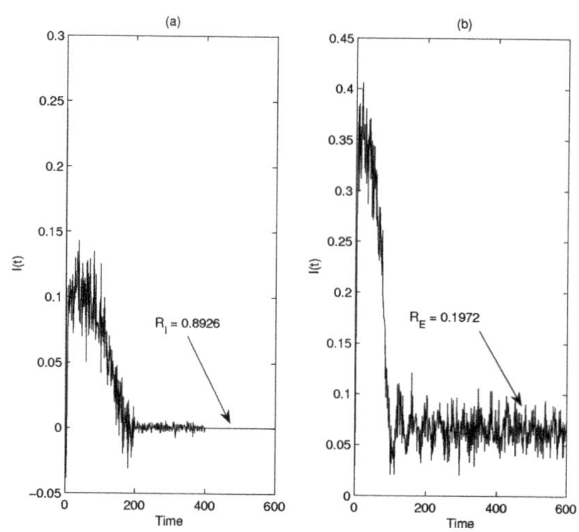

Figure 5. If $\sigma_{i1}(n) = 0.09 + 0.04 \sin n, \sigma_{i2}(n) = \sqrt{0.2 + 0.09 \sin n}$ $(i = 1,2,3,4)$, $c_1(u) = 0.04, c_2(u) = 0.3, c_3(u) = 0.06, c_4(u) = 0.03$ and initial value $(0.35, 0.15, 0.04, 5)$, dynamical responses of the infected individual $I(t)$ are shown in (**a**,**b**), respectively. It can be obtained that $R_I = 0.8926 < 1$, which follows that infectious disease becomes extinct exponentially. In other words, if $c_1(u) = 0.4, c_2(u) = 0.03, c_3(u) = 0.02, c_4(u) = 0.05$, then it can be obtained that $R_E = 0.1972 > 0$, which follows that infectious disease persists in mean.

Remark 3. *From the above two numerical experiments under two different values of* $\sigma_{i1}(n), \sigma_{i2}(n)$ $(i = 1,2,3,4)$, *it reveals that the Gaussian white noises performing on* $I(t)$ *play sufficiently effective roles in reducing the spread of infectious disease. For finite state spaces, although the infectious* *disease only persists within one certain state, there still exists opportunity for infectious disease to persist eventually. Furthermore, Lévy jumps may act double roles in directly controlling the infectious disease based on the values of $c_i(u)$ $(i = 1,2,3,4)$.*

Based on the numerical simulations in Figures 1 and 2, the dynamical responses fluctuate with larger amplitudes under comparatively strong disturbances depicting by Lévy jumps. In the real world, the strong disturbances usually lead to oscillations in the real world, which highly relevant to the vivid phenonomena, i.e., contemporary controlled state—recurrence—re-controlled state within transmission of epidemics.

When the amplitudes of white noises maintain some certain levels, the dynamic changes of total variation norm due to variations of Lévy jumps are discussed, which are indicated in Table 1 and Figure 3. It reveals that the transmission of infectious disease becomes severe with large stochastic fluctuations from surrounding environment in the real world. However, it follows from Figures 4 and 5 that infectious disease may tend to extinction when the stochastic fluctuations from surrounding environment decrease, and the transmission will be controlled within certain duration.

4. Conclusions

Media coverage, random disturbances and time-varying periodic function parameters are important disciplines in the modeling and dynamical analysis of infectious disease transmission. One of the key themes in epidemiology is the study of the stochastic dynamics of infectious disease system. Current field observations of the public health alerts and stochastic perturbations in stochastic nonautonomous infectious disease dynamics has highlighted the necessity of improving related systems that do not consider the joint dynamic impacts of Lévy jumps and media coverage.

In the last few years, scholars have introduced a media coverage feedback mechanism in mathematical model formulation to account for the constructive effects of public health alerts. Stochastic perturbations are usually represented by linear form perturbation of white noise, and the influences of linear noises perturbations on nonautonomous epidemic models were studied in [13–17]. However, in order to accurately depict some stochastic phenomena arising from infectious disease transmission in the real world, it is more constructive to introduce nonlinear noise perturbations into nonautonomous epidemic model.

Furthermore, stochastic models have been established to discuss the prevalence mechanism of infectious disease [23–31] without Lévy jumps. A SIS infectious disease model with regime-switching and driven by Lévy jumps was investigated in [32], while combined dynamic impacts of media coverage and Lévy jumps on random dynamics of infectious disease system are rarely reported.

Hybrid dynamic effects of media coverage and stochastic perturbations in the threshold dynamics of random epidemic system have been investigated in [33–36], while Lévy jumps and periodic function parameters were not considered in [33–36]. The dynamic behavior of infectious disease systems in [37–39] were investigated under nonlinear noise perturbations and Lévy jumps, while all parameters were assumed to be constant values in [37–39], periodicity factors during transmission within the infectious disease regimes were not considered.

Although the stochastic infectious disease model and its dynamic analysis have attracted wide attention, as far as the authors know, the hybrid dynamic impacts of Lévy jumps and media coverage on random dynamics of the nonautonomous SIAM epidemic model with Markov chain and nonlinear noise perturbations have not been reported in previous related studies.

In order to depict the impact of public health alerts and stochastic dynamics of nonautonomous SIAM epidemic model, we extend the work done in [12] by incorporating Lévy jumps, nonlinear noise perturbations and periodic function parameters into the epidemic model. The existence of a stochastically ultimate upper bound and a uniform lower bound of a positive solution of the proposed SIAM epidemic model was studied in Lemma 1.

The existence and uniqueness of globally positive solution to the proposed SIAM epidemic model was studied in Lemma 2. Based on defining certain fitted stochastic

Lyapunov functions, sufficient conditions for existence of a nontrival positive T-periodic solution were discussed in Theorem 1. By verifying a Foster–Lyapunov condition, some sufficient conditions for the exponential ergodicity were investigated in Theorem 2. Furthermore, several conditions were derived in Theorem 3, which were utilized to discuss the persistence in an average sense and the extinction of the epidemic system.

Finally, numerical simulations were provided to support the theoretical findings. The main analytical findings are theoretically beneficial to reveal the transmission mechanism of infectious disease under a stochastic surrounding environment. Furthermore, by utilizing the findings associated with the elimination mechanism of infectious disease, it is also constructive for agencies to formulate policies and measures to control the spread of infectious disease.

Author Contributions: The detailed contributions are as follows. Conceptualization, C.L.; methodology, C.L. and P.C.; software, L.C.; validation, C.L., P.C. and L.C.; investigation, C.L., P.C. and L.C.; resources, L.C.; writing—original draft preparation, C.L. and P.C.; writing—review and editing, C.L. and L.C.; supervision, C.L.; funding acquisition, C.L., P.C. and L.C. All authors have read and approved to the submitted version of the manuscript.

Funding: This work is supported by National Natural Science Foundation of China, grant No. 12171074. Hebei Natural Science Foundation, grant No. A2020501005. Natural Sciences and Engineering Research Council of Canada (NSERC).

Data Availability Statement: Not applicable.

Acknowledgments: Authors would like to express their gratitude to editor and anonymous reviewers for valuable comments and suggestions, and the time and efforts they have spent in the review. Without the expert comments made by editor and anonymous reviewers, the paper would not be of this quality.

Conflicts of Interest: Authors declare that they have no competing interest. All authors of this manuscript declare that there is no conflict of interest regarding the publication of this manuscript. We have no proprietary, financial, professional, or other personal interest of any nature or kind in any product, service, and/or company that could be construed as influencing the position presented in, or review of this manuscript.

References

1. Liu, R.; Wu, J.; Zhu, H. Media/psychological impact on multiple outbreaks of emerging infectious diseases. *Comput. Math. Methods Med.* **2007**, *8*, 153–164. [CrossRef]
2. Cui, J.; Tao, X.; Zhu, H. An SIS infection model incorporating media coverage. *Rocky Mt. J. Math.* **2008**, *38*, 1323–1334. [CrossRef]
3. Funk, S.; Gilad, E.; Jansen, V.A.A. Endemic disease, awareness and local behavioral response. *J. Theor. Biol.* **2010**, *264*, 501–509. [CrossRef] [PubMed]
4. Nyabadza, F.; Chiyaka, C.; Mukandavire, Z.; Musekwa, S.D.H. Analysis of an HIV/AIDS model with public health information campaigns and individual withdrawal. *J. Biol. Syst.* **2010**, *18*, 357–375. [CrossRef]
5. Misra, A.K.; Sharma, A.; Shukla, J.B. Modeling and analysis of effects of awareness programs by media on the spread of infectious diseases. *Math. Comput. Model.* **2011**, *53*, 1221–1228. [CrossRef]
6. Kim, M.; Yoo, B.K. Cost-effectiveness analysis of a television campaign to promote seasonal influenza vaccination among the elderly. *Value Health* **2015**, *18*, 622–630. [CrossRef] [PubMed]
7. Wang, L.W.; Liu, Z.J.; Zhang, X.A. Global dynamics for an age-structured epidemic model with media impact and incomplete vaccination. *Nonlinear Anal. Real World Appl.* **2016**, *32*, 136–158. [CrossRef]
8. Agaba, G.O.; Kyrychko, Y.N.; Blyuss, K.B. Mathematical model for the impact of awareness on the dynamics of infectious disease. *Math. Biosci.* **2017**, *286*, 22–30. [CrossRef]
9. Wang, L.W.; Liu, Z.J.; Xu, D.S.; Zhang, X.A. Global dynamics and optimal control of an influenza model with vaccination, media coverage and treatment. *Int. J. Biomath.* **2017**, *10*, 1750068. [CrossRef]
10. Xiao, Y.; Tang, S.; Wu, J. Media impact switching surface during an infectious disease outbreak. *Sci. Rep.* **2015**, *5*, 7838. [CrossRef]
11. Pharaon, J.; Bauch, C.T. The influence of social behavior on competition between virulent pathogen strains. *J. Theor. Biol.* **2018**, *455*, 47–53. [CrossRef] [PubMed]
12. Misra, A.K.; Rai, R.K.; Takeuchi, Y. Modelling the control of infectious diseases: Effects of TV and social media advertisements. *Math. Biosci. Eng.* **2018**, *15*, 1315–1343. [CrossRef] [PubMed]
13. Greenhalgh, D.; Moneim, I.A. SIRS epidemic model and simulations using different types of seasonal contact rate. *Syst. Anal. Model. Simul.* **2003**, *43*, 573-600. [CrossRef]

14. Bai, Z.; Zhou, Y. Existence of two periodic solutions for a non-autonomous SIR epidemic model. *Appl. Math. Model.* **2011**, *35*, 382–391. [CrossRef]
15. Wang, F.; Wang, X.; Zhang, S.; Ding, C. On pulse vaccine strategy in a periodic stochastic SIR epidemic model. *Chaos Solitons Fractals* **2014**, *66*, 127–135. [CrossRef]
16. Liu, Q.; Jiang, D.Q.; Shi, N.Z.; Hayat, T.; Alsaedi, A. Periodic solution for a stochastic nonautonomous SIR epidemic model with logistic growth. *Physica A* **2016**, *462*, 816–826. [CrossRef]
17. Zhang, W.W.; Meng, X.Z. Stochastic analysis of a novel nonautonomous periodic SIRI epidemic system with random disturbances. *Physica A* **2018**, *492*, 1290–1301. [CrossRef]
18. Liu, Q. Asymptotic properties of a stochastic n-species Gilpin-Ayala competitive model with Lévy jumps and Markovian switching. *Commun. Nonlinear Sci. Numer. Simul.* **2015**, *26*, 1–10. [CrossRef]
19. Zhao, Y.; Yuan, S.L. Stability in distribution of a stochastic hybrid competitive Lotka Volterra model with Lévy jumps. *Chaos Solitons Fractals* **2016**, *85*, 98–109. [CrossRef]
20. Liu, M.; Zhu, Y. Stationary distribution and ergodicity of a stochastic hybrid competition model with Lévy jumps. *Nonlinear Anal. Hybrid Syst.* **2018**, *30*, 225–239. [CrossRef]
21. Liu, L.; Shen, Y. New criteria on persistence in mean and extinction for stochastic competitive Lotka Volterra systems with regime switching. *J. Math. Anal. Appl.* **2015**, *430*, 306–323. [CrossRef]
22. Liu, M.; He, X.; Yu, J.Y. Dynamics of a stochastic regime switching predator prey model with harvesting and distributed delays. *Nonlinear Anal. Hybrid Syst.* **2018**, *28*, 87–104. [CrossRef]
23. Liu, Q.; Jiang, D.Q. Stationary distribution and extinction of a stochastic SIR model with nonlinear perturbation. *Appl. Math. Lett.* **2017**, *73*, 8–15. [CrossRef]
24. Berrhazi, B.E.; Fatini, M.E.; Laaribi, A.; Petterssonc, R.; Taki, R. A stochastic SIRS epidemic model incorporating media coverage and driven by Lévy noise. *Chaos Solitons Fractals* **2017**, *105*, 60–68. [CrossRef]
25. Cortés, J.C.; El-Labany, S.K.; Navarro-Quiles, A.; Selim, M.M.; Slama, H. A comprehensive probabilistic analysis of approximate SIR-type epidemiological models via full randomized discrete-time Markov chain formulation with applications. *Math. Methods Appl. Sci.* **2020**, *43*, 8204–8222. [CrossRef]
26. Lazebnik, T.; Mendrazitsky, S.B.; Shaikhet, L. Novel Method to Analytically Obtain the Asymptotic Stable Equilibria States of Extended SIR-Type Epidemiological Models. *Symmetry* **2021**, *13*, 1120. [CrossRef]
27. Song, M.Y.; Zuo, W.J.; Jiang, D.Q.; Hayat, T. Stationary distribution and ergodicity of a stochastic cholera model with multiple pathways of transmission. *J. Frankl. Inst.* **2020**, *357*, 10773–10798. [CrossRef]
28. Zeb, A.; Kumar, S.; Tesfay, A.; Kumar, A. A stability analysis on a smoking model with stochastic perturbation. *Int. J. Numer. Methods Heat Fluid Flow* **2021**, *15*, 915–930. [CrossRef]
29. Tomé, T.; Silva, A.T.C.; Oliveira, M.J.D. Effect of immunization through vaccination on the SIS epidemic spreading model. *Braz. J. Phys.* **2021**, *51*, 1853. [CrossRef]
30. Lima, L.S. Dynamics based on analysis of public data for spreading of disease. *Sci. Rep.* **2021**, *11*, 12177. [CrossRef] [PubMed]
31. Lima, L.S. Fractional Stochastic Differential Equation Approach for Spreading of Diseases. *Entropy* **2022**, *24*, 719. [CrossRef] [PubMed]
32. Lin, Y.G.; Zhao, Y.N. Exponential ergodicity of a regime-switching SIS epidemic model with jumps. *Appl. Math. Lett.* **2019**, *94*, 133–139. [CrossRef]
33. Guo, W.J.; Cai, Y.L.; Zhang, Q.M.; Wang, W.M. Stochastic persistence and stationary distribution in an SIS epidemic model with media coverage. *Physica A* **2018**, *492*, 2220–2236. [CrossRef] [PubMed]
34. Zhang, Y.; Fan, K.G.; Gao, S.J.; Liu, Y.F.; Chen, S.H. Ergodic stationary distribution of a stochastic SIRS epidemic model incorprating media coverage and saturated incidence rate. *Phys. A* **2019**, *514*, 671–685. [CrossRef]
35. Liu, Q.; Jiang, D.Q.; Hayat, T.; Alsaedi, A.; Ahmad, B. Dynamical behavior of a higher order stochastically perturbed SIRI epidemic model with replapse and media coverage. *Chaos Solitons Fractals* **2020**, *139*, 110013. [CrossRef] [PubMed]
36. Zhu, B.Q.; Jiang, D.Q.; Han, B.T.; Hayat, T. Threshold dynamics and density function of a stochastic epidemic model with media coverage and mean-reverting Ornstein-Uhlenbeck process. *Math. Comput. Simul.* **2022**, *196*, 15–44. [CrossRef]
37. Li, M.L.; Ouyang, Z.Y.; Deng, F.Q.; Wu, Z.H. Dissipativity theory and applications of nonlinear stochastic systems with Markov jump and Lévy noise. *Commun. Nonlinear Sci. Numer. Simul.* **2021**, *99*, 105796 [CrossRef]
38. Lu, C.; Liu, H.H.; Zhang, D. Dynamics and simulations of a second order stochastically perturbed SEIQV epidemic model with saturated incidence rate. *Chaos Solitons Fractals* **2021**, *152*, 111312. [CrossRef]
39. Sun, J.G.; Gao, M.M.; Jiang, D.Q. Threshold Dynamics and the Density Function of the Stochastic Coronavirus Epidemic Model. *Fractal Fract.* **2022**, *6*, 245. [CrossRef]
40. Khasminskii, R. *Stochastic Stability of Differential Equations*, 2nd ed.; Springer: Berlin/Heidelberg, Germany, 2012.
41. Khasminskii, R.Z.; Zhu, C.; Yin, G. Stability of regime-switching diffusions. *Stoch. Process. Their Appl.* **2007**, *117*, 1037–1057. [CrossRef]
42. Meyn, S.P.; Tweedie, R.L. Stability of Markovian processes III: Foster–Lyapunov criteria for continuous-time processes. *Adv. Appl. Probab.* **1993**, *25*, 518–548. [CrossRef]
43. Xi, F. On the stability of jump-diffusions with Markovian switching. *J. Math. Anal. Appl.* **2008**, *341*, 588–600 [CrossRef]

fractal and fractional

MDPI

Article

Modelling the Frequency of Interarrival Times and Rainfall Depths with the Poisson Hurwitz-Lerch Zeta Distribution

Carmelo Agnese [1], Giorgio Baiamonte [1], Elvira Di Nardo [2,*], Stefano Ferraris [3] and Tommaso Martini [2]

[1] Department of Agricultural, Food and Forest Sciences, Università di Palermo, Viale delle Scienze, 13, 90128 Palermo, Italy

[2] Department of Mathematics "G. Peano", Università degli Studi di Torino, Via Carlo Alberto, 10, 10123 Torino, Italy

[3] Interuniversity Department of Regional and Urban Studies and Planning, Università degli Studi di Torino, Politecnico di Torino, Viale Pier Andrea Mattioli, 39, 10125 Torino, Italy

* Correspondence: elvira.dinardo@unito.it

Abstract: The Poisson-stopped sum of the Hurwitz–Lerch zeta distribution is proposed as a model for interarrival times and rainfall depths. Theoretical properties and characterizations are investigated in comparison with other two models implemented to perform the same task: the Hurwitz–Lerch zeta distribution and the one inflated Hurwitz–Lerch zeta distribution. Within this framework, the capability of these three distributions to fit the main statistical features of rainfall time series was tested on a dataset never previously considered in the literature and chosen in order to represent very different climates from the rainfall characteristics point of view. The results address the Hurwitz–Lerch zeta distribution as a natural framework in rainfall modelling using the additional random convolution induced by the Poisson-stopped model as a further refinement. Indeed the Poisson contribution allows more flexibility and depiction in reproducing statistical features, even in the presence of very different climates.

Keywords: Hurwitz-Lerch zeta distribution; log-concavity; compound poisson distribution; one inflated model; moment; simulated annealing

Citation: Agnese, C.; Baiamonte, G.; Di Nardo, E.; Ferraris, S.; Martini, T. Modelling the Frequency of Interarrival Times and Rainfall Depths with the Poisson Hurwitz-Lerch Zeta Distribution. *Fractal Fract.* **2022**, *6*, 509. https://doi.org/10.3390/fractalfract6090509

Academic Editors: Mario Abundo and Viorel-Puiu Paun

Received: 25 July 2022
Accepted: 7 September 2022
Published: 11 September 2022

1. Introduction

Analysis of rainfall data, and the subsequent modelling of the many variables concerning rainfall, is fundamental to many areas such as agricultural, ecological and engineering disciplines. From assessing hydrological risk to both crop and hydropower plannings, rainfall modelling is of the utmost importance. Moreover, being able to provide reliable rainfall modelling is essential in the well known issue of climate change. Due to the complexity of hydrological systems, their analysis and modelling rely heavily on historical records. Rainfall historical records are of various time scales, from hourly data to annual data. However, daily rainfall series are arguably the most used information in environmental, climate, hydrological, and water resources studies [1]. Rainfall manifests one peculiar characteristic which is common to many other geophysical processes: intermittence [2]. Intermittence is found in variables which are related to the internal and external structure of rainfall. The most commonly seen for the external structure are the Wet Spells (*WS*) and Dry Spells (*DS*), meaning the sequences of rainy days and non-rainy days, respectively. A way of studying the alternance of *WS* and *DS* is through the Interarrival Times (*IT*), that is the time elapsed between two consecutive days of rain. If we suppose that *IT* observations are independent and identically distributed (i.i.d.), one natural way to model them is through the well known renewal processes [3]. Many examples can be found in the literature. The simplest renewal process, the Bernoulli process, has been used in [4] for example. In this case, the *IT*'s are geometrically distributed. Its continuous counterpart, the Poisson process, has been used for its simpler mathematical tractability, but requires dealing with the *IT* random variable (r.v.) as continuous, despite its discrete nature. The need to suppose a non-constant probability of rain requires slightly more sophisticated models.

The challenge of this paper is to propose a suitable discrete distribution to fit IT at the daily scale. It is on this time scale that the intermittent character of precipitation can be appreciated and at the same time most practical applications are possible. The proposed distribution must be able to model both the numerous occurrences of the value equal to one, which represent the sequence of rainy days, and some large values scattered over time and responsible for drought phenomena. Our starting point is the three parameter family Hurwitz–Lerch Zeta distribution (HLZD), successfully proposed in [5]. Such a distribution represents a step forward with respect to other commonly used IT modelling distributions, such as the logarithmic one. In Section 3, we summarize the main properties of the HLZD and state new results on its log-concavity and convolution. As a step forward, in this paper we propose to model the IT r.v. using the Poisson-stopped HLZD (PSHLZD). This discrete distribution presents excess zeroes (paralleling the excess of $IT = 1$) and a long tail [6]. The PSHLZD has been used in [6] for comparisons with the negative binomial distribution, a popular model for fitting over-dispersed data. Indeed the PSHLZD can be seen both as a Poisson-stopped sum of HLZD's as well as a generalization of a negative binomial distribution. The Poisson contribution allows us to model the superposition of i.i.d. HLZD's in the observed time series as rare event. In Section 4, we summarize its main properties using the combinatorics of exponential Bell polynomials. It is noteworthy to mention that Bell polynomials are used within fractional calculus, see for example [7,8] and within fractal models [9]. Moreover, new results are added on the the PSHLZD, as for example on log-concavity.

A second goal of this paper is to show that the PSHLZD is also a suitable model for a different feature strictly related to the internal structure of rainfall: the depth (or the intensity) of the rainy days [10]. In the literature, refs. [11,12] rainfall depths are more often treated as continuous despite that sometimes these models fail to account for the time discreteness of the sample process [13]. Daily rainfall depth measurements are almost always performed by automatically counting how many times a small bucket corresponding to 0.2 mm is filled. This led use to treat them as discrete, because of the abundance of ties in the data. Finally, in Section 5 we have also considered a third modelling distribution: the One Inflated HLZD (OIHLZD). Such a distribution mixes two generating processes: the first generates one's and the latter is governed by a HLZD. This stochastic structure takes into account the dominance of one's in the rainfall depth or interarrival time series.

In Section 6, we discuss the results for fitting all these models to rainfall data, proving that the PSHLZD provides a very general framework for rainfall modelling. Indeed the PSHLZD replicates the fitting features of the OIHLZD and outperforms the fitted HLZD in some cases. The PSHLZD has a limited number of parameters and at the same time can adapt very well to data collected in very different climates, from England to Sicily. Let us underline that the analyzed dataset has never been considered in the literature and consists of measures sampled along 70 years at 5 different stations. These stations were chosen in order to represent different climates from the rainfall characteristics point of view. In fact, the interarrival data examined are very different from each other, with a regular pattern of many rainy days in England, and a winter rainy season alternating with long periods in summer without rain, typical of the Sicily Mediterranean climate. The same is for the rainfall depth, namely many small depths in England, and few big storms in Sicily. This made it possible to confirm the great utility of the proposed statistical models within rainfall modelling. Some concluding remarks and future developments are addressed at the end of the paper.

2. Bell Polynomials in a Nutshell

The partial exponential Bell polynomials are usually written as [14]

$$B_{n,j}(z_1,\ldots,z_{n-j+1}) = \sum \frac{n!}{(1!)^{r_1} r_1! (2!)^{r_2} r_2! \cdots} z_1^{r_1} \cdots z_{n-j+1}^{r_{n-j+1}} \quad n \in \mathbb{N}, j \leq n \tag{1}$$

where the summation is over all the solutions in non-negative integers $r_1,\ldots,r_{n-j+1}$ of $r_1+2r_2+\cdots+(n-j+1)r_{n-j+1}=n$ and $r_1+r_2+\cdots+r_{n-j+1}=j$. A lighter expression is obtained using partitions of the integer n with length j. Recall that a partition of an integer n is a sequence $\pi=(\pi_1,\pi_2,\ldots)$ of weakly decreasing positive integers, named parts of π, such that $\pi_1+\pi_2+\cdots=n$. A different notation is $\pi=(1^{r_1},2^{r_2},\ldots)$, where $r_1,r_2,\ldots,$ named multiplicities of π, are the number of parts of π equal to $1,2,\ldots$ respectively. The length of the partition is $l(\pi)=r_1+r_2+\cdots$ and the vector of multiplicities is $m(\pi)=(r_1,r_2,\ldots)$. We write $\pi\vdash n$ to denote that π is a partition of n. Thus the partial exponential Bell polynomials (1) can be rewritten as [15]

$$B_{n,j}(z_1,\ldots,z_{n-j+1})=\sum_{\pi\vdash n,l(\pi)=j} d_\pi z_\pi \tag{2}$$

where the sum is over all the partitions $\pi\vdash n$ with length $l(\pi)=j$ and

$$z_\pi=z_1^{r_1}z_2^{r_2}\cdots \qquad d_\pi=\frac{i!}{(1!)^{r_1}r_1!(2!)^{r_2}r_2!\cdots}. \tag{3}$$

Using integer partitions, the explicit expression of the partial exponential polynomials can be recovered in R using the `kStatistics` package [16]. A useful property used in the following is

$$B_{n,j}(abz_1,\ldots,ab^{n-j+1}z_{n-j+1})=a^j b^n B_{n,j}(z_1,\ldots,z_{n-j+1}) \tag{4}$$

with a,b constants. Equation (4) follows from (2) since from (3) we have

$$(abz_1)^{r_1}(ab^2z_2)^{r_2}\cdots=a^{r_1+r_2+\cdots}b^{r_1+2r_2+\cdots}z_\pi=a^j b^n z_\pi$$

taking into account that $l(\pi)=r_1+r_2+\cdots=j$ and $r_1+2r_2+\cdots=n$.

The n-th complete exponential Bell polynomials in the indeterminates $z_1,\ldots,z_n$ is defined as [14]

$$B_n(z_1,\ldots,z_n)=\sum_{j=0}^{n} B_{n,j}(z_1,\ldots,z_{n-j+1}) \tag{5}$$

with $\{B_{n,j}\}$ the partial exponential Bell polynomials (1). Note that n is the positive integer corresponding to the maximum degree of the monomials in (5). This polynomial sequence satisfies the following recurrence [14]

$$B_{n+1}(z_1,\ldots,z_{n+1})=\sum_{j=0}^{n}\binom{n}{j}z_{j+1}B_{n-j}(z_1,\ldots,z_{n-j}) \tag{6}$$

with the initial value $B_0=1$. The generating function of $\{B_n\}$ is the formal power series composition [14]

$$\exp[h_z(t)-z_0]=\sum_{n\geq 0}\frac{t^n}{n!}B_n(z_1,\ldots,z_n)\in\mathbb{R}[[t]] \tag{7}$$

where $\mathbb{R}[[t]]$ is the ring of formal power series in t and $h_z(t)$ is the generating function of $\{z_k\}_{k\geq 0}$, that is

$$h_z(t)=\sum_{k\geq 0}\frac{t^k}{k!}z_k.$$

A different expression of the n-th complete exponential Bell polynomial involves integer partitions [15] as follows

$$B_n(z_1,\ldots,z_n)=\sum_{\pi\vdash n} d_\pi z_\pi \tag{8}$$

where the sum is over all the partitions $\pi \vdash n, d_\pi$ and z_π are given in (3). In particular we have

$$B_n(\lambda z_1, \dots, \lambda z_n) = \sum_{\pi \vdash n} \lambda^{l(\pi)} d_\pi z_\pi \tag{9}$$

with λ a constant. Now, suppose to replace $\lambda^{l(\pi)}$ in (9) with a numerical sequence $\{a_{l(\pi)}\}$. Thanks to this device, the complete exponential Bell polynomials results as a special case of a wider class of polynomial families, the generalized partition polynomials [16]

$$\mathcal{B}_n(a_1, \dots, a_n; z_1, \dots, z_n) = \sum_{\pi \vdash n} d_\pi a_{l(\pi)} z_\pi \tag{10}$$

where the sum is again over all the partitions $\pi \vdash n$. A different expression of (10) involves the partial exponential Bell polynomials $\{B_{n,j}\}$ in (1)

$$\mathcal{B}_n(a_1, \dots, a_n; z_1, \dots, z_n) = \sum_{j=1}^{n} a_j B_{n,j}(z_1, \dots, z_{n-j+1}). \tag{11}$$

An example of a well known polynomial family, arising from (11) is the logarithmic one [14]

$$\mathcal{L}_n(z_1, \dots, z_n) = \sum_{j=0}^{n} (-1)^{j-1} (j-1)! B_{n,j}(z_1, \dots, z_{n-j+1}). \tag{12}$$

3. The Hurwitz-Lerch Zeta Distribution

Definition 1. *A discrete random variable* $Y \stackrel{d}{=} \mathrm{HLZD}(a, \theta, s)$ *if*

$$q_y := \mathbb{P}(Y = y) = \frac{\theta^y}{\mathcal{T}(\theta, s, a)(y + a)^{s+1}}, \qquad \theta \in (0,1), a > -1, s \in \mathbb{R} \tag{13}$$

for $y = 1, 2, \dots,$ *with* $\mathcal{T}(\theta, s, a) = \theta \Phi(\theta, s+1, a+1)$, *where*

$$\Phi(\theta, s, a) = \sum_{n=0}^{\infty} \frac{\theta^n}{(n+a)^s} \tag{14}$$

is the Lerch Transcendent function.

The probability generating function (pgf) of $Y \stackrel{d}{=} \mathrm{HLZD}(a, \theta, s)$ is

$$G_Y(z) = \frac{\theta \Phi(z\theta, s+1, a+1)}{\Phi(\theta, s, a)}, \quad |z| \le 1 \tag{15}$$

with $G_Y(0) = 0$.

3.1. Moments and Cumulants

HLZD moments have a closed form expression involving the Lerch Transcendent function. Differently from [17], we find this closed form expression using (13).

Proposition 1. *If* $Y \stackrel{d}{=} \mathrm{HLZD}(a, \theta, s)$, *then*

$$\xi_k := \mathbb{E}[Y^k] = \sum_{j=0}^{k} (-a)^{k-j} \binom{k}{j} \frac{\Phi(\theta, s+1-j, a)}{\Phi(\theta, s+1, a+1)}. \tag{16}$$

Proof. Using the binomial expansion of $y^k = (y-a+a)^k$, we have

$$\xi_k = \sum_{y=1}^{\infty} y^k \mathbb{P}(Y=y) = \sum_{y=1}^{\infty} y^k \frac{\theta^{y-1}}{\Phi(\theta,s+1,a+1)(y+a)^{s+1}}$$
$$= \sum_{y=1}^{\infty} \left(\sum_{j=0}^{k} \binom{k}{j} (y+a)^j (-a)^{k-j} \right) \frac{\theta^{y-1}}{\Phi(\theta,s+1,a+1)(y+a)^{s+1}}$$
$$= \sum_{j=0}^{k} \binom{k}{j} (-a)^{k-j} \frac{1}{\Phi(\theta,s+1,a+1)} \sum_{y=1}^{\infty} \frac{\theta^{y-1}}{(y+a)^{s+1-j}}$$

from which (16) follows by taking into account (14). □

As a corollary, the mean and the variance are respectively:

$$\mathbb{E}[Y] = \frac{\mathcal{T}(\theta,s-1,a)}{\mathcal{T}(\theta,s,a)} - a \quad \mathrm{Var}[Y] = \frac{\mathcal{T}(\theta,s-2,a)}{\mathcal{T}(\theta,s,a)} - \left(\frac{\mathcal{T}(\theta,s-1,a)}{\mathcal{T}(\theta,s,a)} \right)^2.$$

More generally, the k-th central moment can be recovered as

$$\xi'_k := \mathbb{E}[(Y-\xi_1)^k] = \sum_{j=0}^{k} \binom{k}{j} \xi_1^j \xi_{k-j}$$

and the factorial moments as

$$(\xi)_k = \mathbb{E}[Y(Y-1)\cdots(Y-k+1)] = \sum_{j=0}^{k} s(k,j)\xi_k \tag{17}$$

with $s(k,j)$ Stirling numbers of the first kind [14]. HLZD cumulants are such that [14]

$$\kappa_n(Y) = \mathcal{L}_n(\xi_1,\ldots,\xi_n) \quad n=1,2,\ldots$$

where $\{\xi_j\}$ are the moments of $Y \stackrel{d}{=} \mathrm{HLZD}(a,\theta,s)$, given in (16), and $\mathcal{L}_n$ is the n-th logarithmic polynomial (12). Let us recall that, if the moment generating function (mgf) $M_Y(t)$ of Y is well defined in a suitable neighborhood of 0, then the coefficients $\{\kappa_n(Y)\}$ in the expansion

$$M_Y(t) = \exp\left(\sum_{n\geq 1} \frac{t^n}{n!} \kappa_n(Y) \right)$$

are the cumulants of Y. The first cumulant is the mean $\mathbb{E}[Y]$, the second cumulant is the variance $\mathrm{Var}(Y)$, the skewness and the kurtosis of Y can be recovered using the third and the fourth cumulant of Y respectively.

3.2. Mode

The HLZ distribution is a particular case of a wider class of distributions called the Modified Power Series Distributions (MPSD) [18].

Definition 2. *A discrete random variable* $Y \stackrel{d}{=} \mathrm{MPSD}(a,g,f)$ *if*

$$p_y := \mathbb{P}(Y=y) = \frac{a(y)g(\theta)^y}{f(\theta)}, \qquad y \in T \subset \mathbb{N} \tag{18}$$

where $a(y)$, $g(\theta)$ *and* $f(\theta)$ *are positive, bounded, and differentiable functions of* y *and* θ *respectively with* $f(\theta) = \sum_{y\in T} a(y)g(\theta)^y$.

Using this wider class of distributions, we will prove that $Y \stackrel{d}{=} \text{HLZD}(a,\theta,s)$ is unimodal for all $s \in \mathbb{R}$. To this aim, let us recall that a discrete distribution with support $T \subset \mathbb{Z}$ is said to be strongly unimodal if and only if the sequence $\{p_y\}_{y\in T}$, with $p_y := \mathbb{P}(Y = y)$, is a logarithmically concave sequence [19], that is if and only if

$$p_y^2 \geq p_{y-1}\, p_{y+1}, \; \forall y \in T. \tag{19}$$

Proposition 2. *Suppose* $Y \stackrel{d}{=} \text{HLZD}(a,\theta,s)$.
(i) *If* $s \geq -1$, *the sequence* $\{q_y\}_{y\geq 1}$ *is monotonically decreasing and the mode is* $y = 1$.
(ii) *If* $s < -1$, Y *is strongly unimodal.*

Proof. Similarly to what stated in Section 2.3 of [20], we have

$$\frac{q_y}{q_{y-1}} = \theta\left(1 - \frac{1}{a+y}\right)^{s+1} \quad y = 2,3,\ldots. \tag{20}$$

Since $\theta \in (0,1), a > -1$ and $s \geq -1$, the rhs of (20) is always between 0 and 1, thus *(i)* follows. For *(ii)* we have to prove that $\{q_y\}_{y\geq 1}$ is log-concave, that is it satisfies (19). Using (13), (19) reduces to $(1-(y+a)^{-2})^{s+1} \geq 1$, which is always true if $s < -1$. □

3.3. Convolution

The family of HLZ distributions is not closed under convolution. Nevertheless, as a subclass of MPS distributions, the HLZD convolution still returns a MPSD. Indeed we will prove that the family of MPS distributions is closed under convolution.

Theorem 1. *If* $Y_1,\ldots,Y_j$ *are independent r.v.'s identically distributed to* $Y \stackrel{d}{=} \text{HLZD}(a,\theta,s)$, *then* $Y_1 + \cdots + Y_j \stackrel{d}{=} \text{MPSD}(a_j, g, f_j)$ *with* $f_j(\theta) = [\mathcal{T}(\theta,s,a)]^j$ *and*

$$a_j(y) = \frac{j!}{y!} \sum_{\pi \vdash y, l(\pi)=j} d_\pi (a_\pi)^{s+1} \quad \textit{with} \quad a_\pi = (a+1)^{-r_1}(a+2)^{-r_2}\cdots \tag{21}$$

and d_π *given in* (3).

Proof. Observe that if $Y_1,\ldots,Y_j$ are r.v.'s i.i.d. to $Y \stackrel{d}{=} \text{MPSD}(a,g,f)$, then $Y_1 + \cdots + Y_j \stackrel{d}{=} \text{MPSD}(a_j, g, f_j)$ with $f_j(\theta) = f^j(\theta)$ and

$$a_j(y) = \begin{cases} \dfrac{j!}{y!} B_{y,j}[a(1),\ldots,a(y-j+1)], & y \in T_j \\ 0, & y \notin T_j \end{cases} \tag{22}$$

with $T_j = \{y_1 + \cdots + y_j \in \mathbb{N} | y_1,\ldots,y_j \in T\}$. Indeed in (18), set $a(y) = 0$ if $y \notin T$ and consider the sequence $\{p_y\}_{y\geq 1}$ such that $p_y = 0$ if $y \notin T$. By using Lemma 1 in [21], we have

$$\mathbb{P}(Y_1 + \cdots + Y_j = y) = \frac{j!}{y!} B_{y,j}(p_1,\ldots,p_{y-j+1}) \tag{23}$$

where $\{B_{y,j}\}$ are the partial exponential Bell polynomials (1). From (23) with p_i replaced by $a(i)g(\theta)^i / f(\theta)$ for $i = 1,\ldots,y-j+1$ and using (4) we have

$$\mathbb{P}(Y_1 + \cdots + Y_j = y) = \frac{j!}{y!} \frac{g(\theta)^y}{f^j(\theta)} B_{y,j}[a(1),\ldots,a(y-j+1)]. \tag{24}$$

Thus $Y_1+\cdots+Y_j \stackrel{d}{=} \text{MPSD}(a_j,g,f_j)$ with $f_j(\theta)=f^j(\theta)$ and $a_j(y)$ given in (22). From (24) note that $a_j(y)=0$ if $1,2,\ldots,y-j+1 \notin T$. By replacing $g(\theta)=\theta, f(\theta)=\mathcal{T}(\theta,s,a)$ and $a(k)=(k+a)^{-(s+1)}$ for $k=1,\ldots,y-j+1$ in (24) we have

$$a_j(y)=\frac{j!}{y!}B_{y,j}\left[(a+1)^{-(s+1)},\ldots,(a+y-j+1)^{-(s+1)}\right] \qquad y=1,2,\ldots.$$

The result follows after some manipulations, rewriting the partial Bell exponential polynomials as in (2). □

3.4. Maximum Likelihood Estimation

Consider a vector $\boldsymbol{y}=(y_1,\ldots,y_n)$ of independent observations of $Y \stackrel{d}{=} \text{HLZD}(a,\theta,s)$. The maximum likelihood estimation (MLE) of (θ,s,a) is

$$(\hat{\theta},\hat{s},\hat{a})=\underset{(\theta,s,a)\in\Theta}{\text{argmax}}\,\ell_n(\theta,s,a,\boldsymbol{y}), \tag{25}$$

with $\Theta=(0,1)\times(-\infty,+\infty)\times(-1,\infty)$, $\ell_n(\theta,s,a,\boldsymbol{y})=\ln L_n(\theta,s,a,\boldsymbol{y})$ the log-likelihood function and

$$L_n(\theta,s,a,\boldsymbol{y})=\prod_{i=1}^{n}\mathbb{P}(Y=y_i).$$

The MLE of the HLZD parameters (θ,s,a) has been studied by Gupta in [20]. He showed that the three likelihood equations arising from maximizing the log-likelihood correspond to the equations of the method of moments. In particular we have

$$\sum_{i=1}^{n}\frac{y_i}{n}=\mathbb{E}[Y], \quad \sum_{i=1}^{n}\frac{\log(a+y_i)}{n}=\mathbb{E}[\log(a+Y)], \quad \sum_{i=1}^{n}\frac{1}{n(a+y_i)}=\mathbb{E}\left[\frac{1}{a+Y}\right].$$

Unfortunately, closed form solutions of the above equations are not available and also the moments $\mathbb{E}[\log(a+Y)]$ and $\mathbb{E}[1/(a+Y)]$ must be numerically approximated. As noted in [20], the likelihood equations may be solved by standard numerical methods to obtain the MLE. However, it is well known that this does not guarantee that global maxima of the likelihood have been achieved. In order to avoid this problem, a global optimization method can be employed to solve (25). The global optimization method takes advantage of the bounds of the parameters. More specifically, the MLE of the parameters can be obtained through a global optimization algorithm known as *Simulated Annealing* [22]. Simulated annealing is a stochastic global optimisation technique applicable to a wide range of discrete and continuous variable problems. It makes use of Markov Chain Monte Carlo samplers, to provide a means to escape local optima by allowing moves which worsen the objective function, with the aim of finding a global optimum. Technical details can be found in [22], a variant of which is the algorithm implemented in the `Optim` function in the base `Stats` R-package.

4. The Poisson-Stopped Hurwitz-Lerch Zeta Distribution

Definition 3. *A discrete random variable* $X \stackrel{d}{=} \text{PSHLZD}(\lambda,a,\theta,s)$ *if its pgf is*

$$G_X(t)=\exp\left(\lambda\left[\frac{\theta\Phi(z\theta,s+1,a+1)}{\Phi(\theta,s,a)}-1\right]\right) \quad \lambda>0, \tag{26}$$

where Φ *is the Lerch Transcendent function* (15).

According to Definition 3, $X \stackrel{d}{=} \text{PSHLZD}(\lambda,a,\theta,s)$ takes non-negative integer values and belongs to the class of generalized r.v.'s [23]. Indeed given two independent r.v.'s Z and Y, with pgf $G_Z(t)$ and $G_Y(t)$ respectively, the generalized r.v. X has pgf

$$G_X(t)=G_Z[G_Y(t)]. \tag{27}$$

The composition (26) matches (27) when $Y \stackrel{d}{=} \text{HLZD}(a,\theta,s)$ and Z is a Poisson r.v. (PS) of parameter $\lambda > 0$, independent of Y, since $G_Z(t) = \exp[\lambda(t-1)]$.

In the following we analyse in detail the properties of the PSHLZD using the complete exponential Bell polynomials. Some of the properties given in [6] will also be briefly recalled.

Proposition 3. *If* $X \stackrel{d}{=} \text{PSHLZD}(\lambda,a,\theta,s)$ *then*

$$p_x := \mathbb{P}(X=x) = \begin{cases} e^{-\lambda}, & x=0, \\ \frac{e^{-\lambda}}{x!} B_x(\lambda q_1,\ldots,\lambda x! q_x), & x=1,2,\ldots \end{cases} \tag{28}$$

where B_x *is the complete exponential Bell polynomial* (5) *of degree* x.

Proof. Observe that $G_Y(0) = 0$ and $G_Y(t) = \sum_{y\geq 1} y! q_y t^y / y!$. The result follows from (7) with $z_0 = 0$ and $h_z(t) = G_Y(t)$, since from (26) we have

$$\exp(-\lambda)\exp[\lambda G_Y(t)] = \sum_{x\geq 0} \frac{t^x}{x!} e^{-\lambda} B_x(\lambda 1! q_1,\ldots,\lambda x! q_x).$$

□

Corollary 1. *If* $X \stackrel{d}{=} \text{PSHLZD}(\lambda,a,\theta,s)$ *then* $p_0 = P(X=0) = e^{-\lambda}$ *and*

$$p_x := \mathbb{P}(X=x) = \theta^x e^{-\lambda} \sum_{\pi \vdash x} \left(\frac{\lambda}{T(\theta,s,a)}\right)^{l(\pi)} \frac{(a_\pi)^{s+1}}{m_\pi!} \qquad x=1,2,\ldots. \tag{29}$$

where the sum is over all the partitions $\pi \vdash x$, $m_\pi! = r_1! r_2! \cdots$ *and* a_π *is given in* (21).

Proof. From (28), using (8) and (3), we have

$$\begin{aligned} p_x &= \frac{e^{-\lambda}}{x!} \sum_{\pi \vdash x} \frac{x!}{(1!)^{r_1} r_1! (2!)^{r_2} r_2! \cdots} (\lambda 1! q_1)^{r_1} (\lambda 2! q_2)^{r_2} \\ &= e^{-\lambda} \sum_{\pi \vdash x} \frac{\lambda^{r_1+r_2+\cdots}}{r_1! r_2! \cdots} \frac{[(a+1)^{-(s+1)}]^{r_1} [(a+2)^{-(s+1)}]^{r_2} \cdots \theta^{r_1+2r_2+\cdots}}{T(\theta,s,a)^{r_1+r_2+\cdots}} \end{aligned}$$

by which (29) follows observing that $r_1 + r_2 + \cdots = l(\pi)$ and $r_1 + 2r_2 + \cdots = x$. □

As a corollary of Proposition 3 and recursion (6), the sequence $\{p_x\}$ in (28) satisfies the following equations.

Corollary 2. *If* $X \stackrel{d}{=} \text{PSHLZD}(\lambda,a,\theta,s)$ *then*

$$p_{x+1} = \frac{\lambda}{x+1} \sum_{j=0}^{x} (j+1) q_{j+1} p_{x-j}, \; x=1,2,\ldots \quad \text{with } p_0 = e^{-\lambda}.$$

Proof. The result follows using (6) since we have

$$\begin{aligned} p_{x+1} &= \frac{e^{-\lambda}}{(x+1)!} B_{x+1}(\lambda q_1,\ldots,\lambda (x+1)! q_{x+1}) \\ &= \frac{e^{-\lambda}}{(x+1)!} \sum_{j=0}^{x} \binom{x}{j} B_{x-j}(\lambda q_1,\ldots,\lambda (x-j)! q_{x-j}) (j+1)! q_{j+1} \\ &= \frac{e^{-\lambda}}{(x+1)!} \sum_{j=0}^{x} \frac{x!}{(x-j)! j!} p_{x-j} \frac{(x-j)!}{e^{-\lambda}} (j+1)! q_{j+1}. \end{aligned}$$

□

The PSHLZD is unimodal if $s \geq 0$ and $-1 < a \leq 0$ ([6], Property 1).

4.1. Log-Concavity

Under suitable conditions, the PSHLZD is log-concave.

Proposition 4. *If $X \stackrel{d}{=} \text{PSHLZD}(\lambda,\theta,a,s)$ and $s \geq -1$, then X has a log-concave cumulative distribution function (cdf), that is*

$$[\mathbb{P}(X \leq x)]^2 \geq \mathbb{P}(X \leq x-1)\mathbb{P}(X \leq x+1), \qquad x = 0,1,2,\ldots.$$

Proof. According to ([24], Theorem 1), a random sum $\sum_{i=1}^{Z} Y_i$ of i.i.d. r.v.'s has a log-concave cdf if Z is strongly unimodal and the distribution of $\{Y_i\}$ has a decreasing pdf. Thus, the result follows as $X \stackrel{d}{=} \sum_{i=1}^{Z} Y_i$ with $Z \stackrel{d}{=} \text{PS}(\lambda)$, which has a log-concave pdf (strongly unimodal), and $Y \stackrel{d}{=} \text{HLZD}(\theta,a,s)$ with a decreasing pdf when $s \geq -1$ (see Proposition 2). □

Proposition 4 gives a sufficient condition to get cdf log-concavity. A different way is to consider the sequence $\{p_x\}$. Indeed, if X has a log-concave pdf (19), then its cdf is also log-concave [24]. In the more general setting of generalized r.v.'s, X has a log-concave pdf if and only if the sequence

$$p_x := \mathbb{P}(X = x) = \frac{1}{x!}\mathcal{B}_x(1!\tilde{q}_1,\ldots,x!\tilde{q}_x;1!q_1,\ldots,x!q_x) \qquad x = 1,2,\ldots \tag{30}$$

is log-concave with $p_0 = \mathbb{P}(X=0) = G_Z[G_Y(0)]$ and $\tilde{q}_x = \mathbb{P}(Z=x), q_x = \mathbb{P}(Y=x)$. Equation (30) follows from Equation (2.3) in [23] using the general partition polynomials (8). When $Z \stackrel{d}{=} \text{PS}(\lambda)$ a necessary and sufficient condition to recover strong unimodality is related to the magnitude of q_1 and q_2, as the following theorem shows.

Theorem 2. *If X is a generalized r.v. with Y strongly unimodal and $Z \stackrel{d}{=} PS(\lambda)$, then X is strongly unimodal if and only if $\lambda q_1 \geq 2q_2$.*

Note that a similar result is proved in ([25], Theorem 4). We provide a different proof using the following lemma.

Lemma 1. *If $\{z_j\}_{j\geq 1} \in [0,\infty)$ is a log-concave sequence, then the sequence $\{\frac{1}{n!}B_n(z_1,\ldots,z_n)\}_{n\geq 1}$ is log-concave if and only if $z_1 \geq 2z_2$, with $\{B_n\}$ given in (5).*

Proof. If $\{z_j\}_{j\geq 0}$ with $z_0 = 1$ is a log-concave sequence of non-negative real numbers and the sequence $\{a(n)\}_{n\geq 0}$ is defined by

$$\sum_{n=0}^{\infty} \frac{a(n)}{n!} y^n = \exp\left(\sum_{j=1}^{\infty} \frac{z_j}{j!} y^j\right) \tag{31}$$

then the sequence $\{\frac{a(n)}{n!}\}_{n\geq}$ is log-concave [26]. Equation (31) parallels (7). Therefore, the sequence $\{\frac{1}{n!}B_n(z_1,\ldots,z_n)\}_{n\geq 1}$ results as log-concave if the sequence $\{z_j\}_{j\geq 0}$ is log-concave. Note that for $j \geq 2$ we have

$$\frac{z_j^2}{[(j-1)!]^2} \geq \frac{z_{j-1}}{(j-2)!}\frac{z_{j+1}}{j!},$$

which easily reduces to $jz_j^2 \geq (j-1)z_{j-1}z_{j+1}$ always satisfied when $\{z_j\}_{j\geq 1}$ is log-concave. Now let $j = 1$. We have $\{z_j\}_{j\geq 0}$ is log-concave if and only if $z_1 \geq 2z_2$ and the result follows. □

Proof of Theorem 2. Following the same arguments of Proposition 3, for a generalized r.v. with $Z \stackrel{d}{=} PS(\lambda)$, (30) reduces to

$$\mathbb{P}(X=x) = \frac{e^{-\lambda}}{x!} B_x(\lambda 1! q_1, \ldots, \lambda x! q_x)$$

with $q_x = \mathbb{P}(Y=x)$ for $x=1,2,\ldots$. The sequence $\{\frac{e^{-\lambda}}{x!} B_x(\lambda 1! q_1, \ldots, \lambda x! q_x)\}_{x \geq 1}$ is log-concave if and only if the sequence $\{\frac{1}{x!} B_x(\lambda 1! q_1, \ldots, \lambda x! q_x)\}_{x \geq 1}$ is log-concave. The result follows using Lemma 1. □

Corollary 3. *If $s < -1$, $X \stackrel{d}{=}$ PSHLZD(λ, θ, a, s) is strongly unimodal if and only if $\lambda q_1 \geq 2 q_2$.*

4.2. Moments and Cumulants

PSHLZD moments and cumulants have closed form expressions in terms of moments of $Y \stackrel{d}{=}$ HLZD(a, θ, s).

Proposition 5. *If $X \stackrel{d}{=}$ PSHLZD(λ, a, θ, s) then*

$$\mu_k := \mathbb{E}[X^k] = B_k(\lambda \xi_1, \ldots, \lambda \xi_k), \qquad k=1,2,\ldots, \tag{32}$$

with B_k the k-th complete exponential Bell polynomial (5) *and $\xi_1, \ldots, \xi_k$ the first k moments of $Y \stackrel{d}{=}$ HLZD(a, θ, s) given in* (16).

Proof. If $M_X(t)$ and $M_Y(t)$ are the mgf's of $X \stackrel{d}{=}$ PSHLZD(λ, a, θ, s) and $Y \stackrel{d}{=}$ HLZD(a, θ, s) respectively, then

$$M_X(t) = G_X(e^t) = e^{\lambda [G_Y(e^t) - 1]} = e^{\lambda [M_Y(t) - 1]} \tag{33}$$

from (27). Equation (32) follows as the rhs of (33) can be written as (7), with $h_z(t) = \lambda M_Y(t)$ and $z_0 = \lambda$. □

Remark 1. *Taking into account* (33), *if $X \stackrel{d}{=}$ PSHLZD(λ, a, θ, s) then $X \stackrel{d}{=} Y_1 + \cdots + Y_Z$ with $Y \stackrel{d}{=}$ HLZD(a, θ, s) and $Z \stackrel{d}{=} PS(\lambda)$, that is X is a compound Poisson r.v. Therefore the PSHLZD is an infinitely divisible distribution [27].*

Moments (32) can be explicited written using (9). A straightforward corollary of recursion (6) is the following.

Corollary 4. $\mu_{k+1} = \lambda \sum_{j=0}^{k} \binom{k}{j} \mu_{k-j} \xi_{j+1}, \quad k=1,2,\ldots$.

If $\mu'_k := \mathbb{E}[(X - \mu_1)^k]$ denotes the k-th central moment of $X \stackrel{d}{=}$ PSHLZD(λ, a, θ, s) then

$$\mu'_k = \sum_{k=0}^{n} \binom{n}{k} (-\lambda \xi_1)^{n-k} B_k(\lambda \xi_1, \ldots, \lambda \xi_k), \; k=1,2,\ldots$$

Proposition 6. *If $X \stackrel{d}{=}$ PSHLZD(λ, a, θ, s) then*

$$(\mu)_k := \mathbb{E}[X(X-1)\cdots(X-k+1)] = B_k(\lambda(\xi)_1, \ldots, \lambda(\xi)_k), \qquad k=1,2,\ldots \tag{34}$$

where $(\xi)_1, \ldots, (\xi)_k$ are the first k factorial moments of $Y \stackrel{d}{=}$ HLZD(a, θ, s) given in (17).

Proof. Let us recall that, if $Q_X(t)$ is the factorial mgf of $\{(\mu)_k\}$, then $Q_X(t) = G_X(t+1)$ with G_X the pgf of X. Therefore we have

$$Q_X(t) = G_X(t+1) = \exp\left(\lambda[G_Y(t+1)-1]\right) = \exp\left(\lambda[Q_Y(t)-1]\right), \tag{35}$$

with $Q_Y(t)$ the generating function of the factorial moments $\{(\xi)_k\}$. Equation (34) follows as the rhs of (35) can be written as (7), with $z_0 = \lambda$ and $h_z(t) = \lambda Q_Y(t)$. □

Proposition 7. *If $\kappa_n(X)$ is the n-th cumulant of $X \overset{d}{=} \mathrm{PSHLZD}(\lambda,a,\theta,s)$ then $\kappa_n(X) = \lambda\xi_n$, for $n = 1,2,\ldots$ where ξ_n is the n-th moment of $Y \overset{d}{=} \mathrm{HLZD}(a,\theta,s)$ given in (16).*

Proof. The result follows since

$$\log M_Y(t) = \log[e^{\lambda(M_X(t)-1)}] = \lambda[M_x(t)-1] = \sum_{n\geq 1} \frac{t^n}{n!}\lambda\mathbb{E}[X^k].$$

□

4.3. Maximum Likelihood Estimation

Suppose to have $\boldsymbol{x} = (x_1,\ldots,x_n)$ independent observations of $X \overset{d}{=} \mathrm{PSHLZD}(\lambda,a,\theta,s)$. The MLE of (λ,θ,s,a) is

$$(\hat{\lambda},\hat{\theta},\hat{s},\hat{a}) = \underset{(\lambda,\theta,s,a)\in\Theta}{\operatorname{argmax}}\ \ell_n(\lambda,\theta,s,a,\boldsymbol{x}),$$

with $\Theta = (0,\infty)\times(0,1)\times(-\infty,+\infty)\times(-1,\infty)$, $\ell_n(\lambda,\theta,s,a,\boldsymbol{x}) = \ln L_n(\lambda,\theta,s,a,\boldsymbol{x})$ the log-likelihood function and

$$L_n(\lambda,\theta,s,a,\boldsymbol{x}) = \prod_{i=1}^{n}\mathbb{P}(X = x_i).$$

The MLE of the PSHLZD parameters in this case must be directly tackled with the global optimization method described in Section 3.4, since $\ell_n(\lambda,\theta,s,a)$ is not analytically tractable referring to (29).

5. The One Inflated Hurwitz-Lerch Zeta Distribution

Definition 4. *A discrete random variable $Z \overset{d}{=} \mathrm{OIHLZD}(p,a,\theta,s)$ if*

$$\begin{aligned} \mathbb{P}(Z=1) &= p + (1-p)\mathbb{P}(Y=1), \\ \mathbb{P}(Z=x) &= (1-p)\mathbb{P}(Y=x), \quad x = 2,3,\ldots \end{aligned} \tag{36}$$

with $p \in [0,1]$ and $Y \overset{d}{=} \mathrm{HLZD}(a,\theta,s)$.

This definition parallels the definition of the Zero Inflated Modified Power Series Distribution given by Gupta [28]. If $G_Z(t)$ denotes the pgf of $Z \overset{d}{=} \mathrm{OIHLZD}(p,a,\theta,s)$ then

$$G_Z(t) = pt + (1-p)G_Y(t) \tag{37}$$

and the HLZD is retrieved by setting $p = 0$.

5.1. Moments and Cumulants

If $Z \overset{d}{=} \mathrm{OIHLZD}(p,a,\theta,s)$ then $M_Z(t) = G_Z(e^t) = pe^t + (1-p)G_Y(e^t) = pe^t + (1-p) M_Y(t)$ from (37). Thus

$$\nu_k := \mathbb{E}[Z^k] = p + (1-p)\xi_k \qquad k = 0,1,\ldots \tag{38}$$

with ξ_k the k-th moment of $Y \overset{d}{=} \mathrm{HLZD}(a,\theta,s)$ given in (16). For example, we have $\mathbb{E}[Z] = p + (1-p)\mathbb{E}[Y]$ and $\mathrm{Var}[Z] = (1-p)\left[\mathrm{Var}[Y] + p(1+\mathbb{E}[Y]^2 - 2\mathbb{E}[Y])\right]$.

Similarly, if $Q_Z(t)$ is the factorial mgf of $Z \stackrel{d}{=} \text{OIHLZD}(p,a,\theta,s)$, since $Q_Z(t) = G_Z(t+1) = p(t+1) + (1-p)G_Y(t+1) = p(t+1) + (1-p)Q_Y(t)$, with $Q_Y(t)$ the factorial mgf of $Y \stackrel{d}{=} \text{HLZD}(a,\theta,s)$, then

$$(\nu)_k := \mathbb{E}[Z(Z-1)\cdots(Z-k+1)] = \begin{cases} p+(1-p)(\xi)_1 & k=1 \\ (1-p)(\xi)_k & k=2,3,\ldots \end{cases}$$

with $(\xi)_k$ the k-th factorial moment of $Y \stackrel{d}{=} \text{HLZD}(a,\theta,s)$ given in (17).

OIHLZD cumulants are $\kappa_n(Z) = \mathcal{L}_n(\nu_1,\ldots,\nu_n), n=1,2,\ldots$ with $\{\nu_j\}$ moments of $Z \stackrel{d}{=} \text{OIHLZD}(p,a,\theta,s)$ given in (38), and $\mathcal{L}_n$ the n-th logarithmic polynomial (12).

5.2. Maximum Likelihood Estimation

To estimate the OIHLZD parameters using the MLE, let us first rewrite (36) using (18), that is

$$\begin{aligned} \mathbb{P}(Z=1) &= 1-w, \\ \mathbb{P}(Z=x) &= (1-p)\frac{a(x)g(\theta)^x}{f(\theta)}, x=1,2\ldots \end{aligned} \tag{39}$$

where $w = (1-p)[1-\mathbb{P}(Y=1)], g(\theta) = \theta, a(x) = (a+x)^{-(s+1)}$ and $f(\theta) = T(\theta,a,s)$. Rewrite (39) as

$$\begin{aligned} \mathbb{P}(Z=1) &= 1-w \\ \mathbb{P}(Z=x) &= w\mathbb{P}(W=x) \end{aligned}$$

where W has a One Truncated Hurwitz-Lerch Zeta Distribution (OTHLZD) [29], that is

$$\mathbb{P}(W=x) := \frac{1}{1-\frac{a(1)g(\theta)}{f(\theta)}}\frac{a(x)g(\theta)^x}{f(\theta)}, \qquad x=1,2\ldots \tag{40}$$

Suppose $\mathbf{z} = (z_1,\ldots,z_n)$ is a vector of independent observations of $Z \stackrel{d}{=} \text{OIHLZD}(p,a,\theta,s)$ and $l(\theta,a,s,w,\mathbf{z}) = \ln L_n(\theta,a,s,w,\mathbf{z})$ is the log likelihood function with

$$L_n(\theta,a,s,w,\mathbf{z}) = \prod_{i=1}^{n}\mathbb{P}(Z=z_i).$$

If n_j is the number of times the integer j appears in the vector $\mathbf{z}$ for $j=1,2,\ldots$, then the log-likelihood $l(\theta,a,s,w,\mathbf{z})$ can be written as

$$l(\theta,a,s,w,\mathbf{z}) = n_1\log(1-w) + (n-n_1)\log(w) + \sum_{j=2}^{\infty} n_j \log\left(\frac{\mathbb{P}(Y=j)}{1-\mathbb{P}(Y=1)}\right).$$

Now set

$$l_1(w,\mathbf{z}) = n_1\log(1-w) + (n-n_1)\log(w) \tag{41}$$

and

$$l_2(\theta,a,s,\mathbf{z}) = \sum_{j=2}^{\infty} n_j\left(\frac{\mathbb{P}(Y=j)}{1-\mathbb{P}(Y=1)}\right). \tag{42}$$

From (41) and (42), the parameters (θ,a,s,w) can be estimated separately, that is the estimation $\hat{w}$ can be recovered from $l_1(w,\mathbf{z})$ and the estimations $(\hat{\theta},\hat{a},\hat{s})$ from $l_2(\theta,a,s,\mathbf{z})$. The latter ones give the MLE of the parameters of $W \stackrel{d}{=} \text{OTHLZD}(\theta,s,a)$ in (40) using the vector $\mathbf{z}$ restricted to the observations which are greater than 1. As a consequence, the estimation $\hat{p}$ of p can be recovered from $\hat{w}$ as

$$\hat{p} = 1 - \frac{\hat{w}T(\hat{\theta},\hat{a},\hat{s})}{T(\hat{\theta},\hat{a},\hat{s}) - a(1)\hat{\theta}}.$$

6. Data-Fitting

6.1. Rainfall Depths and Interarrival Times

With rainfall depth we indicate to what depth liquid precipitation would cover a horizontal surface in an observation period if nothing could drain, evaporate or percolate from this surface. Let a time series of rainfall data be defined as $h = \{h_1, \ldots, h_n\}$, where h (mm) is the rainfall depth recorded at a fixed uniform unit τ of time (e.g., a day). A day k is considered rainy if the rainfall depth $h_k \geq h^*$, where h^* is a fixed rainfall threshold. The sub-series of h of the rainy days can be defined as the event series $E = \{t_1, \ldots, t_{n_r}\}$, where $n_r \leq n$ is an integer multiple of the time-scale τ. The sequence built with the times elapsed between each element of E (except the first one) and the immediately preceding one is defined as the interarrival time series $IT = \{IT_2, \ldots, IT_{n_r}\}$. In order to select an appropriate distribution for IT, some statistical characteristics usually observed in IT samples have to be considered: very high variance and skewness, relatively high frequency associated to the observation $IT = 1$, monotonically decreasing frequencies with a slowly decaying tail and a drop in the passage from the frequency at $IT = 1$ to the one at $IT = 2$. The HLZD in (13) has been fitted to rain IT in [5] for stations in Sicily and in [30] for stations in Piedmont, with good results. However, it has not yet been considered for rainfall depths. Recall that in the following we assume to model rainfall depths with a discrete r.v.

6.2. The Data

In this paper, the IT series analyzed were obtained from the recorded rainfall observations, using the rainfall threshold $h^* = 1$ mm, which is the conventional threshold stated by the World Meteorological Organization in order to discriminate between rainy and non rainy days. This dataset has not been previously considered in the literature and consists of both IT and h measured over 70 years at the following five stations: Floresta, Trapani, Torino, Oxford, Ceva. They were chosen in order to represent different climates from the rainfall characteristics point of view. Floresta and Trapani represent the Mediterranean climate with a very wet and a very dry situation respectively. For both stations, the rainfall is concentrated in the colder part of the year, as typical of the Mediterranean climate. Torino and Ceva are more continental, but Ceva is more influenced by the Ligurian sea. Therefore, Torino has its maximum rain in Spring, while that of Ceva is in Autumn, because of the heating of the sea in the Summer. Finally, Oxford is a northern Europe station with rainfall homogeneously spread across the whole year. The recordings start in 1947 and end in 2017, for a total of 70 years. Moreover, the time series were further subdivided. Thus for each station we considered for a total of 33 samples. Note that we did not consider wet and dry seasons for Oxford station due to its climate.

More specifically let *station_name* ∈ {Floresta, Trapani, Torino, Oxford, Ceva} and *season_name* ∈ {wet, dry, spring, summer, autumn, winter}. Then the samples tagged with *station_name year* span the whole length of the series for the *station_name* station, while the samples tagged with *station_name season_name* are the union of all the *season_name* seasons in the whole time series for the *station_name* station, omitting all the other seasons from the dataset. The MLE was used to fit the HLZD (Section 3), the PSHLZD (Section 4) and the OIHLZD (Section 5) to the dataset (Note that the PSHLZD has support $k = 0, 1, \ldots$ and the r.v. IT naturally has support $k = 1, 2, \ldots$ so we had to consider the shifted r.v. $IT' = IT - 1$.). In all cases, the MLE has been tackled with the method described in Section 3.4. The addressed global optimization procedure was further simplified by the previously mentioned statistical characteristics of the data allowing to work on a subset of the parameter space Θ.

- the whole time series, without subdividing the different seasons;
- all the wet seasons and all the dry seasons;
- the standard meteorological seasons,

6.3. Results

In the following we summarize the results of the distribution fitting for IT and h data. The fitting was satisfactory for both the PSHLZD and the HLZD. The assessment of the goodness-of-fit was obtained by following the methodology suggested by [31]. In the case of long tailed distributions, the goodness-of-fit through the classical χ^2 test might be biased, because if there are several small classes, strong asymmetry might occur [31] and some problems of inaccuracy might appear if the classes are grouped [32]. The alternative procedure used to test the goodness-of-fit relies on Monte Carlo simulation to numerically reconstruct the null hypothesis of the χ^2 test to compute the p-values [33].

To further inspect the differences between the distributions, we have measured the fitting errors whose magnitude is strictly related to the discrepancy between the empirical frequencies and the fitted ones. Since many empirical frequencies are zero (in the tail), the cdf has been considered. In particular we considered the mean absolute error (MAE) and the mean relative absolute error (MRAE). Let us recall that, if $x = (x_1, \ldots, x_n)$ is an ordered sample, then $\text{MAE}(x) = \sum_{j=1}^{n} \frac{1}{n} |F_N(x_j) - \hat{F}(x_j)|$ and $\text{MRAE}(x) = \sum_{j=1}^{n} \frac{1}{n} |F_n(x_j) - \hat{F}(x_j)| / F_n(x_j)$ with F_n the empirical cdf and $\hat{F}$ the fitted cdf.

6.3.1. Interarrival Times

We have compared the fitted PSHLZD with the fitted HLZD and the fitted PSHLZD with the fitted OIHLZD. To summarize the results, we have selected 4 of the 33 available samples since they have been considered particularly meaningful with respect to the whole dataset. The selected samples were *Floresta Summer*, *Trapani Wet*, *Trapani Dry* and *Torino Winter*.

Figure 1 is an example of IT empirical frequencies: they usually range from a high peak located at $IT = 1$ to a multitude of rather smaller values in the slow decaying tails. Therefore, to perform comparisons, a log-log scale for all the plots has been adopted.

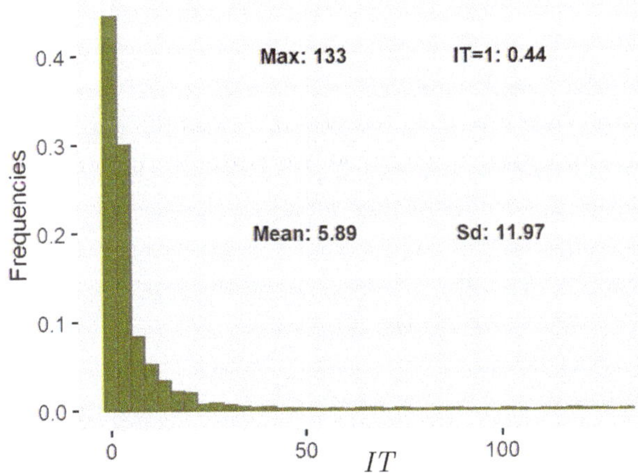

Figure 1. Histogram of the empirical IT frequency for the Trapani station over the whole year. The range is up to 133. The mode is $IT = 1$ with relative frequency 0.44. The mean and the standard deviation are 5.89 and 11.97 respectively.

Figure 2 shows plots of the fitted PSHLZD (solid line) and HLZD (dotted line) compared with the empirical frequencies (dot line) for the 4 selected samples. The fitting in both cases is very good. In particular, in the cases of *Floresta Summer*, *Torino Winter* and *Trapani Dry* the PSHLZD succeeds in fitting the drop from $IT = 1$ to $IT = 2$ whereas the HLZD fails. This happens in the drier periods, where this drop is more prominent.

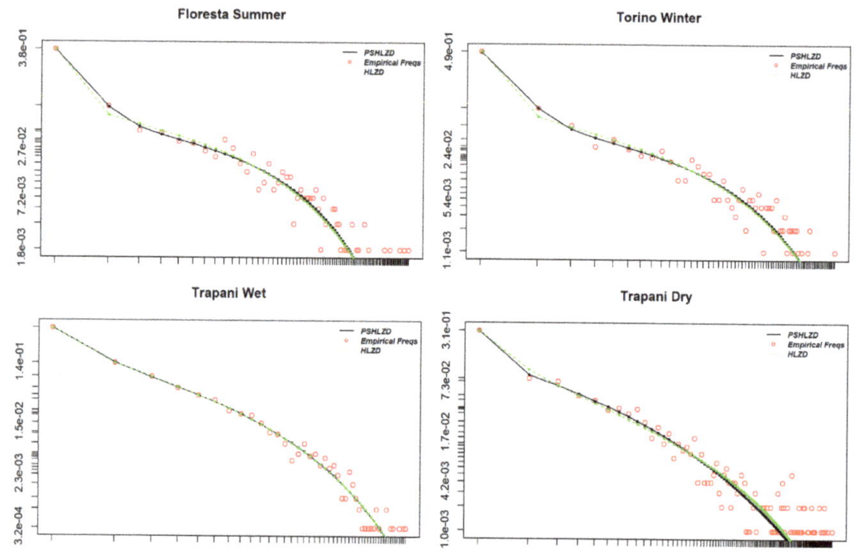

Figure 2. Log-log plots of the fitted HLZD (green dotted line), the fitted PSHLZD (black solid line) and the *IT* empirical frequencies (red dot line) for the 4 selected samples *Floresta Summer*, *Torino Winter*, *Trapani Wet* and *Trapani Dry*.

Moreover, Figure 1 shows the dominance of the frequencies corresponding to $IT = 1$ and $IT = 2$, which are particularly meaningful in hydrology. Figure 3 shows the plots of MAE and MRAE obtained by comparing the fitted cdf's of the PSHLZD (circle) and the HLZD (triangle) with the empirical IT cdf's. Note that the MAE and the MRAE are in general lower for the PSHLZD. Due to the dominance of the frequency corresponding to $IT = 1$, we explored modelling IT with the OIHLZD for all the samples. In all cases, the fitted OIHLZD and the PSHLZD one have minimal differences and are almost indistinguishable (see Figure 4 for an example), confirming the great flexibility of the latter distribution.

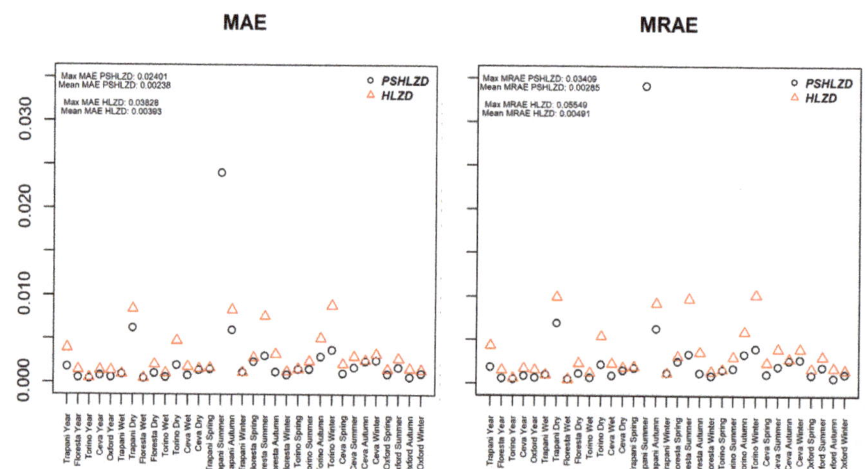

Figure 3. Dot plots of MAE and MRAE taking as reference the cdf of the PSHLZD (black circle) and of the HLZD (red triangle) for all the *IT* samples. The maximum MAE as well as the mean MAE are given in the top left for both the fitting distributions.

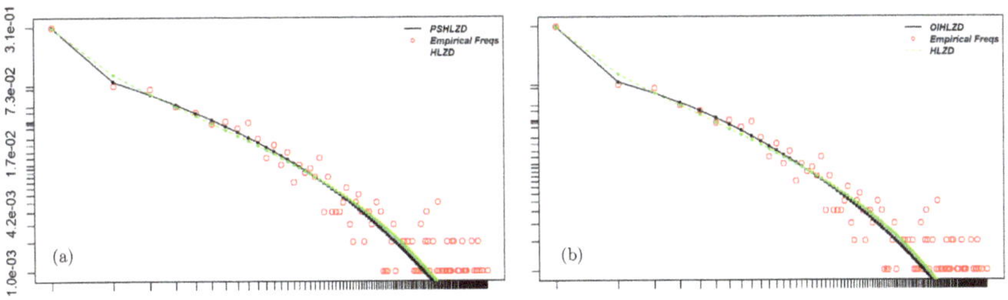

Figure 4. (**a**), the fitted PSHLZD (black solid line) is plotted together with the fitted HLZD (green dotted line) and the IT empirical frequencies (red dot line) for the sample of *Trapani Dry*. (**b**) the fitted OIHLZD (black solid line) is plotted together with the fitted HLZD (green dotted line) and the IT empirical frequencies (red dot line) for the same sample.

To conclude the validation analysis, we compared sample means and sample variances with the same theoretical moments of the HLZD and the PSHLZD computed in Section 3 and 4 respectively. In Table 1, we show the results for the 4 selected samples. In all cases, the fitted distributions agree with the sample means. For the variances, the PSHLZD performs better in many cases. In Table 1, an exception is *Trapani Wet* because the data are highly dispersed.

Table 1. The sample means and the sample variances for the 4 selected samples are given in the first column. The means and the variances of the fitted HLZD and of the fitted PSHLZD are given in the second and in the third column respectively.

		Trapani Dry	
	Sample	**HLZ**	**PSHLZ**
Mean	3.758	3.758	3.758
Var	26.173	26.668	26.588
		Trapani Wet	
	Sample	**HLZ**	**PSHLZ**
Mean	12.788	12.788	12.788
Var	461.938	461.168	529.513
		Floresta Summer	
	Sample	**HLZ**	**PSHLZ**
Mean	8.93	8.93	8.93
Var	161.344	189.699	153.808
		Torino Winter	
	Sample	**HLZD**	**PSHLZD**
Mean	6.594	6.594	6.594
Var	99.561	130.709	107.723

6.3.2. Rainfall Depths

In this section, we summarize the fitting of the rainfall depth time series using both the PSHLZD and the HLZD. We omit the comparison with the OIHLZD since this distribution does not add more insights on the fitting nor what happens for the IT datasets.

Figure 5 shows again an empirical frequency histogram ranging from a high peak in $h = 1$ to a multitude of rather smaller values in the slowly decaying tails. As in the previous section, we employed a log-log scale for all the plots. The selected samples were *Ceva Winter*, *Torino Winter*, *Floresta Dry* and *Trapani Summer*.

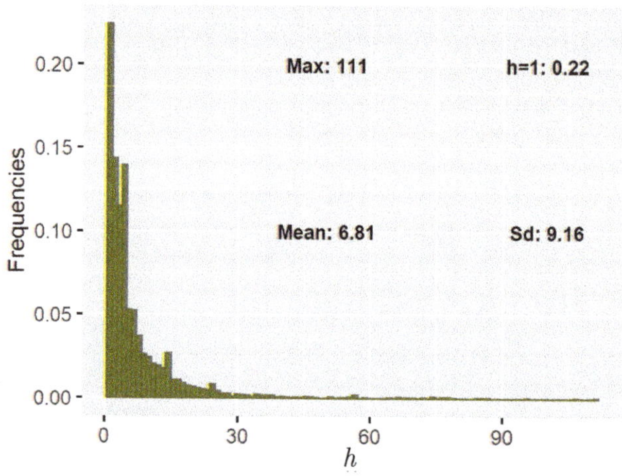

Figure 5. Plot of the empirical h frequency histogram for the Trapani station over the whole year. 111 is the maximum registered depth. The mode is $h = 1$ with relative frequency 0.22. The mean and the standard deviation are 6.81 and 9.16 respectively.

In Figure 6 we have plotted the fitted PSHLZD and HLZD compared with the empirical frequencies. As with IT samples, the fitting is very good, even better that in the IT case. Moreover there is less difference between the performances of the PSHLZD and the HLZD.

Figure 7 shows the plots of the MAE and the MRAE obtained by comparing the fitted cdf's of the PSHLZD (circle) and the HLZD (triangle) with the empirical h cdf's. Even though both errors are smaller for the PSHLZD, there is less difference between the two distributions and they are generally lower than for the IT case.

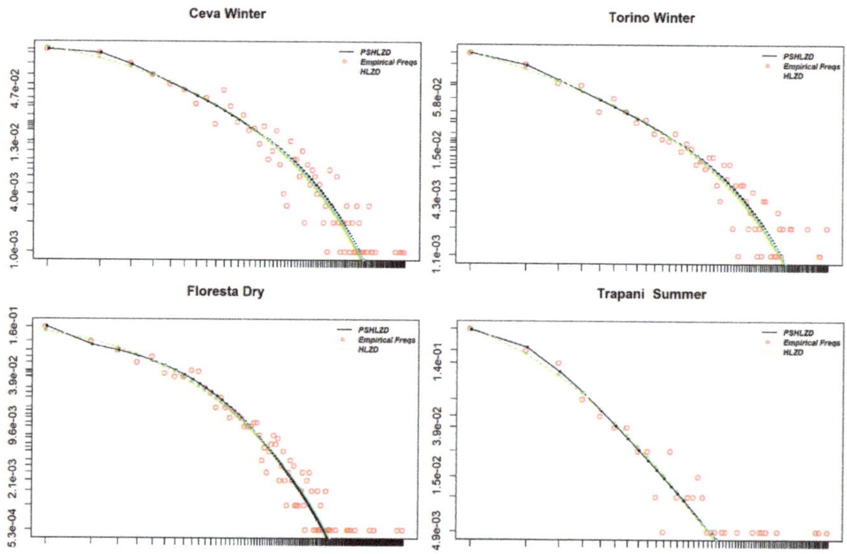

Figure 6. Log-log plots of the fitted HLZD (green dotted line), the fitted PSHLZD (black solid line) and the h empirical frequencies (red dot line) for the 4 selected samples *Ceva Winter*, *Torino Winter*, *Floresta Dry* and *Trapani Summer*.

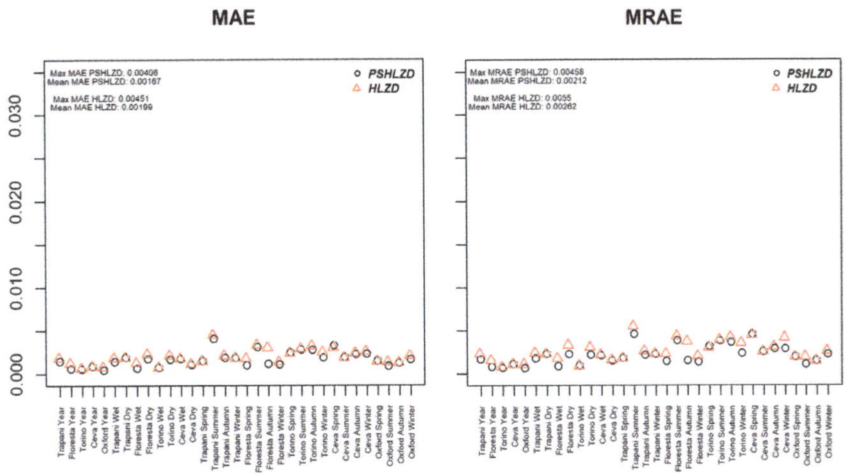

Figure 7. Dot plots of MAE and MRAE taking as reference the cdf of the PSHLZD (black circle) and of the HLZD (red triangle) for all the h samples. The maximum MAE as well as the mean MAE are given in the top left for both the fitting distributions.

7. Conclusions

The first part of this paper focuses on a class of discrete distributions useful to describe very high one counts and long tails. We have reviewed the main properties using the combinatorics of exponential Bell polynomials. This device has permitted the derivation of closed form expressions for the pdf's and their convolutions, as well as moments and cumulants. Moreover, new results on log-concavity have been presented. We have also considered the OIHLZD to compare its features with the HLZD and the PSHLZD. This deep analysis was aimed of investigating how to use these models to find a better fit for rainfall data. Indeed, the PSHLZD and the HLZD were fitted on Interarrival Times IT and rainfall depths h data coming from 5 different stations, which composed a dataset never previously analyzed in the literature. The h data were treated as observations of a discrete r.v., which is not the usual practice in the literature, but seems reasonable when taking into account how they are measured. The fitting was performed with the classical MLE method, but the likelihood was maximized using the Simulated Annealing procedure, which turns out to be fundamental since there are no closed forms of the likelihood equations. The fit was very good for both distributions, with the PSHLZD performing slightly better than the HLZD. This mostly happens for the IT data. Moreover, the PSHLZD was also able to replicate the fit of the OIHLZD further validating its flexibility.

From the modelling point of view, let us underline two final remarks. Firstly, the fit was excellent for both the IT and the h data, suggesting that the PSHLZD can be proposed as a general framework in rainfall modelling. Secondly, it is noteworthy to underline that these models capture the variability of rainfall stochastic phenomena, even though the 5 considered stations represent very different climates: a case study not yet considered in the literature that deals with previous applications of HLZD. Future works will consider modelling the dependence (inter-correlation) between IT and h. Given the remarkable performance of these distribution families in the univariate modelling, a first step would be to consider bivariate modified power series distributions [34] and the methods to estimate their parameters on a rainfall time series. This is in the agenda for our future developments.

Author Contributions: Conceptualization, C.A., G.B., S.F., E.D.N. and T.M.; methodology, E.D.N. and T.M.; software, T.M.; validation, E.D.N. and T.M.; formal analysis, E.D.N. and T.M.; data curation, C.A., G.B. and S.F.; writing—original draft preparation, E.D.N. and T.M.; writing—review and editing, C.A., G.B., S.F., E.D.N. and T.M.; funding acquisition, S.F. All authors have read and agreed to the published version of the manuscript.

Funding: This research was funded by MIUR—Dipartimento di Eccellenza- DIST department funds and by the "PRIN MIUR 2017SL7ABC_005 WATZON Project".

Data Availability Statement: Data are available upon request to the Società Italiana di Meteorologia

Acknowledgments: The authors would like to thank the anonymous referees and the associate editor for carefully reading this manuscript and giving valuable comments to improve the previous version of this paper. The authors would like to thank Nicholas Howden of University of Bristol and Daniele Cat Berro and Luca Mercalli of Società Meteorologica Italiana to provide the English and North-Italy rainfall time series. The authors would like to thank the "Risk Responsible Resilience Interdepartmental Centre" (R3C).

Conflicts of Interest: The authors declare no conflict of interest.

References

1. Serinaldi, F. A multisite daily rainfall generator driven by bivariate copula-based mixed distributions. *J. Geophys. Res. Atmos.* **2009**, *114*, D10103. [CrossRef]
2. Davis, A.; Wiscombe, W.; Cahalan, R.; Marshak, A. Multifractal characterizations of nonstationary and intermittency in geophysical fields: Observed, retrieved, or simulated. *J. Geophys. Res.* **1994**, *99*, 8055–8072. [CrossRef]
3. Lawrance, A.J. Stochastic Modelling of Daily Rainfall Sequences. *J. R. Stat. Soc. Ser. C Appl. Stat.* **1979**, *28*, 84. [CrossRef]
4. Chatfield, C. Wet and dry spells. *Weather* **1966**, *21*, 308–310. [CrossRef]
5. Agnese, C.; Baiamonte, G.; Cammalleri, C. Modelling the occurrence of rainy days under a typical Mediterranean climate. *Adv. Water Resour.* **2014**, *64*, 62–76. [CrossRef]
6. Liew, K.W.; Ong, S.H.; Toh, K.K. The Poisson-stopped Hurwitz–Lerch zeta distribution. *Commun. Stat.—Theory Methods* **2022**, *51*, 5638–5652. [CrossRef]
7. Liu, M.; Zhang, L.; Fang, Y.; Dong, H. Exact solutions of fractional nonlinear equations by generalized bell polynomials and bilinear method. *Therm. Sci.* **2021**, *25*, 1373–1380. [CrossRef]
8. Taghavian, H. The use of partition polynomial series in Laplace inversion of composite functions with applications in fractional calculus. *Math. Methods Appl. Sci.* **2019**, *42*, 2169–2189. [CrossRef]
9. Fathizadeh, F.; Kafkoulis, Y.; Marcolli, M. Bell polynomials and Brownian bridge in spectral gravity models on multifractal Robertson–Walker cosmologies. In *Annales Henri Poincaré*; Springer: Berlin/Heidelberg, Germany, 2020; Volume 21, pp. 1329–1382.
10. Bernardara, P.; De Michele, C.; Rosso, R. A simple model of rain in time: An alternating renewal process of wet and dry states with a fractional (non-Gaussian) rain intensity. *Atmos. Res.* **2007**, *84*, 291–301. [CrossRef]
11. Yang, L.; Franzke, C.L.E.; Fu, Z. Power-law behaviour of hourly precipitation intensity and dry spell duration over the United States. *Int. J. Climatol.* **2020**, *40*, 2429–2444. [CrossRef]
12. Porporato, A.; Vico, G.; Fay, P.A. Superstatistics of hydro-climatic fluctuations and interannual ecosystem productivity. *Geophys. Res. Lett.* **2006**, *33*, L15402. [CrossRef]
13. Foufoula-Georgiou, E.; Lettenmaier, D.P. Continuous-time versus discrete-time point process models for rainfall occurrence series. *Water Resour. Res.* **1986**, *22*, 531–542. [CrossRef]
14. Charalambides, C.A. *Enumerative Combinatorics*; CRC Press Series on Discrete Mathematics and its Applications; Chapman & Hall/CRC: Boca Raton, FL, USA, 2002; Volume 16, p. 609.
15. Di Nardo, E.; Guarino, G.; Senato, D. A unifying framework for k-statistics, polykays and their multivariate generalizations. *Bernoulli* **2008**, *14*, 440–468. [CrossRef]
16. Di Nardo, E.; Guarino, G. kStatistics: Unbiased Estimates of Joint Cumulant Products from the Multivariate Faà Di Bruno's Formula. *arXiv* **2022**, arXiv:2206.15348.
17. Aksenov, S.V.; Savageau, M.A. Some properties of the Lerch family of discrete distributions. *arXiv* **2005**, arXiv:math/0504485.
18. Gupta, R.C. Modified power series distribution and some of its applications. *Sankhyā Ser. B* **1974**, *36*, 288–298.
19. Keilson, J.; Gerber, H. Some Results for Discrete Unimodality. *J. Am. Stat. Assoc.* **1971**, *66*, 386–389. [CrossRef]
20. Gupta, P.L.; Gupta, R.C.; Ong, S.H.; Srivastava, H. A class of Hurwitz–Lerch Zeta distributions and their applications in reliability. *Appl. Math. Comput.* **2008**, *196*, 521–531. [CrossRef]
21. Eger, S. Identities for partial Bell polynomials derived from identities for weighted integer compositions. *Aequationes Math.* **2016**, *90*, 299–306. [CrossRef]
22. Bélisle, C.J.P. Convergence theorems for a class of simulated annealing algorithms on $\mathbf{R}^d$. *J. Appl. Probab.* **1992**, *29*, 885–895. [CrossRef]
23. Charalambides, C.A. On the generalized discrete distributions and the Bell polynomials. *Sankhyā Ser. B* **1977**, *39*, 36–44.
24. Badía, F.G.; Sangüesa, C.; Federgruen, A. Log-concavity of compound distributions with applications in operational and actuarial models. *Probab. Eng. Inf. Sci.* **2021**, *35*, 210–235. [CrossRef]
25. Yu, Y. On the entropy of compound distributions on nonnegative integers. *IEEE Trans. Inform. Theory* **2009**, *55*, 3645–3650. [CrossRef]

26. Bender, E.A.; Canfield, E.R. Log-concavity and related properties of the cycle index polynomials. *J. Comb. Theory Ser. A* **1996**, *74*, 57–70. [CrossRef]
27. Sato, K.i. Lévy processes and infinitely divisible distributions. In *Cambridge Studies in Advanced Mathematics*; Translated from the 1990 Japanese original, Revised edition of the 1999 English translation; Cambridge University Press: Cambridge, UK, 2013; Volume 68, pp. 14, 521.
28. Gupta, P.L.; Gupta, R.C.; Tripathi, R.C. Inflated modified power series distributions with applications. *Commun. Stat.-Theory Methods* **1995**, *24*, 2355–2374. [CrossRef]
29. Conceição, K.S.; Louzada, F.; Andrade, M.G.; Helou, E.S. Zero-modified power series distribution and its Hurdle distribution version. *J. Stat. Comput. Simul.* **2017**, *87*, 1842–1862. [CrossRef]
30. Baiamonte, G.; Mercalli, L.; Berro, D.C.; Agnese, C.; Ferraris, S. Modelling the frequency distribution of inter-arrival times from daily precipitation time-series in North-West Italy. *Hydrol. Res.* **2019**, *50*, 339–357. [CrossRef]
31. Martínez-Rodríguez, A.M.; Sáez-Castillo, A.; Conde-Sánchez, A. Modelling using an extended Yule distribution. *Comput. Stat. Data Anal.* **2011**, *55*, 863–873. [CrossRef]
32. Spierdijk, L.; Voorneveld, M. Superstars without talent? The Yule distribution controversy. *Rev. Econ. Stat.* **2009**, *91*, 648–652. [CrossRef]
33. Hope, A.C. A simplified Monte Carlo significance test procedure. *J. R. Stat. Soc. Ser. B Methodol.* **1968**, *30*, 582–598. [CrossRef]
34. Shoukri, M.M.; Consul, P.C. Bivariate Modified Power Series Distribution Some Properties, Estimation and Applications. *Biom. J.* **1982**, *24*, 787–799. [CrossRef]

fractal and fractional

Article

A Model Based on Fractional Brownian Motion for Temperature Fluctuation in the Campi Flegrei Caldera

Antonio Di Crescenzo *,†, Barbara Martinucci † and Verdiana Mustaro †

Dipartimento di Matematica, Università degli Studi di Salerno, 84084 Fisciano, Italy; bmartinucci@unisa.it (B.M.); vmustaro@unisa.it (V.M.)
* Correspondence: adicrescenzo@unisa.it
† These authors contributed equally to this work.

Abstract: The aim of this research is to identify an efficient model to describe the fluctuations around the trend of the soil temperatures monitored in the volcanic caldera of the Campi Flegrei area in Naples (Italy). This study focuses on the data concerning the temperatures in the mentioned area through a seven-year period. The research is initially finalized to identify the deterministic component of the model given by the seasonal trend of the temperatures, which is obtained through an adapted regression method on the time series. Subsequently, the stochastic component from the time series is tested to represent a fractional Brownian motion (fBm). An estimation based on the periodogram of the data is used to estabilish that the data series follows an fBm motion rather than fractional Gaussian noise. An estimation of the Hurst exponent H of the process is also obtained. Finally, an inference test based on the detrended moving average of the data is adopted in order to assess the hypothesis that the time series follows a suitably estimated fBm.

Keywords: fractional Brownian motion; stochastic model; regression; time series; residuals analysis; Hurst exponent; Campi Flegrei caldera; temperature fluctuation

Citation: Di Crescenzo, A.; Martinucci, B.; Mustaro, V. A Model Based on Fractional Brownian Motion for Temperature Fluctuation in the Campi Flegrei Caldera. *Fractal Fract.* **2022**, *6*, 421. https://doi.org/10.3390/fractalfract6080421

Academic Editor: Mario Abundo

Received: 4 July 2022
Accepted: 27 July 2022
Published: 30 July 2022

1. Introduction

In several applied contexts, one of the main problems in the description of a phenomenon that evolves over time is the individuation of the probabilistic laws by which the phenomenon itself is driven. To this aim, schemes based on the superposition of deterministic and random components are often chosen, in which the deterministic part describes the phenomenon's trend, while the random one describes the fluctuations determined by unpredictable exogenous factors. An example of such a model is presented in the work by Sebastiani and Malagnini [1], where a physical model with non-constant variance is proposed to describe the phenomenon of coupling erosion that caused the 2011 earthquake in Tohoku-Oki (Japan). Analogously, another compound model can be found in the work of Giordano and Morale [2]. Here, the authors propose a mathematical model for the prices of electricity in the Italian market through the years, whose stochastic component is given by the sum of a fractional Ornstein–Uhlenbeck process and the solution of a mean-reverting SDE driven by a Hawkes-marked process.

The aim of the present study is to implement an analogous scheme to describe the soil temperatures observed on the surface of the Campi Flegrei caldera. This volcanic area is located near the city of Naples in Italy and is famous for the ground deformations that have been registered in the area since ancient times. Recent observations, carried out from 2011 to 2017, were led to monitor the presence of radon in two sites of the area (cf. Sabbarese et al. [3]). In [3], the presence of radon in the soil was tracked, along with its dependence on physical quantities such as temperature, pressure and humidity, in order to prove how the seismic activity in the Campi Flegrei area was influenced by the presence of the element.

Starting from the research mentioned above, in this work, we aim to build a model able to describe the fluctuations in soil temperatures in the caldera through the identification of the temporal trend and the corresponding random component.

What emerged from the present study is that the deterministic component is described well by a piecewise linear model, where the segments' slopes are alternating in sign. The slopes can be determined by applying the ordinary least squares (OLS) method on the observed data. The endpoints for each segment of the piecewise curve are chosen through an emphirical method. The stochastic model describing the random fluctuations component is instead recognized to be the fractional Brownian motion (fBm). For a description of the most relevant probabilistic and statistical aspects of fBm, see, for example, the works of Nourdin [4] and Prakasa-Rao [5]. This stochastic process has been proposed in several studies for modeling various geophysical phenomena. In the work by Mattia et al. [6], fBm is proposed as an approximation of the process describing the trend of ground data involving soil roughness collected over three different European sites. In the paper by Yin and Ranalli [7], the authors show that the event of earthquake rupturing in a fault is related to factors such as static shear stress and static frictional strength through a potential dynamic stress drop. The latter is regarded as a one-dimentional stochastic process, and it is modeled as an fBm with a Hurst exponent close to zero.

The evidence of the reasonableness of the model first emerges as consequence of a statistical test, in which the hypothesis that the random fluctuations of the model can be described by Brownian motion is rejected. The subsequent phase of the work involves the study of the periodogram of the data. It allows verifying that the random fluctuations are described well by fBm, in opposition to a fractional Gaussian noise (fGn). Such analysis also leads to the estimation of the involved parameters, namely the Hurst exponent H and the scale parameter D of the fBm. Many estimation techniques for the former parameters have been proposed in the literature, such as the methods presented in [8]. Other works often provide modified versions of the already known algorithms.

Subsequently, this study moves toward analyzing the residual series emerging from the considered stochastic process and simulated fBm with the previously estimated parameters. First of all, such a residual series is plotted with respect to the temperatures series, which suggests the Gaussianity of the residuals. Then, the procedure is repeated through suitable simulations. Two statistical tests are performed on the obtained values, with both leading to the acceptance of the null hypothesis of Gaussianity.

The goodness of the considered model is finally confirmed through an inference test, in which the test statistics involve the detrending moving averages obtained from the data.

The main results of this article are as follows:

- The development of a suitable stochastic model $X(t)$ describing a geophysical phenomenon;
- The estimation of the parameters involved in the model based on real observed data;
- The design of revised statistical algorithms to construct the model and to perform appropriate testing for confirming its validity.

Here is the plan of the article. In Section 2, we introduce the starting equation for modeling the soil temperature trend (cf. Equation (1)). Section 3 focuses on the data analysis directed toward identifying the deterministic component of the temperature trend. The remaining part of the article is dedicated to the analysis of the stochastic component of the model. More specifically, Section 4 is aimed at studying the periodogram obtained from the data observations, from which emerges the eligibility of fBm to describe the random component of the temperature compared to fGn. An estimation of the relevant parameters of the process (i.e., the Hurst exponent and the scale parameter) is also performed. Testing on the residuals of the model is then performed by resorting to the Shapiro–Wilk test and the robust Jarque–Bera test. The work is completed by Section 5, which contains the statistical test leading to the acceptance of the model, and Section 6, which collects the conclusions of the study.

2. The Stochastic Model

The time series considered throughout this study describes the temperatures of the surface soil of the Campi Flegrei caldera, obtained via daily observations in the time period between 1 July 2011 and 31 December 2017. The main aim of the research is to develop a model able to describe the fluctuations of the data. As usually observed in geophysical studies, the trajectory of the time series is seen as the superposition of a deterministic component and a stochastic one, where the latter represents the fluctuations from the trend. As a consequence, the process $\{X(t),\ t \geq 0\}$ that represents the daily observed soil temperatures is described by the following equation:

$$X(t) = r(t) + B_H(t), \tag{1}$$

where $r(t)$ constitutes the deterministic component of the process, while $B_H(t)$ is a stochastic process. We will show that a suitable assumption is to identify $B_H(t)$ as fBm. A similar model has been proposed for the vertical motion of the soil in the Campi Flegrei area (cf. Travaglino et al. [9]).

Under the assumption that $B_H(t)$ is indeed fBm, it is easy to observe that the model shown in Equation (1) is such that $\mathbb{E}(X(t)) = r(t)$. We recall, in fact, that fractional Brownian motion is a continuous-time Gaussian process with zero mean and covariance:

$$\mathrm{Cov}[B_H(t), B_H(s)] = D(t^{2H} + s^{2H} - |t-s|^{2H}), \quad t, s \geq 0, \tag{2}$$

where $D > 0$ is a scale parameter named the diffusion constant and $0 < H < 1$ is a parameter known as the Hurst exponent. The process was first introduced by Mandelbrot and Van Ness in [10]. It is a generalization of Brownian motion, since for $D = 1/2$ and $H = 1/2$ it reduces to a standard Wiener process. We remark that fBm has the property of self-similarity; that is, for any choice of $a > 0$, the following is true:

$$\{B_H(at),\ t \geq 0\} \stackrel{d}{=} \{a^H B_H(t),\ t \geq 0\},$$

meaning that the two processes are equal in distribution. The parameter H is therefore also referred to as the scaling exponent or fractal index of the process.

The model in Equation (1) is not stationary, since fBm is not a stationary process itself. The process of the increments of $B_H(t)$, defined by $Z(t) = B_H(t+1) - B_H(t)$ for all $t \geq 0$, is, however, stationary. This is known as fractionary Gaussian noise (fGn).

Introducing the process $Z(t)$ leads to another well-known property of fBm, which is long-range dependency (LRD). As pointed out in [11], a stationary process $X(t)$ has long-range dependence (or is a long memory process) if its autocorrelation function, defined as

$$\rho(k) = \frac{\mathrm{Cov}(X(t), X(t+k))}{\mathrm{Var}[X(1)]},$$

satisfies the condition

$$\sum_{k=-\infty}^{+\infty} \rho(k) = \infty.$$

This definition implies that the autocorrelation of the process decays slowly over time, making the sum of the autocorrelations divergent. Therefore, if $X(t)$ is an LRD process, then

$$\lim_{k \to +\infty} \frac{\rho(k)}{ck^{-\alpha}} = 1, \tag{3}$$

where $c > 0$ and $0 < \alpha < 1$ are constants. This definition states that the decay of the autocorrelation function is power-like and hence slower than exponential decay. In the case of fBm, long-range dependence can be seen when looking at the increments in $Z(t)$. Moreover, the parameter α is related to the Hurst exponent through the equation $\alpha = 2H - 2$, evidencing that the value of the Hurst exponent can be used to determine the nature of the process. We recall the following (see also [11,12]):

- If $\frac{1}{2} < H < 1$, then the increments of the process are positively correlated, making the process persistent (i.e., likely to keep the trend exhibited in the previous observations);
- If $0 < H < \frac{1}{2}$, the increments of the process are negatively correlated, and the process is counter-persistent (i.e., likely to break the trend followed in the past).

A relation similar to Equation (3) but concerning the frequency domain of the time series is shown in Section 3, where it is used for the estimation of a parameter of the process. It is worth mentioning that recent investigations showed that fBm with a random Hurst exponent can be employed to describe biological phenomena such as particle motion and intracellular transport, as reported by Balcerek et al. [13] and Han et al. [14], respectively.

3. The Deterministic Component

The model in Equation (1) was adopted to analyze the time series of the recorded temperatures of the Campi Flegrei caldera soil (cf. Sabbarese et al. [3]). The initial dataset consisted of $\hat{N} + 1 = 2005$ observations collected at the times indicated by $t_0, t_1, \ldots, t_{\hat{N}}$ and shown in the scatter plot in Figure 1.

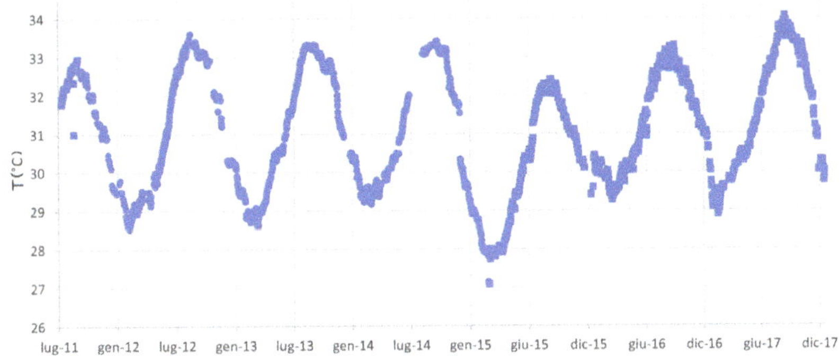

Figure 1. Scatter plot of the data series $X(t)$ for the temperatures of the surface soil of the Campi Flegrei caldera at the Monte Olevano (NA) site, collected during the observation period.

The observations are not equally spaced time-wise, since the data series was collected daily with the lack of a quite large number of observations. As shown in the plot in Figure 2, a seasonal trend is clearly visible in the dataset, with the temperatures increasing during spring and summer in each year and then starting to decrease around September. For this reason, the deterministic trend $r(t)$ is constructed by alternating segments with opposite slopes, whcih were obtained from the data through the OLS method. The first and the last measurement of the dataset, which are also the extremes of the curve, are given by

$$t_0 = 07/29/2011, \qquad X(t_0) = 31.78;$$
$$t_{\hat{N}} = 12/31/2017, \qquad X(t_{\hat{N}}) = 29.94.$$

To obtain the points in which the curve changes its slope, a heuristic method based on linear regression was used, which was developed upon the methodologies illustrated by Hudson in [15]. First, for the dataset, we detected the local minima m_i, $i = 1, \ldots, 6$ and maxima M_i, $i = 1, \ldots, 7$, as shown in Table 1.

Table 1. Local maxima M_i and minima m_i for each subset of the dataset and the corresponding dates τ_i (day, month, and year).

τ_i	$M_i = X(\tau_i)$ (°C)	τ_i	$m_i = X(\tau_i)$ (°C)
16 September 2011	31.78	14 February 2012	28.68
25 August 2012	33.6	22 March 2013	28.63
11 September 2013	33.31	4 March 2014	29.17
22 September 2014	33.4	27 February 2015	27.08
5 September 2015	32.4	11 March 2016	29.27
19 September 2016	33.29	03 February 2017	28.88
30 August 2017	34.04		

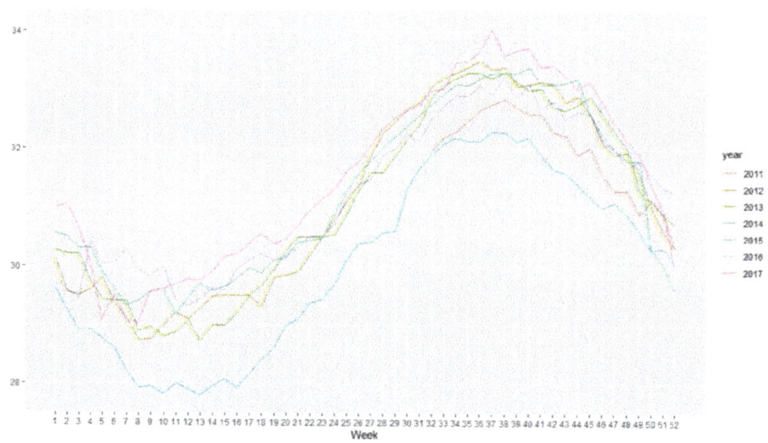

Figure 2. Seasonal plot of the temperatures described by $X(t)$. This was obtained by taking weekly spaced values from the time series $X(t)$ and representing each year of observation (2011–2017) separately.

The time t_0 was used as the first extreme of the first segment of the piecewise linear curve. Subsequently, the slope α_j and the intercept of the regression line between τ_0 and the first local maxima M_1 were estimated using the OLS method. The corresponding coefficient of determination R^2 was also calculated. The second step consisted of repeating the operation and choosing the second extreme among the five values that preceded and the five ones that followed M_1. Hence, among the 11 possible choices, we chose the regression line that maximized R^2. Table 2 shows the values obtained for each endpoint.

Table 2. Slopes of the segments obtained by linear regression between initial time t_0 and a neighborhood of M_1 with their corresponding coefficients of determination. The slope was chosen such that R^2 was the highest, as underlined.

t_j	α_j	R^2
11 September 2011	0.0219	0.9302
12 September 2011	0.0217	0.9326
13 September 2011	0.0215	0.9341
14 September 2011	0.0213	0.9369
15 September 2011	0.0213	0.9405
16 September 2011	0.0213	0.9442
23 September 2011	0.0199	0.9021
24 September 2011	0.0190	0.8787
25 September 2011	0.0182	0.8574
26 September 2011	0.0174	0.8369
27 September 2011	0.0162	0.7880

Then, the same method was applied for the next 12 segments, in which the first initial time was determined by the previous step. The last regression line was finally calculated using $t_{\hat{N}}$ as the endpoint. We denote by $c_i > 0$ and $v_i < 0$ the alternating slopes obtained through the regression model. The endpoints θ_i for each interval, as well as the estimated slopes and the coefficients of determination, are reported in Table 3.

The values obtained by means of this procedure allowed us to identify the deterministic component $r(t)$ of the model in Equation (1). The values of R^2 confirm that the regression lines provided a good fitting of the data in each interval. The scatter plot and the resulting linear model are shown in Figure 3.

Table 3. Estimated slopes c_i and v_i and corresponding coefficients of determination evaluated for each subset $[\theta_{i-1}, \theta_i]$ of the data series, with $\theta_0 = t_0 =$ 29 July 2011.

θ_i	c_i	R^2	θ_i	v_i	R^2
16 September 2011	0.0213	0.9442	27 February 2012	−0.0264	0.9709
13 September 2012	0.0262	0.9356	7 March 2013	−0.0311	0.9722
6 September 2013	0.0287	0.9818	7 March 2014	−0.0256	0.9573
24 September 2014	0.0237	0.9701	1 March 2015	−0.0418	0.9635
23 August 2015	0.0294	0.9659	19 March 2016	−0.0145	0.9287
8 September 2016	0.0238	0.9619	16 February 2017	−0.0261	0.9357
14 September 2017	0.0228	0.9674	31 December 2017	−0.0346	0.9153

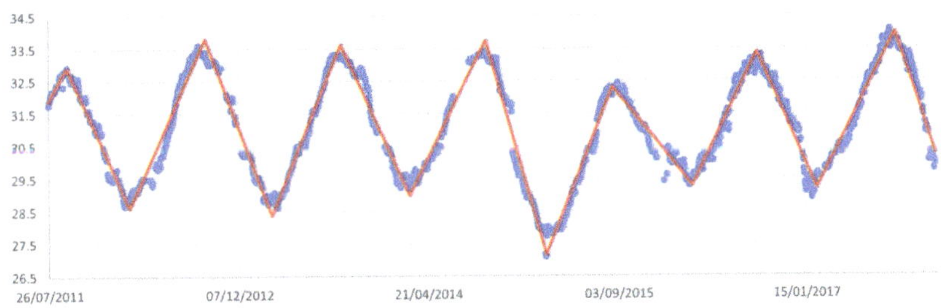

Figure 3. The scatter plot of the data series $X(t)$ and the linear approximation of the deterministic trend $r(t)$.

4. Analysis of the Stochastic Component

In this section, we focus on the investigation concerning the nature of the time series $B_H(t) = X(t) - r(t)$. To this aim, some preparatory interpolations were performed on the dataset, as is customary. Since the times of the initial series were not equally spaced, we collected the missing data by performing a linear interpolation covering the whole observation period. The prepared dataset obtained after this phase consisted of $N + 1 = 2348$ observations. A plot of $B_H(t)$ is shown in Figure 4.

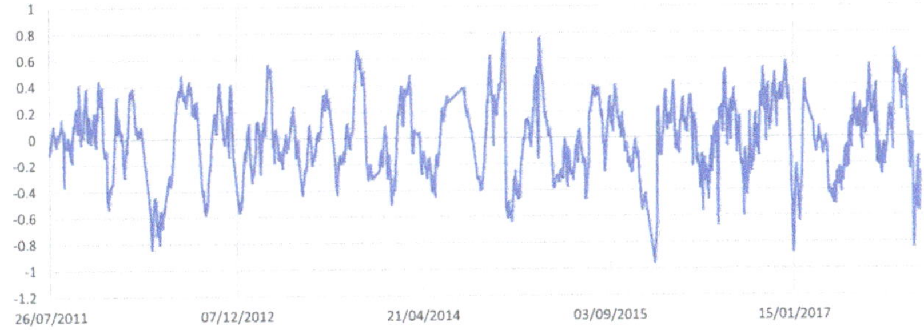

Figure 4. Plot of the time series $B_H(t) = X(t) - r(t)$.

As a preliminary study on the nature of the process $B_H(t)$, we represented the time series through a histogram and a qq-plot, as shown in Figure 5 and Figure 6, respectively. The obtained plots suggest classifying the stochastic component among Gaussian processes, thus justifying the following testing.

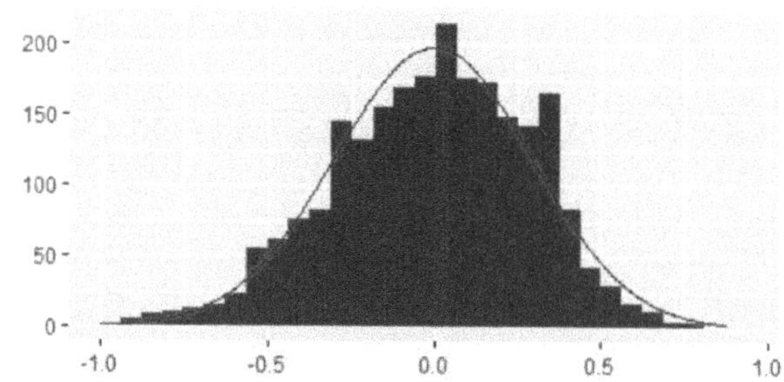

Figure 5. Histogram of the discrete time series $B_H(t)$.

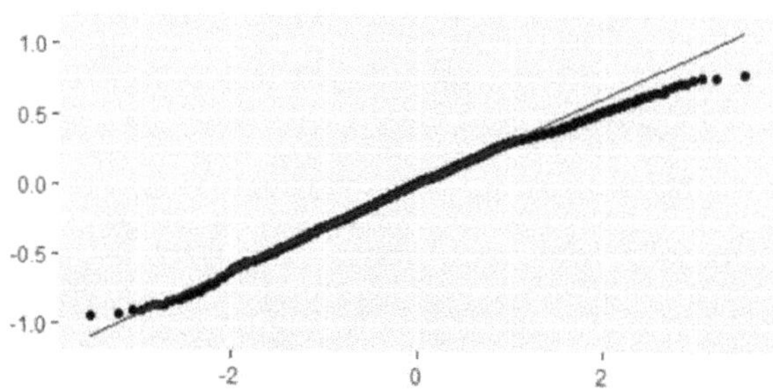

Figure 6. Qq-plot of the discrete time series $B_H(t)$.

4.1. Testing for Brownian Motion

With reference to the model in Equation (1), hereafter, we perform a test on the process $B_H(t)$, aiming to assess its suitable nature. With reference to the Brownian motion $B(t)$, we consider the following hypothesis:

Hypothesis 1. $B_H(t) = \sigma B(t)$.

Hypothesis 2. *$B_H(t)$ is a confined or directed diffusion.*

By adopting the test proposed by Briane et al. in [16], the asymptotic region of acceptance for $\mathcal{H}_0$ is given by

$$\left\{ q\left(\frac{\alpha}{2}\right) \le \frac{S_D^N}{\hat{\sigma}\sqrt{t_N}} \le q\left(1-\frac{\alpha}{2}\right) \right\},$$

where

$$S_D^N = \max_{j=1,\dots,N} |B_H(t_j) - B_H(t_0)|,$$

$$\hat{\sigma} = \left\{ \frac{1}{N} \sum_{j=1}^{N} \frac{[B_H(t_j) - B_H(t_{j-1})]^2}{t_j - t_{j-1}} \right\}^{1/2},$$

and $q(\alpha)$ is the quantile of

$$\sup_{0 \le s \le 1} |B(s) - B(0)|.$$

The extremes of the acceptance interval for the significance level $\alpha = 0.05$ are given by

$$q(0.025) = 0.834, \qquad q(0.975) = 2.940,$$

while the value obtained for the test statistic is 0.0787. The null hypothesis of the process $B_H(t)$ being Brownian motion is therefore rejected. This justifies the further analysis conducted hereafter.

Let us now conduct an investigation on $B_H(t)$ to verify whether it might follow an fBm or fGn trend. Consequently, we also estimate the Hurst exponent H of the process. To this aim, we apply the theoretical procedure consisting of evaluating the Fourier spectrum $S(f)$ of the data and studying its behavior with respect to the frequencies f. In fact, a relation analogous to Equation (3) for long memory processes can be presented in the frequency domain of the time series. In particular, for an fBm or a fGn process, the estimated Fourier spectrum $S(f)$ and the frequencies f are asymptotically evaluated as

$$S(f) = S(f_0)\, f^{-\beta}, \tag{4}$$

where both $S(f_0)$ and β are constants. The coefficient β is linked to the Hurst exponent H by different equations, depending on the nature of the underlying process. Indeed, for an fGn (fBm) process, Equation (4) holds with $\beta = 2H - 1$ ($\beta = 2H + 1$) (cf. [17–19]). Recalling that $0 < H < 1$, we find the following:

(i) If $-1 < \beta < 1$, then the time series can be identified as a realization of fractional Gaussian noise;
(ii) If $1 < \beta < 3$, then the data series represents a sample path of fractional Brownian motion.

From Equation (4), we have

$$\log S(f) = \log S(f_0) - \beta \log f,$$

so that an estimation of the coefficient β can be obtained by plotting on a log-log scale the frequencies and the values of $S(f)$ and then taking the opposite value of the slope of the least squares line as an estimation.

To attain an estimation of the function $S(f)$, we evaluate the periodogram of the data series, which for $B_H(t)$ is defined by

$$I(f) = \left| \frac{1}{2\pi N} \sum_{i=0}^{N} B_H(t_i) e^{ift_i} \right|^2.$$

In order to achieve the best possible approximation for the Fourier spectrum, many modified versions of the periodogram have been devised. In particular, we chose the estimation of $S(f)$ using the modified periodogram method suggested by Welch in [20]. The technique consists of dividing the signal into overlapping segments and averaging the modified periodograms calculated in each window to obtain the final result. The modified Welch's periodogram is shown in Figure 7.

Hence, by applying a log-log transformation on the frequencies and the estimated periodogram, we obtained the plot shown in Figure 8. The estimate of β obtained as the slope of the regression line is given by

$$\beta = 1.853,$$

with the corresponding $R^2 = 0.907$. Since the coefficient β belongs to the interval $(1, 3)$, we could conclude that the data series followed an fBm trend. This also led to an estimation of the Hurst parameter as follows:

$$\hat{H} = \frac{\beta - 1}{2} = 0.427. \tag{5}$$

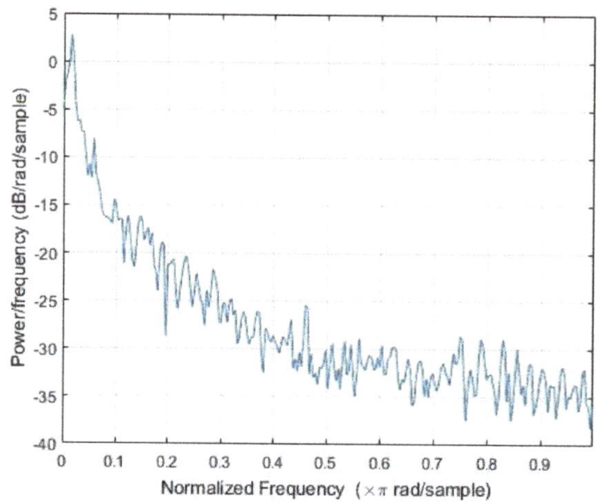

Figure 7. Modified Welch periodogram for the data series $B_H(t)$ plotted for different values of the frequency.

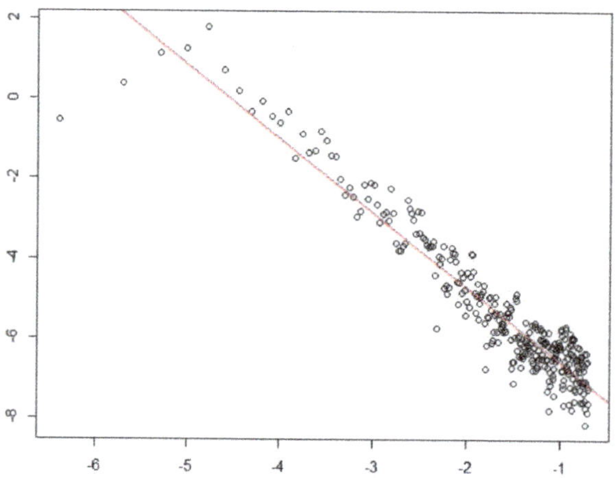

Figure 8. Log-log plot for the Welch's periodogram of the data series with respect to frequency. The slope of the least squares line estimates the coefficient β.

The estimation obtained in Equation (5) shows that the Hurst exponent of the process was less than 0.5, implying anti-persistence in the data trend. This result has already emerged in the literature in investigations related to soil temperatures. In the work by Zhang et al. [21], surveys of the soil temperature at different levels of depth showed that a weak anti-persistence ($H \sim 0.4$) was observed near the soil surface, consistent with our result.

In order to complete the estimation of the parameters for $B_H(t)$, we also evaluated the diffusion constant D of the process. This was performed directly from the data, calculating the sample variance of the increments $B_H(t) - B_H(t-1)$ of the data series. The estimate obtained to this point was

$$\hat{D} = 0.0846. \tag{6}$$

To further confirm the hypothesis suggested from the periodogram, we performed analysis of the residuals of the time series making use of simulations of fBm. To this aim, we denoted with $B_{\hat{H},\hat{D}}(t)$ a simulated sample path of a fractional Brownian motion process with a Hurst exponent and scale parameter estimated as in Equations (5) and (6). Then, the residual

$$z(t) = X(t) - r(t) - B_{\hat{H},\hat{D}}(t)$$

was evaluated by means of the residuals plot. Moreover, the normality of the residuals was assessed hereafter through two tests: the Shapiro–Wilk test and robust Jarque–Bera test for Gaussianity.

4.2. The Shapiro–Wilk Test

Since the test statistic W of the Shapiro–Wilk test tended to detect even small departures from the null hypothesis when the sample size was sufficiently large, it was convenient for reducing the sample to a smaller size. Hence, from now on, we consider a subset $\tilde{X}(t)$ of the main dataset which contains weekly spaced observations. After simulating a sample path $\tilde{B}_{\hat{H},\hat{D}}$ of fBm of a length $M = \lfloor (N+1)/7 \rfloor$, we plotted the residual series

$$\tilde{z}(t) = \tilde{X}(t) - \tilde{r}(t) - \tilde{B}_{\hat{H},\hat{D}}(t)$$

with respect to the values of $\tilde{X}(t)$ (see Figure 9). Notice that the residuals do not seem to follow a specific path; rather, they appear to be distributed randomly with respect to $\tilde{X}(t)$.

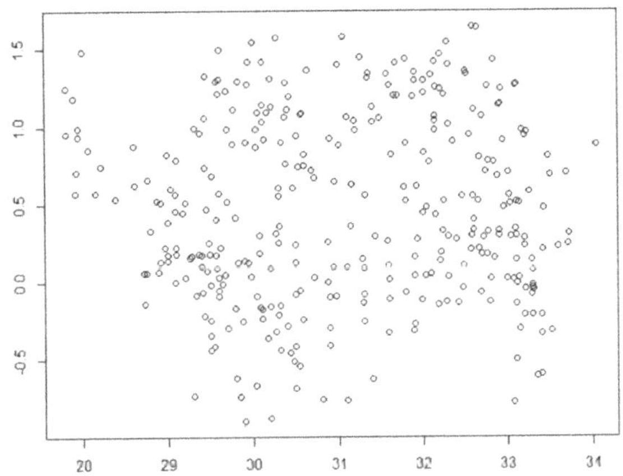

Figure 9. Plot of the residuals series $\tilde{z}(t)$ with respect to the temperatures $\tilde{X}(t)$, equally spaced with one observation for each 7 days.

Let us now describe the adopted testing strategy. Since we were interested in studying the behavior of the residual series, we first simulated the fBm path $\tilde{B}_{\hat{H},\hat{D}}(t)$ for a total of $n = 10^4$ iterations. For each simulation, we evaluated the residuals series $\tilde{z}_i(t)$ and first performed a Shapiro–Wilk test for $i = 1, \ldots, n$. The procedure was necessarily based on the repetition of the Shapiro–Wilk test since we dealt with simulated fBm paths. Hence, we considered the collection of test statistics

$$W_i = \frac{\sum_{j=1}^{M} a_j \tilde{z}_i(t_{(j)})}{\sum_{j=1}^{M} (\tilde{z}_i(t_j) - \mu_i)^2}, \qquad i = 1, \ldots, n, \tag{7}$$

where W_i refers to the ith simulation and where μ_i represents the sample mean of $\tilde{z}_i(t)$. The coefficients a_j, as is well known, are related to some moments of the order statistics of i.i.d. random variables sampled from the standard normal distribution. The values of W_i, as in Equation (7), ranged between 0 and 1. The ith test rejects the Gaussian hypothesis when W_i is close to 0, and hence we accepted the null hypothesis if $W_i \geq 0.98$ for $i = 1, \ldots, n$. As shown in Figure 10, a large proportion of the values of W_i was greater than 0.98. This result was observed for more than 70% of the values of W_i even for various repetitions of the n-simulation procedure (cf. [22,23]). This was also confirmed by the cumulative histogram of W_i shown in Figure 11. The results of the test suggest the acceptance of the Gaussianity hypothesis.

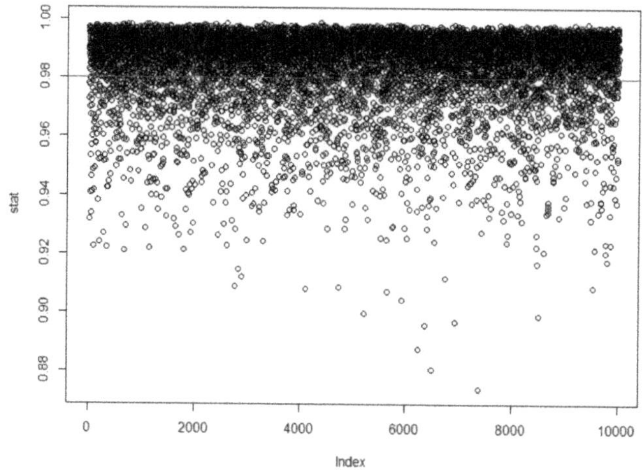

Figure 10. The values of the statistic W_i evaluated according to Equation (7).

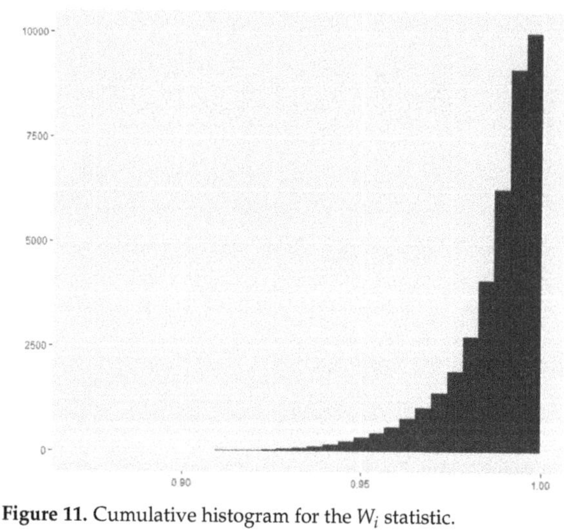

Figure 11. Cumulative histogram for the W_i statistic.

4.3. Robust Jarque–Bera Test

Let us now perform a modified version of the Jarque–Bera test. The statistic for the original Jarque–Bera test for a sample of a size n is given by (cf. [24])

$$JB = \frac{n}{6}\left(\frac{\hat{\mu}_3}{\hat{\mu}_2^{3/2}}\right)^2 + \frac{n}{24}\left(\frac{\hat{\mu}_4}{\hat{\mu}_2} - 3\right)^2,$$

which is a combination of the sample skewness $\left(\frac{\hat{\mu}_3}{\hat{\mu}_2^{3/2}}\right)$ and the sample kurtosis $\left(\frac{\hat{\mu}_4}{\hat{\mu}_2} - 3\right)$ obtained from the data through the estimates $\hat{\mu}_i$ of the central moments $i = 2, 3, 4$. If the data series is distributed normally, one has that $JB \sim \chi^2(2)$ asymptotically. Thus, under the null hypothesis of the time series being Gaussian, both the expected skewness and the expected kurtosis are zero. Therefore, higher values of the JB statistic lead to the rejection of the Gaussianity hypothesis.

For our testing procedure, we followed the modified version of the test proposed by Gel and Gastwirth in [25], which is known as the robust Jarque–Bera (RJB) test. The modification consisted of substituting the estimate of the spread, formerly $\hat{\mu}_2$, with the average absolute deviation from the sample median, henceforth denoted as Med, given by

$$J_M = \frac{\sqrt{\pi/2}}{M} \sum_{j=1}^{M} |z(\tilde{t}_j) - \text{Med}|,$$

where M refers to the sample size. This substitution makes the statistics less influenced by outliers. Therefore, the statistic becomes

$$JB_M = \frac{M}{C_1}\left(\frac{\hat{\mu}_3}{J_M^3}\right)^2 + \frac{M}{C_2}\left(\frac{\hat{\mu}_4}{J_M^4} - 3\right)^2,$$

where C_1 and C_2 are positive constants estimating moments of J_M, which is obtained through Monte Carlo simulations. Moreover, the p-values for the RJB test are obtained by use of k = 10,000 Monte Carlo simulations instead of the $\chi^2(2)$ distribution, thus obtaining a better approximation (cf. [26]).

To perform the RJB test, we repeated 10 blocks of n simulations with 3 different numbers of iterations ($n = 100, 500$ and 1000). For each simulation, the residual series

$$\{\tilde{z}_i(t_j),\ j = 1, \ldots, M\}_{i=1,\ldots,n} \tag{8}$$

was tested. The choice of performing blocks of simulations arose from the need for reducing the computational times. The results, reported in Table 4, showed compliance with the Shapiro–Wilk test and led to the acceptance of the Gaussianity hypothesis for the residuals series.

Table 4. Percentages of acceptable values for the residuals series in Equation (8) tested with the RJB method and repeated for $h = 1, \ldots, 10$, with significance level $\alpha = 0.02$ and different values of n.

h	$n = 100$	$n = 500$	$n = 1000$
1	64%	66.8%	68.0%
2	76%	72.2%	69.9%
3	68%	70.0%	66.1%
4	66%	69.4%	71.3%
5	70%	71.0%	70.5%
6	68%	71.4%	71.1%
7	68%	70.8%	70.5%
8	71%	67.0%	69.9%
9	70%	73.0%	68.1%
10	70%	69.4%	71.5%

5. Statistical Test for the Model

To further confirm the results obtained in Section 4, we performed a statistical test on the whole N-term data series for the hypothesis of $B_H(t)$ being fractional Brownian motion with parameters $\hat{H} = 0.427$ and $\hat{D} = 0.0846$. To this aim, we used the inference test presented in Sikora [27] based on the detrending moving average (DMA) of the data. The hypotheses for the test were

$$\mathcal{H}_0 : \{B_H(t_1), B_H(t_2), \ldots, B_H(t_N)\} \text{ is a trajectory of fBm with parameters } \hat{H} \text{ and } \hat{D},$$

$$\mathcal{H}_1 : \{B_H(t_1), B_H(t_2), \ldots, B_H(t_N)\} \text{ is not a trajectory of fBm with parameters } \hat{H} \text{ and } \hat{D}.$$

The corresponding test statistic for a fixed value $m > 1$ was given by

$$S^2(m) = \frac{1}{N-m} \sum_{j=m}^{N} (B_H(t_j) - \overline{B}_H^m(t_j))^2, \tag{9}$$

where $\overline{B}_H^m(t_j)$ denotes the moving average of the m observations $B_H(t_{j-i})$, $i = 0, 1, \ldots, m-1$.

The p-value for this test was defined by an infinite series, so it needed to be computed as an empirical quantile obtained from series of generalized chi-squared random variables. Recalling Step 5 in Section 3 of [27], the p-value of the test was calculated as

$$p = \frac{2}{L} \min\{\#(\sigma_l^2(m) < S^2(m)), \#(\sigma_l^2(m) > S^2(m))\}. \tag{10}$$

In this formula, L is the number of samples generated for the chi-squared distribution. For a good estimate of p, we set $L = 1000$. Moreover, in Equation (10), we considered

$$\sigma_l^2(m) = \frac{1}{N-m} \sum_{j=1}^{N-m+1} \lambda_j(m) U_j^l, \qquad l = 1, 2, \ldots, L,$$

where $\mathbf{U}^l = \{U_1^l, U_2^l, \ldots, U_{N-m+1}^l\}$ is the lth chi-square sample and $\lambda_j(m)$ is the jth eigenvalue of the sample matrix

$$\tilde{\Sigma} = \{\mathbb{E}[B_H(t_j)B_H(t_k)];\ j, k = 1, 2, \ldots, N-m+1\}.$$

Recalling Equation (9), since the search for an optimal value for m was still an open question, we divided the data series into 10 subsets of about 250 observations each in order to adapt the procedure even to cases with large numbers of observed data. Then, we ran the test in every subset for each suitable value of m.

After that, we collected the p-values and created a box plot with jitter for each subset in order to determine the outlier values for p (cf. Figure 12).

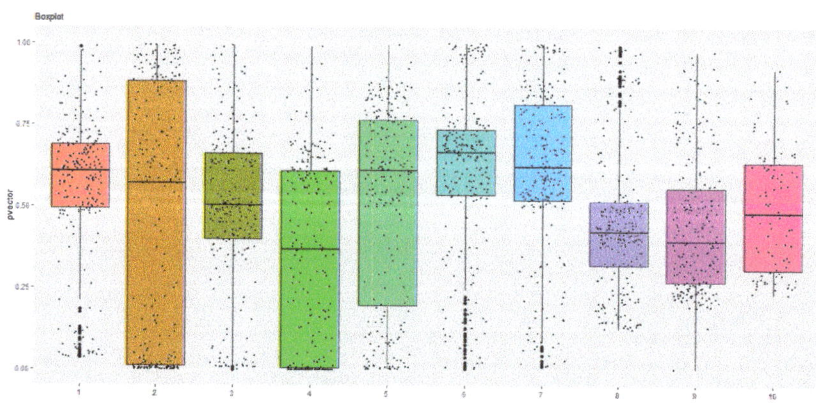

Figure 12. Box plot for the values of p obtained in the 10 subsets of the data.

The significance level considered for this test was $\alpha = 0.02$. Therefore, the results in each subset were collected and divided into three categories:

(i) For $p \leq 0.02$, the null hypothesis is rejected;
(ii) For $0.02 < p \leq 0.05$, the values are considered as a "warning";
(iii) For $p > 0.05$, the hypothesis $\mathcal{H}_0$ cannot be rejected.

Table 5 shows the results, with the percentages of the number of rejected, warning and acceptable values over the total p-values calculated in the 10 datasets. It can be observed that the null hypothesis could not be rejected in almost 90% of the cases covered, considering all the subsets and all the different choices of m for the test statistic. This allowed stating the validity of the hypothesis that the time series $B_H(t)$ followed an fBm trend.

Table 5. Number of cases in which the DMA statistic test had different outcomes for the data series $B(t)$.

Set	Reject Values ($p \leq 0.02$)	Perc.	Warning Values ($0.02 \leq p \leq 0.05$)	Perc.	Acceptable Values ($p > 0.05$)	Perc.
1	0	0%	12	4.86%	235	95.14%
2	70	28.23%	12	4.84%	154	62.10%
3	17	6.85%	6	2.42%	219	88.31%
4	70	28.23%	6	2.42%	166	66.94%
5	23	9.27%	10	4.03%	205	82.66%
6	4	1.61%	4	1.61%	236	95.16%
7	3	1.21%	3	1.21%	239	96.37%
8	0	0%	0	0%	248	100%
9	1	0.4%	0	0%	247	99.6%
10	0	0%	0	0%	95	100%

6. Conclusions

The performed analysis shows that the proposed stochastic model is suitable to fit the observed temperatures. Indeed, the obtained results suggest that the coefficients obtained for the process $X(t)$ display high values for R^2, thus ensuring the goodness of the fit. In conclusion, this study reveals that the evolution of the temperatures in the Campi Flegrei caldera can be modeled by the sum of a deterministic component, which represents the seasonal trend, and fractional Brownian motion.

Possible future developments of the performed analysis can be oriented toward (1) the prevision of the trend of the seasonal terms and (2) the study of other quantities of interest, such as leaked radon.

Author Contributions: Conceptualization, A.D.C., B.M. and V.M.; Investigation, A.D.C., B.M. and V.M.; Methodology, A.D.C., B.M. and V.M.; Writing, review and editing, A.D.C., B.M. and V.M. All authors have read and agreed to the published version of the manuscript.

Funding: This work was supported in part by the Italian MIUR-PRIN 2017, project "Stochastic Models for Complex Systems", No. 2017JFFHSH.

Institutional Review Board Statement: Not applicable.

Informed Consent Statement: Not applicable.

Data Availability Statement: Not applicable.

Acknowledgments: The authors are members of the research group GNCS of INdAM (Istituto Nazionale di Alta Matematica).

Conflicts of Interest: The authors declare no conflict of interest.

Abbreviations

The following abbreviations are used in this manuscript:

OLS	ordinary least squares
fBm	fractional Brownian motion
fGn	fractional Gaussian noise
RJB	robust Jarque–Bera test
DMA	detrending moving average
i.i.d.	independent and identically distributed
LRD	long-range dependency

References

1. Sebastiani, G.; Malagnini, L. Forecasting the Next Parkfield Mainshock on the San Andreas Fault (California). *J. Ecol. Nat. Resour.* **2020**, *4*, 1–45. [CrossRef]
2. Giordano, L.M.; Morale, D. A fractional Brownian–Hawkes model for the Italian electricity spot market: Estimation and forecasting. *J. Energy Mark.* **2021**,*14*, 1–44. [CrossRef]
3. Sabbarese, C.; Ambrosino, F.; Chiodini, G.; Giudicepietro, F.; Macedonio, G.; Caliro, S.; De Cesare, W.; Bianco, F.; Pugliese, M.; Roca, V. Continuous radon monitoring during seven years of volcanic unrest at Campi Flegrei caldera (Italy). *Sci. Rep.* **2020**, *10*, 9551. [CrossRef] [PubMed]
4. Nourdin, I. *Selected Aspects of Fractional Brownian Motion*; Bocconi & Springer Series; Springer: Berlin/Heidelberg, Germany, 2012.
5. Prakasa Rao, B.L.S. *Statistical Inference for Fractional Diffusion Processes;* Wiley: Hoboken, NJ, USA, 2010. [CrossRef]
6. Mattia, F.; Le Toan, T.; Davidson, M.; Borgeaud, M. On the scattering from natural rough surfaces. *IGARSS* **1999**, *5*, 2413–2415. [CrossRef]
7. Yin, Z.M.; Ranalli, G. Modelling of earthquake rupturing as a stochastic process and estimation of its distribution function from earthquake observations. *Geophys. J. Int.* **1995**, *123*, 838–848. [CrossRef]
8. Taqqu, M.S.; Teverovsky, V.; Willinger, W. Estimators for longrange dependence: An empirical study. *Fractals* **1996**, *3*, 785–798. [CrossRef]
9. Travaglino, F.; Di Crescenzo, A.; Martinucci, B.; Scarpa, R. A New Model of Campi Flegrei Inflation and Deflation Episodes based on Brownian Motion Driven by Telegraph Process. *Math. Geosci.* **2018**, *50*, 961–975. [CrossRef]
10. Mandelbrot, B.B.; Van Ness, J.W. Fractional Brownian Motions, Fractional Noises and Applications. *SIAM Rev.* **1968**, *10*, 422–437. [CrossRef]
11. Dieker, T. Simulation of Fractional Brownian Motion. Ph.D. Thesis, University of Twente, Enschede, The Netherland, 2004.
12. Li, M. On the Long-Range Dependence of Fractional Brownian Motion. *Math. Probl. Eng.* **2013**, *2013*, 842197. [CrossRef]
13. Balcerek, M.; Burnecki, K.; Thapa, S.; Wyłomańska, A.; Chechkin, A. Fractional Brownian motion with random Hurst exponent: Accelerating diffusion and persistence transitions. *arXiv* **2022**, arXiv:2206.03818.
14. Han, D.; Korabel, N.; Chen, R.; Johnston, M.; Gavrilova, A.; Allan, V.J.; Fedotov, S.; Waigh, T.A. Deciphering anomalous heterogeneous intracellular transport with neural networks. *eLife* **2019**, *9*, e52224.[CrossRef] [PubMed]
15. Hudson, D.J. Fitting Segmented Curves Whose Join Points Have to Be Estimated. *J. Am. Stat. Assoc.* **1966**, *61*, 1097–1129. [CrossRef]
16. Briane, V.; Vimond, M.; Kervrann, C. An adaptive statistical test to detect non Brownian diffusion from particle trajectories. In Proceedings of the 2016 IEEE 13th International Symposium on Biomedical Imaging (ISBI), Prague, Czech Republic, 13–16 April 2016.
17. Roume, C.; Ezzina, S.; Blain, H.; Delignieres, D. Biases in the Simulation and Analysis of Fractal Processes. *Comput. Math. Methods Med.* **2019**, *2019*, 4025305. [CrossRef] [PubMed]
18. Montanari, A.; Taqqu, M.S.; Teverovsky, V. Estimating Long-Range Dependence in the Presence of Periodicity: An Empirical Study. *Math. Comput. Model.* **1999**, *29*, 217–228. [CrossRef]
19. Flandrin, P. On the spectrum of fractional Brownian motions. *IEEE Trans. Inf. Theory* **1989**, *35*, 197–199. [CrossRef]
20. Welch, P. The use of fast Fourier transform for the estimation of power spectra: A method based on time averaging over short, modified periodograms. *IEEE Trans. Audio Electroacoust.* **1967**, *15*, 70–73. [CrossRef]
21. Zhang, T.; Shen, S.; Cheng, C.; Song, C.; Ye, S. Long-Range Correlation Analysis of Soil Temperature and Moisture on A'rou Hillsides, Babao River Basin. *J. Geophys. Res. Atmos.* **2018**, *123*, 12–606. [CrossRef]
22. Royston, P. Approximating the Shapiro-Wilk W-test for non-normality. *Stat. Comput.* **1992**, *2*, 117–119. [CrossRef]
23. Royston, P. Remark AS R94: A Remark on Algorithm AS 181: The W-test for Normality. *J. R. Stat. Soc.* **1995**, *44*, 547–551. [CrossRef]
24. Jarque, C.M.; Bera, A.K. A test for normality of observations and regression residuals. *Int. Stat. Rev.* **1987**, *55*, 163–172.[CrossRef]
25. Gel, Y.R.; Gastwirth, J.L. A robust modification of the Jarque–Bera test of normality. *Econ. Lett.* **2008**, *99*, 30–32. [CrossRef]
26. Deb, P.; Sefton, M. The distribution of a Lagrange multiplier test of normality. *Econ. Lett.* **1996**, *51*, 123–130. [CrossRef]
27. Sikora, G. Statistical test for fractional Brownian motion based on detrending moving average algorithm. *Chaos Solitons Fractals* **2018**, *116*, 54–62. [CrossRef]

fractal and fractional

MDPI

Article

Threshold Dynamics and the Density Function of the Stochastic Coronavirus Epidemic Model

Jianguo Sun *, Miaomiao Gao and Daqing Jiang

School of Science, China University of Petroleum (East China), Qingdao 266580, China;
s18090014@upc.edu.cn (M.G.); daqingjiang2010@upc.edu.cn (D.J.)
* Correspondence: sunjg616@upc.edu.cn

Abstract: Since November 2019, each country in the world has been affected by COVID-19, which has claimed more than four million lives. As an infectious disease, COVID-19 has a stronger transmission power and faster propagation speed. In fact, environmental noise is an inevitable important factor in the real world. This paper mainly gives a new random infectious disease system under infection rate environmental noise. We give the existence and uniqueness of the solution of the system and discuss the ergodic stationary distribution and the extinction conditions of the system. The probability density function of the stochastic system is studied. Some digital simulations are used to demonstrate the probability density function and the extinction of the system.

Keywords: stochastic epidemic model; threshold dynamics; infection rate; extinction; ergodic stationary distribution

Citation: Sun, J.; Gao, M.; Jiang, D. Threshold Dynamics and the Density Function of the Stochastic Coronavirus Epidemic Model. *Fractal Fract.* **2022**, *6*, 245. https://doi.org/10.3390/fractalfract6050245

Academic Editor: Mario Abundo

Received: 2 April 2022
Accepted: 27 April 2022
Published: 29 April 2022

1. Introduction

So far, infectious diseases have become one of the important factors endangering human health. Medical research shows that there are three outcomes of any infectious disease: the first is that the infectious virus is eliminated by human drugs; the second is that the virus exists only in a small area, such as Ebola, SARS (severe acute respiratory syndrome), and so on; the third is the long-term coexistence of viruses and humans, such as influenza, AIDS (acquired immunodeficiency syndrome), and so on [1].

In order to describe the dynamic behavior of the epidemic, mathematical modeling is considered to be an important tool [2]. According to the occurrence, transmission, and development law of infectious disease in the population, Mathematicians and ecologists have established several epidemic models to study and control various epidemics [3–5]. The authors gave the dynamics and stationary distribution of the hepatitis E model. Meanwhile, the authors obtained the optimal control analysis and the Atangana–Baleanu derivative for the dynamical analysis of the hepatitis E model in [6]. Through the qualitative analysis and numerical simulation of the dynamic behavior of the infectious disease model [7], the authors gave an exact expression of the probability density function of the stochastic model SVI (susceptible, vaccinated, infectious) around the unique endemic equilibrium of the deterministic system by solving the corresponding Fokker–Planck equation, which is guaranteed by a new critical value R_0^s in [8], and other models, such as SIR (susceptible, infectious, recovered), SIRS (susceptible, infectious, recovered, susceptible), SEIR (susceptible, exposed, infectious, recovered), etc. [9,10].

Since November 2019, the world has been enveloped in COVID-19 (coronavirus disease 2019). As a contagious virus, COVID-19 is highly infectious [11]. Since April 2021, only half a year, the number of newly confirmed cases in the world has increased by 100 million [12]. At the same time, Griffin B.D., etc. [13], found that the new coronavirus could infect and spread among North American deer rats, which has increased the difficulty in controlling COVID-19. Many countries are working hard to prevent the spread of

COVID-19. Mathematicians use simulated mathematical models to predict the future behavior of coronavirus disease transmission in 2019. So far, several typical 2019 coronavirus disease transmission models have been proposed and are being used in some decisions. Recently, people have grasped valuable inferences through mathematical modeling and obtained in-depth understanding of the novel coronavirus (COVID-19) [14]. A. Atangana confirmed the effect of lockdown as a possible adequate measure to help flatten the curve of deaths and infections with the epidemic model as follows: [15]:

$$\begin{cases} dS(t) = [\Lambda - \frac{\beta S(t)D(t)}{N} - (\delta+\mu)S(t) + \eta R(t)]dt, \\ dC(t) = [(\frac{\beta S(t)D(t)}{N} + \delta S(t) - (\beta+\mu+\pi)C(t)]dt, \\ dI(t) = [\delta(1-\theta)S(t) + \pi C(t) - (\tau+\mu+\sigma)I(t)]dt, \\ dR(t) = [\beta C(t) + \tau I(t) - (\mu+\eta)R(t)]dt, \\ dD(t) = [\sigma I(t)]dt, \end{cases} \quad (1)$$

where the parameters are in Table 1.

Table 1. The definitions of the parameters.

Parameter	Definitions
$S(t)$	The susceptible class
$I(t)$	Infected people
$C(t)$	Carriers (dead corpse)
$R(t)$	Recovered persons
$D(t)$	Total number of deaths
μ	Rate of natural death recruitment rate into $S(t)$
θ	Probability of an $S(t)$ class to join $C(t)$ class
σ	Death rate induced by COVID-19
β	Recovery rate of $C(t)$ class
δ	Force of infection of class $S(t)$
τ	Recovery rate of $I(t)$ class
π	Rate at which an $C(t)$ class is recovered
η	Rate at which treated persons become $C(t)$ class

However, in the real world, due to the influence of various factors, such as the environment, a random model is constructed by random components with some distribution. Through the addition of some white noise, these distributions may reflect the uncertainty of the input content or random process [16]. Meanwhile, the quarantined measures play a very important role in fighting and preventing the increase in COVID-19. The authors found that the dynamic system with the external source was more reliable than the suspected people travelling, and that the rate of isolation is extremely important for controlling the increase in the cumulative confirmed people of COVID-2019 [17]. The authors in [18] put forward the stochastic coronavirus epidemic model with the parameter disturbance by the natural mortality rate by translating the quarantined factor as follows:

$$\begin{cases} dS(t) = [\Lambda - \frac{\beta S(t)I(t)}{N} - \mu_0 S(t)]dt + \eta_1 S(t)dB_1(t), \\ dI(t) = [(\frac{\beta S(t)I(t)}{N} - (\gamma_1+\mu_1+\mu_0)I(t) + \sigma Q(t)]dt + \eta_2 I(t)dB_2(t), \\ dQ(t) = [\gamma_1 I(t) - (\mu_0+\mu+\sigma)Q(t)]dt + \eta_3 Q(t)dB_3(t), \end{cases} \quad (2)$$

where the definitions of the parameters are in Table 2 and $N = S(t) + I(t) + Q(t)$.

In the system (2), the important role of isolation in COVID-19 is pointed out, and the stable distribution of the model under extinction conditions is obtained. However, in the real world, according to the COVID-19 data in Pakistan [19], the disturbance of the infection rate coefficient plays a very important role in the spread of COVID-19. Meanwhile, vaccination and isolation measures can also affect the infection rate. In the present paper, we give the stochastic coronavirus epidemic model with the stochastic disturbance of the infection rate coefficient. The system is the following:

$$\begin{cases} dS(t) = [\Lambda - \frac{\beta S(t)I(t)}{N} - \mu_0 S(t)]dt - \frac{\eta S(t)I(t)}{N}dB(t), \\ dI(t) = [(\frac{\beta S(t)I(t)}{N} - (\gamma_1 + \mu_1 + \mu_0)I(t) + \sigma Q(t)]dt + \frac{\eta S(t)I(t)}{N}dB(t), \\ dQ(t) = [\gamma_1 I(t) - (\mu_0 + \mu + \sigma)Q(t)]dt, \end{cases} \quad (3)$$

The main composition of the present paper is as follows. The second section gives the basic lemma and basic concepts of this paper. The existence and uniqueness of the global positive solution of the system (3) are obtained in the third section, and the fourth section gives the ergodic stationary distribution of the system (3). In order to better understand the degree of control of the virus, we consider the extinction condition of the system (3) in the fifth section. Meanwhile, in the sixth section, the probability density function of the system (3) is given to understand the trend of the coronavirus viruses in the system (3). In the last section, by numerical simulation, two examples give the extinction, long-term persistence, and the probability density function with the corresponding conditions.

Table 2. The definitions of the parameters.

Parameter	**Definitions**
$S(t)$	The susceptible class
$I(t)$	Infected people
$Q(t)$	Quarantined people
N	Total population
Λ	Capita constant fecundity rate
β	Infection rate
μ_0	Infected natural mortality rate
μ_1	Quarantined natural mortality rate
μ	Disease-related mortality rate
γ_1	The constant rate of quarantining infected
σ	The quarantined rate from infected people
$B_i(t), i = 1,2,3$	Brownian motion
$\eta_i, i = 1,2,3$	The intensity of $B_i(t)$

2. Preliminaries

We give some basic conceptions as in [3,5,16,20]. Suppose $(\Omega, \mathscr{F}, \{\mathscr{F}_t\}_{t\geq 0}, \mathbb{P})$ is a complete probability space with a filtration $\{\mathscr{F}_t\}_{t\geq 0}$; we define $\mathbb{R}^3_+ = \{x \in \mathbb{R}^3 : x_i > 0$ for all $1 \leq i \leq 3\}$ and $\overline{\mathbb{R}}^3_+ = \{x \in \mathbb{R}^3 : x_i \geq 0$ for all $1 \leq i \leq 3\}$. In addition, if $f(t)$ is an integral function on $t \in [0, \infty)$, we define $f^u = sup\{f(t) \mid t \geq 0\}$, $f^l = inf\{f(t) \mid t \geq 0\}$. In the following, we give the Itô′s formula.

Lemma 1 ([3]). *Let $x(t)$ be an Itô′ process with the stochastic differential*

$$dx(t) = f(t)dt + g(t)dB_t, \ for \ t \geq t_0, \quad (4)$$

where $f \in L^1(\mathbb{R}_+, \mathbb{R})$ and $g \in L^1(\mathbb{R}_+, \mathbb{R})$. Let $V \in C^{2,1}(\mathbb{R} \times \mathbb{R}_+, \mathbb{R})$. Then, $V(x(t), t)$ is again an Itô′ process with the stochastic differential given by

$$dV(x(t), t) = [V_t(x(t), t) + V_x(x(t), t)f(t) + \frac{1}{2}V_{xx}(x(t), t)g^2(t)]dt + V_x(x(t), t)g(t)dB_t \ a.s.$$

Firstly, we consider the general three-dimensional stochastic differential equation

$$dx(t) = f(x(t), t)dt + g(x(t), t)dB(t), \ \text{for} \ t \geq t_0 \quad (5)$$

with initial value $x(t_0) = x_0 \in \mathbb{R}^3$, where $B(t)$ denotes three-dimensional standard Brownian motion defined on the above probability space $(\Omega, \Gamma, \{\Gamma_t\}_{t\geq 0}, \mathbb{P})$. Define the differential operator L by Mao [3] as

$$L = \frac{\partial}{\partial t} + \Sigma f_i(x,t)\frac{\partial}{\partial x_i} + \frac{1}{2}\Sigma[g^T(x,t)g(x,t)]_{ij}\frac{\partial^2}{\partial x_i \partial x_j}.$$

If L acts on a function $V \in C^{2,1}(\mathbb{R}^3 \times \bar{\mathbb{R}}_+; \bar{\mathbb{R}}_+)$, where $\overline{\mathbb{R}}_+ = \{x \in \mathbb{R} : x \geq 0\}$, then

$$LV(x,t) = V_t(x,t) + V_x(x,t) + \frac{1}{2}trac[g^T(x,t)V_{xx}(x,t)g(x,t)],$$

where $V_t = \frac{\partial V}{\partial t}$, $V_x = (\frac{\partial V}{\partial x_1}, \cdots, \frac{\partial V}{\partial x_3})$ and $V_{xx} = (\frac{\partial^2 V}{\partial x_i \partial x_j})_{3\times 3}$.

By Lemma 1, we obtain

$$dV(x(t),t) = LV(x(t),t)dt + V_x(x(t),t)g(x(t),t)dB(t).$$

The diffusion matrix is defined as follows:

$$A(x) = (a_{ij}(x)),\ a_{ij} = \sum_{r=1}^{3} g_r^i(x)g_r^j(x), 1 \leq i,j \leq 3.$$

3. Existence and Uniqueness of the Global Positive Solution

The problem where the solution is global and positive is important in studying the dynamical behavior of the system (3). The coefficients of the system (3) are not the linear growth, and the solutions of the system (3) may explode at a fixed time. The main theorem is as follows.

Theorem 2. *There is a unique positive solution $(S(t), I(t), Q(t))$ of system (3) on $t \geq 0$ by the initial value $(S(0), I(0), Q(0)) \in \mathbb{R}^3_+$, and the solution $(S(t), I(t), Q(t)) \in \mathbb{R}^3_+$ for all $t \geq 0$ almost surely (a.s.).*

Proof. Based on [5], we obtain the fact that there is a unique solution $(x(t), y(t), z(t), w(t))$ on $[0, \tau_0)$ for the reason that the coefficients of the system (3) are the locally Lipschitz continuous, where τ_0 is an explosion time. We can obtain the fact that the local solution is global when $\tau_0 = \infty$ a.s. By the definitions in [3], we define a fundamental C^2-function $U : \mathbb{R}^3_+ \to \bar{\mathbb{R}}_+$, which is

$$U(S,I,Q) = (S(t) - 1 - \ln S(t)) + a(I(t) - 1 - \beta \ln I(t)) + b(Q(t) - 1 - \ln Q(t)), \quad (6)$$

where a, b are positive constants, which will be determined in the following text. The non-negativity of the function U can be seen from $x - 1 - \ln x \geq 0$ for any $x > 0$.

Applying Itô's formula [3], we obtain

$$dU(S,I,Q) = LUdt + \frac{\eta(S(t)-1)I(t)}{N^2}dB(t) - \frac{\eta(I(t)-1)S(t)}{N^2}dB(t), \quad (7)$$

where

$$
\begin{aligned}
LU &= (1-\tfrac{1}{S})(\Lambda - \tfrac{\beta SI}{N} - \mu_0 S) - \tfrac{\eta^2}{2}(\tfrac{I}{N})^2 + (1-\tfrac{1}{I})[(\tfrac{\beta SI}{N} - (\gamma_1+\mu_1+\mu_0)I + \sigma Q] + \tfrac{\eta^2}{2}(\tfrac{S}{N})^2 \\
&\quad +(1-\tfrac{1}{Q})[\gamma_1 I - (\mu_0+\mu+\sigma)Q] \\
&= \Lambda - \tfrac{\beta SI}{N} - \mu_0 S - \tfrac{\Lambda}{S} + \tfrac{\beta I}{N} + \mu_0 - (\mu_0+\mu_1+\gamma_1)I + \tfrac{\beta SI}{N} + \sigma Q - \tfrac{\beta S}{N} + (\mu_0+\mu_1+\gamma_1) - \tfrac{\sigma Q}{I} + \gamma_1 I \\
&\quad -(\mu_0+\mu+\sigma)Q - \tfrac{\gamma_1 I}{Q} + (\mu_0+\mu+\sigma) - \tfrac{\eta^2}{2}(\tfrac{I}{N})^2 + \tfrac{\eta^2}{2}(\tfrac{S}{N})^2 \\
&\le \Lambda - \mu_0 S - \tfrac{\Lambda}{S} + \beta + \mu_0 - (\mu_0+\mu_1+\gamma_1)I + \sigma Q + (\mu_0+\mu_1+\gamma_1) - (\mu_0+\mu+\sigma)Q \\
&\quad -\tfrac{\sigma Q}{I} + \gamma_1 I - \tfrac{\gamma_1 I}{Q} + (\mu_0+\mu+\sigma) + \tfrac{\eta^2}{2} \\
&\le \Lambda - \mu_0 S - \tfrac{\Lambda}{S} + \beta + \mu_0 + (\mu_0+\mu_1+\gamma_1) - \tfrac{\sigma Q}{I} - \tfrac{\gamma_1 I}{Q} + (\mu_0+\mu+\sigma) + \tfrac{\eta^2}{2} \\
&\le \Lambda + \beta + 3\mu_0 + \mu_1 + \gamma_1 + \mu + \sigma - 2\sqrt{\mu_0\Lambda} - 2\sqrt{\sigma\gamma_1} + \tfrac{\eta^2}{2} \\
&= (\sqrt{\Lambda} - \sqrt{\mu_0})^2 + (\sqrt{\sigma} - \sqrt{\gamma_1})^2 + \beta + 2\mu_0 + \mu_1 + \mu + \tfrac{\eta^2}{2},
\end{aligned} \tag{8}
$$

Then, we can obtain

$$
LU \le (\sqrt{\Lambda} - \sqrt{\mu_0})^2 + (\sqrt{\sigma} - \sqrt{\gamma_1})^2 + \beta + 2\mu_0 + \mu_1 + \mu + \tfrac{\eta^2}{2} := K, \tag{9}
$$

where K is a positive constant. The remainder of the proof is similar to Theorem 3.1 in Mao [5]. Hence, we omit it here. □

4. Ergodic Stationary Distribution of the Stochastic Coronavirus Epidemic Model

In this section, the existence of ergodic stationary components of the system (3) is given. Firstly, we define R_0^* as a stochastic reproductive ratio of the system (3), such as

$$
R_0^* = \frac{\mu_0 \beta \sigma \gamma_1}{(\mu_0 + \frac{1}{2}\eta^2)(\gamma_1+\mu_1+\mu_0+\frac{1}{2}\eta^2)^2(\mu_0+\mu+\sigma)},
$$

which is equal to $\frac{\beta\sigma\gamma_1}{(\gamma_1+\mu_1+\mu_0)^2(\mu_0+\mu+\sigma)}$ when $\eta = 0$ [3]. The following is a known result.

Lemma 3 ([3,5]). *The Markov process $X(t)$ has a stationary distribution $\mu(\cdot)$ if there exists a bounded domain $U \subset E_l$ with regular boundary Γ and*
(B.1) there is a positive number M such that $\sum_{i,j-1}^{l} a_{ij}(x)\xi_i\xi_j > M|\xi|^2,\ x \in U,\ \xi \in R^l$;
(B.2) there exists a nonnegative C^2 function V such that LV is negative for any $E_l \backslash U$. Then,

$$
P_x\Big\{ \lim_{T\to\infty} \frac{1}{T}\int_0^T f(X(t))dt = \int_{E_l} f(x)\mu(dx) \Big\} = 1,
$$

for all $x \in E_l$, where $f(\cdot)$ is a function integrable with respect to the measure μ.

Theorem 4. *When $R_0^* > 1$, for the solution $(S(t), I(t), Q(t))$ of the system (3), there exists an ergodic unique stationary distribution.*

Proof. We construct a C^2-function $\widetilde{V} : \mathbb{R}_+^3 \to \mathbb{R}$ as follows:

$$
\widetilde{V} = N(t) - c_1 \ln S(t) - c_2 \ln I(t) - c_3 \ln Q(t).
$$

Applying Itô's formula [3], we obtain

$$
\begin{aligned}
L\widetilde{V} &= (\Lambda - \mu_0 N - \mu_1 I - \mu Q) + c_1[-\tfrac{\Lambda}{S} + \tfrac{\beta I}{N} + \mu_0 + \tfrac{1}{2}(\eta IN)^2] + c_2[-\tfrac{\beta S}{N} + (\gamma_1+\mu_1+\mu_0) - \sigma\tfrac{Q}{I} + \tfrac{1}{2}(\tfrac{\eta S}{N})^2] \\
&\quad + c_3[-\gamma_1\tfrac{I}{Q} + (\mu_0+\mu+\sigma)] \\
&= c_1\tfrac{\beta I}{N} + [-\mu_0 N - c_1\tfrac{\Lambda}{S} - c_2\tfrac{\beta S}{N} - c_2\sigma\tfrac{Q}{I} - c_3\gamma_1\tfrac{I}{Q}] + c_1(\mu_0 + \tfrac{1}{2}\eta^2) + c_2(\gamma_1+\mu_1+\mu_0+\tfrac{1}{2}\eta^2) \\
&\quad + c_3(\mu_0+\mu+\sigma) + \Lambda - \mu_1 I - \mu Q \\
&\le c_1\tfrac{\beta I}{N} - 5(\mu_0 c_1 c_2^2 c_3 \beta\sigma\Lambda\gamma_1)^{\frac{1}{5}} + c_1(\mu_0+\tfrac{1}{2}\eta^2) + c_2(\gamma_1+\mu_1+\mu_0+\tfrac{1}{2}\eta^2) + c_3(\mu_0+\mu+\sigma) + 2\Lambda \\
&\le c_1\tfrac{\beta I}{N} - 5\Lambda[(R_0^*)^{\frac{1}{5}} - 1],
\end{aligned} \tag{10}
$$

and $R_0^* = \frac{\mu_0\beta\sigma\gamma_1}{(\mu_0+\frac{1}{2}\eta^2)(\gamma_1+\mu_1+\mu_0+\frac{1}{2}\eta^2)^2(\mu_0+\mu+\sigma)}$. We choose $c_1 = \frac{\Lambda}{\mu_0+\frac{1}{2}\eta^2}$, $c_2 = \frac{\Lambda}{\gamma_1+\mu_1+\mu_0+\frac{1}{2}\eta^2}$, $c_3 = \frac{\Lambda}{\mu_0+\mu+\sigma}$.

When $R_0^* > 1$, we suppose

$$\overline{V} = M\widetilde{V} - \ln S(t) - \ln I(t) - \ln Q(t) + N(t),$$

and $V := \overline{V}(S, I, Q) - \overline{V}(S_0, I_0, Q_0)$.

Applying Itô's formula to V, we obtain

$$\begin{aligned} LV &= ML\widetilde{V} - L\ln S(t) - L\ln I(t) - L\ln Q(t) + LN(t) \\ &= -5\Lambda M[(R_0^*)^{\frac{1}{5}} - 1] + c_1M\tfrac{\beta I}{N} + (-\tfrac{\Lambda}{S} + \tfrac{\beta I}{N} + \mu_0 + \tfrac{1}{2}\eta^2 + [-\tfrac{\beta S}{N} + (\gamma_1 + \mu_1 + \mu_0) - \sigma\tfrac{Q}{I} - \tfrac{1}{2}\eta^2] \\ &\quad +[-\gamma_1\tfrac{I}{Q} + (\mu_0 + \mu + \sigma)] + \Lambda - \mu_0 N - \mu_1 I - \mu Q \\ &\leq -5\Lambda M[(R_0^*)^{\frac{1}{5}} - 1] + (c_1M + 1)\beta - \tfrac{\Lambda}{S} - \sigma\tfrac{Q}{I} - \gamma_1\tfrac{I}{Q} - \tfrac{\beta S}{N} + 3\mu_0 + \gamma_1 + \mu_1 + \mu + \sigma + \Lambda \\ &\quad -\mu_0 N - \mu_1 I - \mu Q \\ &\leq (c_1M + 1)\beta + 3\mu_0 + \gamma_1 + \mu_1 + \mu + \sigma + \Lambda - \tfrac{\Lambda}{S} - \sigma\tfrac{Q}{I} - \gamma_1\tfrac{I}{Q} - \mu_0 S - (\mu_1 + \mu_0)I - (\mu + \mu_0)Q. \end{aligned} \tag{11}$$

Define

$$f_1(S) = (c_1M + 1)\beta + 3\mu_0 + \gamma_1 + \mu_1 + \mu + \sigma + \Lambda - \frac{\Lambda}{S} - \mu_0 S,$$

$$f_2(I) = -\sigma\frac{Q}{I} - (\mu_1 + \mu_0)I,$$

$$f_3(Q) = -\gamma_1\frac{I}{Q} - (\mu + \mu_0)Q.$$

We can divide $\mathbb{R}_+^3 \setminus D_\varepsilon$ into the following six domains:

$$D_1 = \{(S, I, Q) \in \mathbb{R}_+^3 : 0 < S < \varepsilon\}; \quad D_2 = \{(S, I, Q) \in \mathbb{R}_+^3 : S > \frac{1}{\varepsilon}\};$$

$$D_3 = \{(S, I, Q) \in \mathbb{R}_+^3 : 0 < I < \varepsilon\}; \quad D_4 = \{(S, I, Q) \in \mathbb{R}_+^3 : I > \frac{1}{\varepsilon}\};$$

$$D_5 = \{(S, I, Q) \in \mathbb{R}_+^3 : Q < \varepsilon^2, I > \varepsilon\}; \quad D_6 = \{(S, I, Q) \in \mathbb{R}_+^3 : Q > \frac{1}{\varepsilon}\};$$

Clearly, $D_\varepsilon = \bigcup_{j=1}^{6} D_j$. In the following text, we will show that $LV(S, I, Q) \leq -1$ on $\mathbb{R}_+^3 \setminus D_\varepsilon$.

Case 1. If $(S, I, Q) \in D_1 \bigcup D_2$, one can choose

$$M < -\frac{\mu_0 + \frac{1}{2}\eta^2}{\Lambda},$$

and

$$LV(S, I, Q) \leq 3\mu_0 + \gamma_1 + \mu_1 + \mu + \sigma + \Lambda - \frac{\Lambda}{S} - \mu_0 S \leq -2;$$

Case 2. If $(S, I, Q) \in D_3 \bigcup D_4$,

$$LV(S, I, Q) \leq (c_1M + 1)\beta + 3\mu_0 + \gamma_1 + \mu_1 + \mu + \sigma + \Lambda + f_2(I) \leq -2;$$

Case 3. If $(S, I, Q) \in D_5 \bigcup D_6$,

$$LV(S, I, Q) \leq (c_1M + 1)\beta + 3\mu_0 + \gamma_1 + \mu_1 + \mu + \sigma + \Lambda + f_3(Q) \leq -2;$$

therefore, for all $(S, I, Q)) \in \mathbb{R}_+^3 \setminus D_\varepsilon$, $LV(S, I, Q)) \leq -1$, which indicates that assumption $(B.2)$ holds.

We can know that the system (3) is ergodic and has a unique stationary distribution. This completes the proof. □

5. Extinction of the Stochastic Coronavirus Epidemic Model

It is a very important topic to consider the dynamic behavior of the epidemic virus to obtain the conditions for the virus to be eliminated in a long time. We mainly discuss the extinction conditions of the system (3). According to the results in [10], we can obtain the following lemma.

Lemma 5. *For any initial value, the solution of stochastic model satisfies*

$$\lim_{t\to\infty}\frac{\ln x(t)}{t}\le 0,\ \lim_{t\to\infty}\frac{\ln y(t)}{t}\le 0,\ \lim_{t\to\infty}\frac{\ln z(t)}{t}\le 0,\ \lim_{t\to\infty}\frac{\ln w(t)}{t}\le 0\ a.s. \tag{12}$$

$$\lim_{t\to\infty}\frac{x(t)+y(t)+z(t)+w(t)}{t}=0,\ a.s.. \tag{13}$$

Moreover,

$$\lim_{t\to 0}\frac{1}{t}\int_0^t x(m)dB_1(m)=0\,,\lim_{t\to 0}\frac{1}{t}\int_0^t y(m)dB_2(m)=0,\ \lim_{t\to 0}\frac{1}{t}\int_0^t z(m)dB_3(m)=0\ \ a.s.. \tag{14}$$

Theorem 6. *Let $(S(t),I(t),Q(t))$ be the solution of system (3) with any initial value $(S(0),I(0),Q(0))\in\mathbb{R}^3_+$. If $R_0^S<1$, then the solution $(S(t),I(t),Q(t))$ of system (3) satisfies*

$$\limsup_{t\to\infty}\frac{\ln I(t)}{t}\le\frac{1}{\mu_0+\gamma_1+\mu_1+\mu+\sigma}(R_0^S-1)<0\ \ a.s.,$$

where $R_0^S=\frac{\beta}{2\eta^2}(\mu_0+\gamma_1+\mu_1+\mu+\sigma)$. Namely, the disease will be eradicated in the long term.

Proof. Applying Itô′s formula to $\ln I(t)$, we obtain

$$\begin{aligned} d\ln I(t) &= \tfrac{dI(t)}{I(t)}=(\tfrac{\beta S}{N}-(\gamma_1+\mu_1+\mu_0)+\sigma\tfrac{Q}{I})dt-\tfrac{1}{2}(\tfrac{\eta S}{N})^2dt-\tfrac{\eta S}{N}dB(t)\\ &\le[\tfrac{\beta S}{N}-\tfrac{1}{2}(\tfrac{\eta S}{N})^2-(\gamma_1+\mu_1+\mu_0)+\sigma]dt-\tfrac{\eta S}{N}dB(t)\\ &\le\{-\tfrac{\eta^2}{2}[(\tfrac{S}{N})^2-\tfrac{2\beta}{\eta^2}\tfrac{S}{N}+(\tfrac{\beta}{\eta^2})^2-(\tfrac{\beta}{\eta^2})^2]-(\gamma_1+\mu_1+\mu_0)+\sigma\}dt-\tfrac{\eta S}{N}dB(t)\\ &\le[\tfrac{\beta^2}{2\eta^2}-(\gamma_1+\mu_1+\mu_0-\sigma)]dt-\tfrac{\eta S}{N}dB(t). \end{aligned} \tag{15}$$

Integrating the above formula from 0 to t on both sides, we obtain

$$\ln I(t)-\ln I(0)\le\int_0^t[\frac{\beta^2}{2\eta^2}-(\gamma_1+\mu_1+\mu_0-\sigma)]ds-\int_0^t\frac{\eta S}{N}dB(t).$$

According to the strong law of large numbers [20], we have

$$\lim_{t\to 0}\frac{1}{t}\int_0^t\frac{\eta S}{N}dB(t)=0\ \ a.s..$$

and we can obtain

$$\begin{aligned} \limsup_{t\to\infty}\tfrac{\ln I(t)}{t} &\le\int_0^t[\tfrac{\beta^2}{\eta^4}-(\gamma_1+\mu_1+\mu_0-\sigma)]ds\\ &\le\tfrac{\beta^2}{\eta^4}-(\gamma_1+\mu_1+\mu_0-\sigma)\\ &=(\gamma_1+\mu_1+\mu_0-\sigma)(\tfrac{\beta^2}{\eta^4(\gamma_1+\mu_1+\mu_0-\sigma)}-1)\\ &<0\ \ a.s.. \end{aligned} \tag{16}$$

We choose $R_0^S=\frac{\beta}{2\eta^2}(\mu_0+\gamma_1+\mu_1+\mu+\sigma)<1$, which is equal to $\eta>(\frac{\eta^2}{\mu_0+\gamma_1+\mu_1+\mu+\sigma})^{\frac{1}{4}}$.

Therefore, the above indicates that

$$\lim_{t\to\infty} I(t) = 0 \quad a.\,s..$$

and we can obtain the fact that the viral will be eradicated, which completes the proof. □

6. The Probability Density Function of the Stochastic Coronavirus Epidemic Model

Let $N = S + I + Q$; we can transfer system (3) into the following system:

$$\begin{cases} dS(t) = [\Lambda - \frac{\beta S(t)(N(t)-S(t)-Q(t))}{N} - \mu_0 S(t)]dt - \frac{\eta S(t)(N(t)-S(t)-Q(t))}{N} dB(t) \\ dQ(t) = (\gamma_1 N(t) - \gamma_1 S(t) - (\gamma_1 + \mu_0 + \mu + \sigma)Q(t))dt \\ dN(t) = [\Lambda - (\mu_1 + \mu_0)N(t) + \mu_1 S - (\mu_1 - \mu)Q(t)]dt. \end{cases} \quad (17)$$

Hence, we can consider the probability density function of system (17) in place of system (3).

Theorem 7. *We consider the condition that $R_0^* > 1$, for any initial value $(S(0), Q(0), N(0)) \in \mathbb{R}_+^3$; then, the solution $(S(t), Q(t), N(t))$ of system (3) with a weak kernel will have a normal probability density function $\Phi(S(t), Q(t), N(t))$ around (S^*, Q^*, N^*), which is given by*

$$\Phi(S(t), Q(t), N(t)) = (2\pi)^{-\frac{3}{2}} \mid \Sigma \mid^{-\frac{1}{2}(S(t),Q(t),N(t))\Sigma^{-1}(S(t),Q(t),N(t))^T},$$

where Σ is a positive definite matrix and satisfies

$$\Sigma = \begin{pmatrix} \frac{a_2}{2(a_1a_2-a_3)} & 0 & -\frac{1}{2(a_1a_2-a_3)} \\ 0 & \frac{1}{2(a_1a_2-a_3)} & 0 \\ -\frac{1}{2(a_1a_2-a_3)} & 0 & \frac{a_1}{2a_3(a_1a_2-a_3)} \end{pmatrix},$$

where $a_1 = \gamma_1 + 2\mu_0 + \mu + \sigma + \frac{\beta(N^*-2S^*-Q^*)}{N^*}$, $a_2 = (\mu_0 - \mu_1)(\gamma_1 + 2\mu_0 + \mu + \sigma) + \frac{\beta(N^*-2S^*-Q^*)}{N^*}(2\mu_0 + \mu + \sigma) + \frac{\gamma_1\beta(N^*-S^*)}{N^*}$, $a_3 = \frac{\beta S^*(N^*-S^*-Q^*)}{(N^*)^2} + (\gamma_1\mu + (\mu_0 - \mu_1)(\gamma_1 + \mu_0 + \mu + \sigma))(\mu_0 + \frac{\beta(N^*-2S^*-Q^*)}{N^*}) + \mu_1(2\gamma_1 + \mu_0 + \mu + \sigma)$.

Proof. Firstly, we can obtain the linear system of system (17) at point $(y_1, y_2, y_3) = (S^*, Q^*, N^*)$.

$$\begin{cases} dy_1 = (b_{11}y_1(t) + b_{12}y_2(t) + b_{13}y_3(t))dt - \frac{\eta S^*(N^*-S^*-Q^*)}{N^*} dB(t) \\ dy_2 = (b_{21}y_1(t) + b_{22}y_2(t) + b_{23}y_3(t))dt \\ dy_3 = (b_{31}y_1(t) + b_{32}y_2(t) + b_{33}y_3(t))dt, \end{cases} \quad (18)$$

where

$$A = \begin{pmatrix} b_{11} & b_{12} & b_{13} \\ b_{21} & b_{22} & b_{23} \\ b_{31} & b_{32} & b_{33} \end{pmatrix} = \begin{pmatrix} -(\mu_0 + \frac{\beta(N^*-2S^*-Q^*)}{N^*}) & \frac{\beta S^*}{N^*} & -\frac{\beta S^*(S^*+Q^*)}{(N^*)^2} \\ -\gamma_1 & -(\gamma_1 + \mu_0 + \mu + \sigma) & \gamma_1 \\ \mu_1 & \mu - \mu_1 & -(\mu_1 + \mu_0) \end{pmatrix},$$

$$B = \begin{pmatrix} \frac{\eta S^*(N^*-S^*-Q^*)}{N^*} & 0 & 0 \\ 0 & 0 & 0 \\ 0 & 0 & 0 \end{pmatrix},$$

Let $Y = (y_1, y_2, y_3)^T$, $G = diag(\frac{\eta S^*(N^*-S^*-Q^*)}{N^*}, 0, 0)$; then,

$$dY = AYdt + GdB(t).$$

By the Roozen [21], we can obtain the fact that the density function $\Phi(Y) = \Phi(y_1, y_2, y_3)$ of system (18) nearby the origin point can approximate to the Fokker–Plank equation as follows:

$$\begin{aligned} -\frac{\eta^2}{2}\frac{\partial^2\Phi(t)}{\partial y_1^2} &+\frac{\partial}{\partial y_1}[(b_{11}y_1(t)+b_{12}y_2(t)+b_{13}y_3(t))\Phi(t)] \\ &+\frac{\partial}{\partial y_2}[(b_{21}y_1(t)+b_{22}y_2(t)+b_{23}y_3(t))\Phi(t)] \\ &+\frac{\partial}{\partial y_3}[(b_{31}y_1(t)+b_{32}y_2(t)+b_{33}y_3(t))\Phi(t)] \\ &=0. \end{aligned} \tag{19}$$

By Gaussian distribution,

$$\Phi(Y) = C\exp\{-\frac{1}{2}(Y-Y^*)P(Y-Y^*)^T\}, \tag{20}$$

where P is a real symmetric matrix that satisfies

$$PG^2P + A^TP + PA = 0. \tag{21}$$

Let $P^{-1} = \Sigma$; then, we have

$$G^2 + A\Sigma + \Sigma A^T = 0. \tag{22}$$

We know there exists a matrix

$$\widetilde{M} = \begin{pmatrix} 1 & 0 & 0 \\ 0 & 1 & 0 \\ 0 & \frac{\mu_1}{\gamma_1} & 1 \end{pmatrix},$$

satisfying

$$\widetilde{M}A\widetilde{M}^{-1} = \begin{pmatrix} -a_{11} & a_{12} & a_{13} \\ a_{21} & -a_{22} & a_{23} \\ a_{31} & a_{32} & -a_{33} \end{pmatrix} = \begin{pmatrix} -(\mu_0+\frac{\beta(N^*-2S^*-Q^*)}{N^*}) & \frac{\beta S^*}{N^*} & -\frac{\beta S^*(S^*+Q^*)}{(N^*)^2} \\ -\gamma_1 & -(\gamma_1+\mu_0+\mu+\sigma) & \gamma_1 \\ 0 & \frac{-\mu_1(2\gamma_1+\mu_0+\mu+\sigma)+\gamma_1\mu}{\gamma_1} & -\mu_0 \end{pmatrix}.$$

Hence, we have the characteristic polynomials of A as

$$\varphi_A(\lambda) = \lambda^3 + a_1\lambda^2 + a_2\lambda + a_3.$$

Denote

$$dY = d\begin{pmatrix} y_1 \\ y_2 \\ y_3 \end{pmatrix} = \begin{pmatrix} -a_1 & -a_2 & -a_3 \\ 1 & 0 & 0 \\ 0 & 1 & 0 \end{pmatrix}\begin{pmatrix} y_1 \\ y_2 \\ y_3 \end{pmatrix}dt,$$

where $a_1 = a_{11}+a_{22}+a_{33} = \gamma_1+2\mu_0+\mu+\sigma+\frac{\beta(N^*-2S^*-Q^*)}{N^*} > 0$, $a_2 = a_{11}a_{22}+a_{11}a_{33}+a_{22}a_{33}+a_{23}a_{32}+a_{12}a_{21} = (\mu_0-\mu_1)(\gamma_1+2\mu_0+\mu+\sigma)+\frac{\beta(N^*-2S^*-Q^*)}{N^*}(2\mu_0+\mu+\sigma)+\frac{\gamma_1\beta(N^*-S^*)}{N^*} > 0$, $a_3 = a_{11}a_{22}a_{33}+a_{11}a_{23}a_{32}+a_{12}a_{21}a_{33}+a_{13}a_{21}a_{32} = \frac{\beta S^*(N^*-S^*-Q^*)}{(N^*)^2}+(\gamma_1\mu+(\mu_0-\mu_1)(\gamma_1+\mu_0+\mu+\sigma))(\mu_0+\frac{\beta(N^*-2S^*-Q^*)}{N^*})+\mu_1(2\gamma_1+\mu_0+\mu+\sigma) > 0$. We can easily obtain $a_1a_2 - a_3 > 0$. Therefore, by some transformation, the standard R_1 matrix of A is unique. By the same method of the Lemma 3 in [22], we can obtain a positive matrix.

$$\Sigma = \begin{pmatrix} \frac{a_2}{2(a_1a_2-a_3)} & 0 & -\frac{1}{2(a_1a_2-a_3)} \\ 0 & \frac{1}{2(a_1a_2-a_3)} & 0 \\ -\frac{1}{2(a_1a_2-a_3)} & 0 & \frac{a_1}{2a_3(a_1a_2-a_3)} \end{pmatrix}.$$

Hence, Σ is a positive definite, and we complete the proof. □

7. Examples and Numerical Simulations

In this section, we give the numerical simulation of system (3) by using the discrete equation with the same method as [10]. The equations are as follows:

$$\begin{cases} S(k+1) = S(k) + [\Lambda - \frac{\beta S(t)I(t)}{N(t)} - \mu_0 S(t)]\triangle t - \eta \frac{S(t)I(t)}{N(t)}\sqrt{\triangle t}\xi_k - \frac{\eta^2}{2}\frac{S(t)I(t)}{N(t)}\triangle t(\xi_k^2 - 1), \\ I(k+1) = I(k) + [(\frac{\beta S(t)I(t)}{N} - (\gamma_1 + \mu_1 + \mu_0)I(t) + \sigma Q(t)]\triangle t + \eta \frac{S(t)I(t)}{N(t)}\sqrt{\triangle t}\xi_k + \frac{\eta^2}{2}\frac{S(t)I(t)}{N(t)}\triangle t(\xi_k^2 - 1), \\ Q(k+1) = Q(k) + [\gamma_1 I(t) - (\mu_0 + \mu + \sigma)Q(t)]\triangle t, \end{cases} \quad (23)$$

where the time increment $\triangle t > 0$, and ξ_k is a the Gaussian random variable $(k = 0, 1, 2, \cdots n)$.

Example 1. *Here, in system (3), we use the environmental noise parameter as $\eta = 0.1$. In addition, following the biological feasibility result, the values of the parameters are as shown in Table 3.*

Table 3. Parameters value.

Notation	Value	References
Λ	0.028	[18]
β	0.2	Estimated
μ_0	0.011	[18]
μ_1	0.2	Estimated
γ	0.06	[18]
σ	0.3	Estimated
μ	0.5	[18]

In addition, we can choose the following real data $S(0) = 355{,}250, I(0) = 1453, Q(0) = 51{,}343$, in Pakistan on 7 October 2021 [19]. Then, $R_0^* = 3.208 > 1$, where R_0^* is defined in Section 4. By the results of Theorem 4, we can find that system (3) will persist for a long time by a distribution $\mu(\cdot)$. The numerical simulations (Figure 1) confirm this.

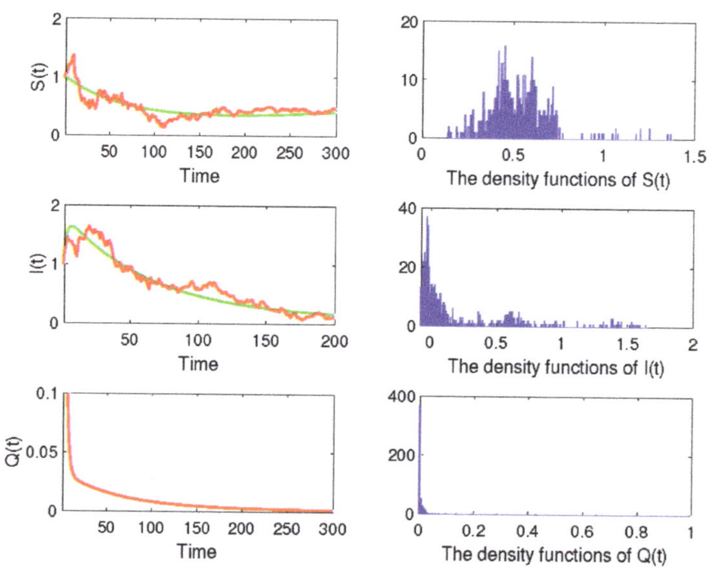

Figure 1. The red lines describe the solution of system (3) and the green lines stand for the solution of the corresponding system (2). The right pictures are the histogram of the density function for S, I, and Q populations.

Example 2. *Here, in system (3), we use the environmental noise parameter $\eta = 0.1$. In addition, following the biological feasibility result, the values of the parameters are as shown in Table 4.*

Table 4. Parameters value.

Notation	Value	References
Λ	0.5	[15]
β	0.6	Estimated
μ_0	0.2	[15]
μ_1	0.2	Estimated
γ	0.3	[15]
σ	0.1	Estimated
μ	0.2	[15]

In addition, we can choose the following real data $S(0) = 355{,}250, I(0) = 1453, Q(0) = 51{,}343$, in Pakistan on 7 October 2021 [19]. Then, $R_0^* = 0.0769 < 1$, where R_0^* is defined in Section 4. We can find that system (3) will be extinct in a long time. The numerical simulations (Figure 2) confirm this.

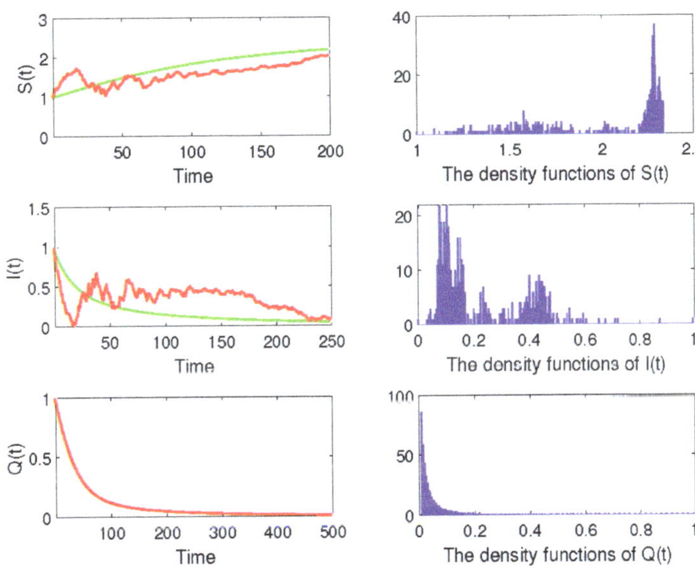

Figure 2. The red lines discribe the solution of system (3) and the green lines stand for the solution of the corresponding system (2). The right pictures are the histogram of the density function for S, I, and Q populations.

In fact, using the statistical data of Pakistan for September to December 2021, it can be seen from the figure (Figures 3–5) that the control of the isolation number will affect the disturbance of the infection rate and control the increase in the infection number. At the same time, when the infection rate is disturbed by other factors, such as vaccine injection, the number of deaths decreases with the decrease in the number of infections. This is basically consistent with the research results of system (3) in this paper.

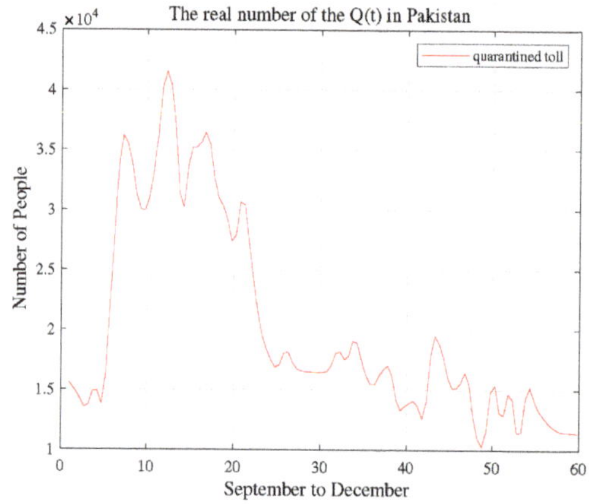

Figure 3. The number of the daily statistics of quarantined people in Pakistan from September to December 2021.

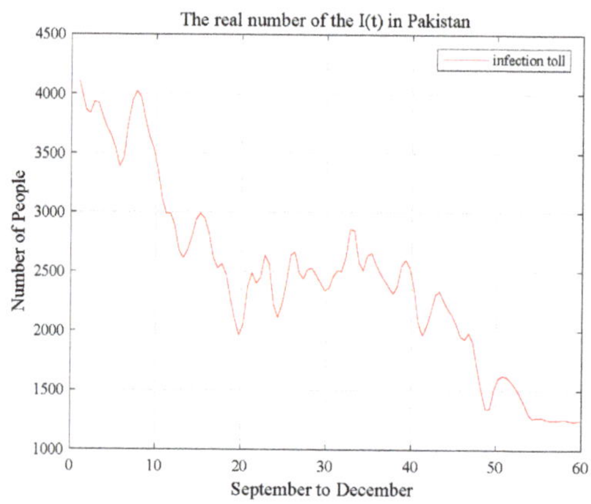

Figure 4. The red curve represents the daily statistics number of infections in Pakistan from September to December 2021.

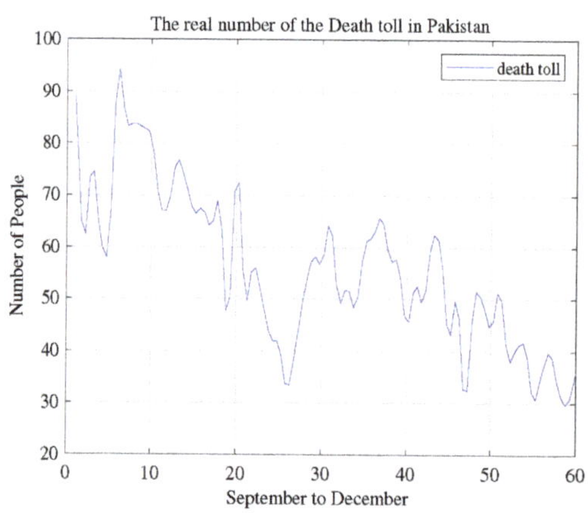

Figure 5. The blue curve represents the daily statistics death toll in Pakistan from September to December 2021.

8. Discussion

So far, the COVID-19 coronavirus disease is still one of the most serious diseases in the world. Until today, there is no appropriate treatment. At the same time, due to the strong transmission of the virus, with the existence of many uncertain factors (human activities, animal activities, express delivery, etc.), it also contains a lot of randomness. With the help of stochastic theory, we developed a model for the new 2019 coronavirus disease, and considered studying the transmission characteristics of the disease and understanding its transmission dynamics in the change in population and environment. The important role of isolation measures in controlling transmission is introduced. By disturbing the infection coefficient, the existence and positivity of Lyapunov function theory are studied. In this paper, in order to further discuss the extinction and stable distribution, we gave a new random infectious disease system under infection rate environmental noise. We give the existence and uniqueness of the solution of the system and discuss the ergodic stationary distribution and the extinction conditions of the system. The probability density function of the stochastic system is studied. Some digital simulations are used to demonstrate the probability density function and the extinction of system (3). Through numerical simulation, we analyzed the above results and drew a conclusion with the support of graphics. This work shows that random analysis is a better method used to study the dynamics of infectious diseases, especially the new 2019 coronavirus disease.

We know that the control of infectious diseases needs to consider a variety of random disturbance factors, and we will consider system (3) in more general random phenomena and their persistence and extinction properties in our future research.

Author Contributions: Writing—original draft preparation, J.S.; writing—review and editing, J.S. and M.G.; funding acquisition, J.S. and D.J. All authors have read and agreed to the published version of this manuscript.

Funding: The authors were supported by Shandong Provincial Natural Science Foundation (No. ZR2021MA052) and the National Natural Science Foundation of China (No. 11601520).

Institutional Review Board Statement: Not applicable.

Informed Consent Statement: Not applicable.

Data Availability Statement: Not applicable.

Conflicts of Interest: The authors declare no conflict of interest.

References

1. Lai, C.C.; Shih, T.P.; Ko, W.C.; Tang, H.J.; Hsueh, P.R. Severe acute respiratory syndrome coronavirus 2 (SARS-CoV-2) and corona virus disease-2019 (COVID-19): The epidemic and the challenges. *Int. J. Antimicrob. Agents* **2020**, *53*, 105924. [CrossRef] [PubMed]
2. Wu, Z.; Mcgoogan, J.M. Characteristics of and important lessons from the coronavirus disease 2019 (COVID-19) outbreak in china: Summary of a report of 72 314 cases from the chinese center for disease control and prevention. *JAMA* **2020**, *323*, 1239–1242. [CrossRef] [PubMed]
3. Mao, X. *Stochastic Differential Equations and Applications*; Horwood: Chichester, UK, 1997.
4. Has'miniskii, R. *Stochastic Stability of Differential Equations*; Horwood: Chichester, UK, 1997.
5. Mao, X.; Marion, G.; Renshaw, E. Environmental Brownian noise suppresses explosions in population dynamics. *Stoch. Process. Appl.* **2002**, *97*, 95–110 [CrossRef]
6. Alzahrani, E.O.; Khan, M.A. Modeling the dynamics of hepatitis E with optimal control. *Chaos Solit. Fract.* **2018**, *116*, 287–301. [CrossRef]
7. Wilder-Smith, A.; Freedman, D.O. Isolation, quarantine, social distancing and community containment: Pivotal role for old-style public health measures in the novel coronavirus (2019-nCoV) outbreak. *J. Travel Med.* **2020**, *27*, taaa020. [CrossRef] [PubMed]
8. Han, B.; Zhou, B.; Jiang, D.; Hayat, T.; Alsaedi, A. Stationary solution, extinction and density function for a high-dimensional stochastic SEI epidemic model with general distributed delay. *Appl. Math. Comput.* **2021**, *405*, 126236. [CrossRef]
9. Thomas, C.G. *Introduction to Stochastic Differential Equations*; Dekker: New York, NY, USA, 1988.
10. Gao, M.M.; Jiang, D.; Hayat, T.; Alsaedi, A. Threshold behavior of a stochastic Lotka-Volterra food chain chemostat model with jumps. *Phys. A* **2019**, *523*, 191–203 [CrossRef]

11. Siqueira, J.D.; Goes, L.R.; Alves, B.M.; da Silva, A.C.P.; de Carvalho, P.S.; Cicala, C.; Arthos, J.; Viola, J.P.; Soares, M.A. Distinguishing SARS-CoV-2 bonafide re-infection from pre-existing minor variant reactivation, Infection. *Genet. Evol.* **2021**, *90*, 104772. [CrossRef] [PubMed]
12. Lipster, R. A strong law of large numbers for local martingales. *Stochastics* **1980**, *3*, 217–218.
13. Griffin, B.D.; Chan, M.; Tailor, N.; Mendoza, E.J.; Leung, A.; Warner, B.M.; Duggan, A.T.; Moffat, E.; He, S.; Garnett, L.; et al. SARS-CoV-2 infection and transmission in the North American deer mouse. *Nat. Commun.* **2021**, *12*, 3612. [CrossRef]
14. Khan, M.A.; Abdon, A. Modeling the dynamics of novel coronavirus (2019-nCoV) with fractional derivative. *Alex. Eng. J.* **2020**, *59*, 2379–2389. [CrossRef]
15. Abdon, A. Modelling the spread of COVID-19 with new fractal-fractional operators: Can the lockdown save mankind before vaccination. *Chaos Solitons Fractals* **2020**, *136*, 109860
16. Zhou, B.; Zhang, X.; Jiang, D. Dynamics and density function analysis of a stochastic SVI epidemic model with half saturated incidence rate. *Chaos Solitons Fractals* **2020**, *137*, 109865. [CrossRef]
17. Chen, Y.; Cheng, J.; Jiang, Y.; Liu, K. A time delay dynamic system with external source for the local outbreak of 2019-nCoV. *Appl. Anal.* **2020**, *12*, 1–12. [CrossRef]
18. Din, A.; Khan, A.; Baleanu, D. Stationary distribution and extinction of stochastic coronavirus (COVID-19) epidemic model. *Chaos Solitons Fractals* **2020**, *139*, 110036. [CrossRef] [PubMed]
19. Available online: http://covid.gov.pk/ (accessed on 10 January 2022).
20. Herwaarden, O.A.; Grasman, J. Stochastic epidemics: Major outbreaks and the duration of the endemic period. *J. Math. Biol.* **1995**, *33*, 581–601. [CrossRef] [PubMed]
21. Roozen, H. An asymptotic solution to a two dimensional exit problem arising in population dynamics. *SIAM J. Appl. Math.* **1989**, *49*, 1973. [CrossRef]
22. Hethcote, H.W. Qualitative analyses of communicable disease models. *Math. Biosci.* **1976**, *28*, 335–356 [CrossRef]

MDPI
St. Alban-Anlage 66
4052 Basel
Switzerland
www.mdpi.com

Fractal and Fractional Editorial Office
E-mail: fractalfract@mdpi.com
www.mdpi.com/journal/fractalfract

www.ingramcontent.com/pod-product-compliance
Lightning Source LLC
Chambersburg PA
CBHW061552210926
51850CB00009B/2083
* 9 7 8 3 7 2 5 8 1 2 4 1 7 *